SCALE INVARIANCE AND BEYOND

Les Houches Workshop, March 10-14, 1997

Editors

B. DUBRULLE, F. GRANER
and D. SORNETTE

EDP Sciences

7, avenue du Hoggar
Parc d'Activités de Courtabœuf, B.P. 112
91944 Les Ulis cedex A, France

Springer-Verlag Berlin Heidelberg GmbH

Suite 81
875 Massachusetts Avenue
Cambridge, MA 02139, USA

Centre de Physique des Houches

Books already published in this series

1 Porous Silicon Science and Technology
Jean-Claude VIAL and Jacques DERRIEN, Eds. 1995

2 Nonlinear Excitations in Biomolecules
Michel PEYRARD, Ed. 1995

3 Beyond Quasicrystals
Françoise AXEL and Denis GRATIAS, Eds. 1995

4 Quantum Mechanical Simulation Methods for Studying Biological Systems
Dominique BICOUT and Martin FIELD, Eds. 1996

5 New Tools in Turbulence Modelling
Olivier MÉTAIS and Joel FERZIGER, Eds. 1997

6 Catalysis by Metals
Albert Jean RENOUPREZ and Hervé JOBIC, Eds. 1997

Book series coordinated by Michèle LEDUC

Editors of "Scale Invariance and Beyond" (N° 7)

B. Dubrulle, *Service d'Astrophysique, L'Orme des Merisiers, 709, 91191 Gif-sur-Yvette cedex, France*

F. Graner, *Laboratoire de Spectrométrie Physique, B.P. 87, 38402 Saint-Martin-d'Hères cedex, France*

D. Sornette, *Laboratoire de Physique de la Matière Condensée, CNRS et Université de Nice - Sophia Antipolis, 06108 Nice, France*

ISBN 978-3-540-64000-4 ISBN 978-3-662-09799-1 (eBook)
DOI 10.1007/978-3-662-09799-1

Originally published by Springer-Verlag Berlin Heidelberg New York in 1997

AUTHORS

Arneodo A., Centre de Recherche Paul Pascal, avenue Schweitzer, 33600 Pessac, France

Audit B., Centre de Recherche Paul Pascal, avenue Schweitzer, 33600 Pessac, France

Bacry E., Centre de Mathématiques Appliquées, École Polytechnique, 91128 Palaiseau, France

Biferale L., University of Rome "Tor Vergata", Department of Physics and INFN, Via della Ricerca Scientifica 1, 00133 Rome, Italy

Bouchaud J.-P., Science & Finance, 109-111 rue Victor Hugo, 92532 Levallois, France

Bouchaud J.-P., Service de Physique de l'État Condensé, Centre d'Études de Saclay, Orme des Merisiers, 91191 Gif-sur-Yvette cedex, France

Bréchet Y., L.T.P.C.M., B.P. 75, I.N.P. Grenoble, 38402 Saint-Martin-d'Hères, France

Butler S.L., Department of Physics, University of Toronto, Toronto, Ontario, Canada, M5S 1A7

Castaing B., Centre de Recherches sur les Très Basses Températures, Associé à l'Université Joseph Fourier, CNRS, B.P. 166, 38042 Grenoble cedex 9, France

Chatagny R., Group of Parallel Computing, CUI, University of Geneva, 24 rue du Général Dufour, 1211 Geneva 4, Switzerland

Chopard B., Group of Parallel Computing, CUI, University of Geneva, 24 rue du Général Dufour, 1211 Geneva 4, Switzerland

Cont R., Service de Physique de l'État Condensé, Centre d'Études de Saclay, 91191 Gif-sur-Yvette, France, and, Science & Finance, 109-111 rue Victor Hugo, 92532 Levallois, France; and Laboratoire de Physique de la Matière Condensée, URA 190 du CNRS, Université de Nice, 06108 Nice cedex 2, France

Dahmen K.A., Department of Physics, Harvard University, Cambridge, MA 02138, U.S.A.

Dubrulle B., CNRS, CEA/DSM/DAPNIA/Service d'Astrophysique, L'Orme des Merisiers 709, 91191 Gif-sur-Yvette, France

Graner F., CNRS, UMR 5588, Laboratoire de Spectrométrie Physique, Université Grenoble I - Joseph Fourier, B.P. 87, 38402 Saint-Martin-d'Hères cedex, France

Harrington S.T., Center for Polymer Studies and Physics Department, Boston University, Boston, MA 02215, U.S.A.

Henriksen R.N., Department of Physics, Queen's University at Kingston, Ontario, K7L 3N6, Canada

Hilfer R., ICA-1, Universität Stuttgart, 70569 Stuttgart, Germany; and Institut für Physik, Universität Mainz, 55099 Mainz, Germany

Knopoff L., Department of Physics and Astronomy and Institute of Geophysics and Planetary Physics, University of California, Los Angeles, U.S.A.

Lachièze-Rey M., C.E.-Saclay, 91191 Gif-sur-Yvette, France

Lebyodkin M., L.T.P.C.M., B.P. 75, I.N.P. Grenoble, 38402 Saint-Martin-d'Hères, France

Manneville S., Centre de Recherche Paul Pascal, avenue Schweitzer, 33600 Pessac, France

Marsili M., Institut de Physique Théorique, Université de Fribourg, 1700 Fribourg, Switzerland

Muzy J.F., Centre de Recherche Paul Pascal, avenue Schweitzer, 33600 Pessac, France

Nottale L., DAEC, Observatoire de Paris-Meudon, CNRS and Université Paris VII, 92195 Meudon cedex, France

Peltier W.R., Department of Physics, University of Toronto, Toronto, Ontario, Canada, M5S 1A7

Perkovic O., Laboratory of Atomic and Solid State Physics, Cornell University, Ithaca, NY 14853-2501, U.S.A.

Pocheau A., IRPHE, Université Aix-Marseille I & II, CNRS, Centre de Saint-Jérôme, Case 252, 13397 Marseille, France

Poole P.H., Department of Applied Mathematics, University of Western Ontario, London, Ontario N6A 5B7, Canada

Potters M., Science & Finance, 109-111 rue Victor Hugo, 92532 Levallois, France

Roux S.G., Centre de Recherche Paul Pascal, avenue Schweitzer, 33600 Pessac, France

Sciortino F., Dipartimento di Fisica, Università di Roma La Sapienza, 00185 Roma, Italy

Sethna J.P., Laboratory of Atomic and Solid State Physics, Cornell University, Ithaca, NY 14853-2501, U.S.A.

Sornette D., Laboratoire de Physique de la Matière Condensée, CNRS and Université de Nice-Sophia Antipolis, Parc Valrose, 06108 Nice, France; and Department of Earth and Space Sciences and Institute of Geophysics and Planetary Physics, University of California, Los Angeles, California 90095-1567, U.S.A.

Stanley H.E., Center for Polymer Studies and Physics Department, Boston University, Boston, MA 02215, U.S.A.

Zajdenweber D., Department of Economics, Paris-X Nanterre University, Paris, France

CONTENTS

FOREWORD

Scale Invariance and Beyond

by B. Dubrulle, F. Graner and D. Sornette

1. Introduction 1
2. Scale Invariance 2
   2.1. Concept 2
      2.1.1. A definition of scale invariance 2
      2.1.2. Open questions 3
      2.1.3. Analogy with translation invariance 3
      2.1.4. Link with dimensional analysis 6
   2.2. Tools to study scale invariance 7
      2.2.1. Wavelets 8
      2.2.2. Fractional derivative 8
      2.2.3. Dimensional similarity 8
      2.2.4. Stable laws 9
      2.2.5. Computation of scaling exponents 10
   2.3. Examples 10
      2.3.1. Condensed matter systems 11
      2.3.2. From cosmology to earthquakes 12
      2.3.3. From biology to finance 14
3. Beyond scale invariance 16
   3.1. From scale invariance to scale covariance 16
   3.2. Scale symmetry breaking 17
4. Conclusion 19

Scale Invariance

Concept

LECTURE 1

Scale Invariance Without Mechanism?

by F. Graner

1. Introduction 25
2. Magic in the solar system? 25
   2.1. A « log versus rank » plot 26
   2.2. Explanations of this geometrical progression 27
   2.3. Comments 28
3. Hand-waving symmetry argument 28
   3.1. Generic equations 28
   3.2. Generic solutions 29
   3.3. Comments 30
4. Example: the flat, cold disk 30
   4.1. Symmetry of the equations 31
   4.2. Linear stability analysis 31
   4.3. Unstable modes 32
5. Discussion 33
   5.1. From continuous to discrete scale invariance 33
   5.2. Warnings 34
6. Summary 35

Tools

LECTURE 2

Scale Invariance and Beyond: What Can We Learn from Wavelet Analysis?

by A. Arneodo, B. Audit, E. Bacry, S. Manneville, J.F. Muzy and S.G. Roux

1. From global to local characterization of the regularity of fractal functions 37
2. Statistical analysis of the regularity of fractal functions: the multifractal formalism 38
3. A thermodynamics of fractal signals based on wavelet analysis 39
   3.1. Singularity detection and processing with wavelets 39
   3.2. The wavelet transform modulus maxima method 40
   3.3. The structure function approach versus the WTMM method 41
4. Numerical and experimental applications of the WTMM method 41
   4.1. Fully developed turbulence 42
   4.2. DNA sequences 45
5. Further developments and perspectives 48

LECTURE 3

Fractional Derivatives in Static and Dynamic Scaling

by R. Hilfer

1. Introduction 53
2. Fractional derivatives 54
3. Scaling in thermodynamics 56
4. Scaling in macroscopic dynamics 58

LECTURE 4

Multi-Dimensional Self-Similarity, and Self-Gravitating N-Body Systems

by R.N. Henriksen

1. Introduction 63
2. CH prescription for multi-dimensional self-similarity 65
3. Rescaling symmetries in N-body self-gravitating systems 68
   3.1. Stationary example 68
   3.2. Asymptotic self-similarity: speculations 69

LECTURE 5

Scaling in Stock Market Data: Stable Laws and Beyond

by R. Cont, M. Potters and J.-Ph. Bouchaud

1. Introduction 75
2. Statistical description of market data 76
3. Scale invariance and stable laws 77
4. Beyond scale invariance: truncated Lévy flights 78
5. Correlation and dependence 81
6. Turbulence and finance 82
7. Conclusion 83

Examples

Condensed Matter Systems

LECTURE 6

Hysteresis, Avalanches, and Barkhausen Noise

by J.P. Sethna, O. Perković and K.A. Dahmen

1. Introduction 87
2. The model 89

3. The critical exponents 89
4. The epsilon expansion 90
5. Widom scaling 93
6. Conclusion 94

LECTURE 7

Burgers Turbulence and the Energy Landscape of Randomly Pinned Objects

by J.-Ph. Bouchaud

LECTURE 8

Power Laws and Scale Invariance in Physical Metallurgy

by Y. Bréchet and M. Lebyodkin

1. Introduction 107
2. The actors and the length scales in physical metallurgy 107
   2.1. The actors of physical metallurgy 107
   2.2. A microstructure viewpoint on mechanical properties 108
   2.3. Changing scales 109
3. Scale invariance and power laws in physical metallurgy: examples 110
   3.1. Examples taken from phase transformations 110
   3.2. Organisation of structural defects 111
   3.3. Scaling laws for mechanical behaviour 112
4. Suggestions for further work and need of new approaches 115
5. Conclusion 117

LECTURE 9

Scale Invariance in Fluids with Anticorrelated Entropy-Specific Volume Fluctuations

by H.E. Stanley, S.T. Harrington, P.H. Poole and F. Sciortino

1. Introduction 119
2. The hypothesis 120
3. Why care? 120
4. Why believe? 121
   4.1. Plausibility arguments 121
   4.2. Evidence from simulations 123
   4.3. Does $1/K_T^{max}$ extrapolate to zero at $(T_{C'}, P_{C'})$? 123
   4.4. Is there a « kink » in the P_ρ isotherms for sufficiently low temperature? 124
   4.5. Is there a unique structure of the liquid near the kink point? 124
   4.6. Does the coordination number approach four as C' is approached? 124
   4.7. Is it possible that two apparent « phases » may coexist below125 C'? 125
   4.8. Do fluctuations appear on all time scales? 125
   4.9. Is there « critical slowing down » of a characteristic time scale? 125
   4.10. Is the characteristic dynamics of each « phase » different? 126
   4.11. Is there evidence for a HDL-LDL critical point from independent simulations? 126
   4.12. Evidence from experiments 126
   4.13. Density fluctuations along the P = 0.1 MPa isobar 127
   4.14. Structure along isobars up to P = 600 MPa 127
   4.15. Reversible conversion of LDA to HDA with pressure 128
5. Discussion 128

LECTURE 10

Scale Invariance(s) in the Cosmic Matter Distribution?

by M. Lachièze-Rey

1. Introduction 133
2. The large scale matter distribution 134
   2.1. Typical scales 134
   2.2. Correlation functions 135
      2.2.1. Galaxy - galaxy correlation functions 135
      2.2.2. The cluster - cluster correlation function 136
      2.2.3. Higher order correlations 137
   2.3. Void and count probability functions 137
   2.4. Difficulties with correlation functions 138
      2.4.1. Fractal or multifractal dimensions? 139
3. Scale invariance and dynamics 140
   3.1. The power spectrum of fluctuations 140
   3.2. Gravitational instability: standard views 141
   3.3. A time dependent scale for gravitational clustering 142
   3.4. A global scale invariance? 143
   3.5. The hierarchical assumption 143
4. Scale invariance in local collpase? 144
5. Conclusions 145

LECTURE 11

Scale Invariance of the Avalanche Effect in Phase Transition Modulated Mantle Mixing

by S.L. Butler and W.R. Peltier

1. Introduction 147
2. Mantle convection with phase transitions 148

3. A theoretical model 149
4. Results 151
5. Discussion and conclusions 155

LECTURE 12

Scale Invariance of Earthquakes

by L. Knopoff

1. Earthquake energy distributions 159
2. Aftershocks 162
3. Stress redistribution 164
4. The physics of the clustering of large earthquakes 167
5. The physics of small earthquakes 169
6. Discussion 171

From Biology to Finance

LECTURE 13

Models for Evolution and Extinction

by M. Marsili

1. Introduction 173
2. The Lotka-Volterra equations 174
3. Evolution and extinction from Lotka-Volterra equations 176
4. Conclusions 180

LECTURE 14

Scale Invariance in Economics and in Finance

by D. Zajdenweber

1. Introduction 185
2. Empirical evidence of single-tail scale invariant distributions in economics 185
3. Mathematical background 188
4. Economic analysis 188
   4.1. Absence of characterisric scale in empirical and in theoretical economics 189
   4.2. Non democratic sharing 190
   4.3. Economics of scale 191
   4.4. Stochastic models of gravitational fields 191
   4.5. Theory of ruin and economies of scale 192
   Conclusion 193

LECTURE 15

Models of Artificial Foreign Exchange Markets

by B. Chopard and R. Chatagny

1. Introduction 195
2. The foreign exchange market 196
   2.1. Ther market makers and brokers 196
   2.2. The traders 197
3. Properties of the FX market 198
   3.1. Exchange rates and market price 198
   3.2. Spread 199
   3.3. Change of price 199
   3.4. Risk and volatility 200
   3.5. Return distribution 201
4. Simulation of the FX market 201

5. Results of our microsimulation 203
6. Conclusion 204

Beyond Scale Invariance

From Scale Invariance to Scale Covariance

LECTURE 16

From Scale-Invariance to Scale-Covariance in Out-of-Equilibrium Systems

by A. Pocheau

1. Introduction 209
2. Definitions and concepts 209
3. Turbulent propagative fronts 211
3.1. Scale-invariant regime 211
3.2. Scale-covariance; scale-covariant laws 211
3.3. Experimental evidence 214
3.4. Fractal construction in scale space 216
3.5. Scale-covariance versus scale-similarity 217
4. Intermittency in fully developed turbulence 218
4.1. Scale-invariance as a gauge-invariance 218
4.2. Scale-covariance of single scale functions 219
4.3. Observers, measurement procedure and gauge covariance 219
4.4. Scale-covariant intermittency function 221
4.4.1. Statistical homogeneity ignored 221
4.4.2. Statistical homogeneous state 222
5. Conclusion 223

LECTURE 17

Turbulence: Statistical Approach

by B. Castaing

1. Introduction 225
2. The turbulent velocity profile 226
   2.1. Dimensional analysis 226
   2.2. A first refinement 226
   2.3. Second refinement 227
3. Velocity intermittency in turbulence 227
   3.1. The problem 227
   3.2. The cascade of multipliers 228
   3.3. Experimental check of infinite divisibility 230
   3.4. Behavior of $n(r)$; first and second refinements 231
4. Markovian approaches and conclusion 232

Scale Symmetry Breaking

LECTURE 18

Discrete Scale Invariance

by D. Sornette

1. What is discrete scale invariance (DSI)? 235
2. What are the signatures of DSI? 236
3. Scenarios leading to DSI 238
   3.1. Built-in geometrical hierarchy 238
   3.2. Diffusion in anisotropic quenched random lattices 239
   3.3. Cascade of ultra-violet instabilities: growth processes and rupture 240
   3.4. Cascades of sub-harmonic bifurcations in the transition to chaos ... 241
   3.5. Non-local geometry and non-unitary field theory 242
   3.6. Quenched disordered systems 242

4. Other systems 243
5. Open problems 243

LECTURE 19

Scale Relativity

by L. Nottale

1. Introduction 249
2. Toward a nondifferentiable space-time 251
3. Scale invariance and scale relativity 252
3.1. Least action principle for scale laws 252
3.2. Scale-symmetry breaking 253
3.3. Special scale-relativity 255
4. Generalized scale laws 256
4.1. Scale « dynamics » 256
4.1.1. Constant scale-force 257
4.1.2. Harmonic oscillator 257
4.2. Discrete scale invariance and log-periodic behavior 258
4.3. Scale-motion coupling and gauge invariance 259
5. Discussion and open problems 260

LECTURE 20

Scaling in Turbulent Flows

by L. Biferale

1. Introduction 263
2. Extended self similarity 265
3. Generalized-ESS 266
4. A theoretical interpretation 267
5. Multiscaling 272
6. Conclusions 273

LECTURE 21

Statistical Scale Symmetry Breaking

by B. Dubrulle

1. Introduction 275
2. Scale invariant case 276
3. Building new tools 277
   3.1. The composition rule 277
   3.2. The scale transformations 278
   3.3. Analogy with special relativity 279
4. Scale symmetry breaking 280
   4.1. Lagrangian framework 281
   4.2. An example: the log-normal case 281
      4.2.1. Dynamics 282
      4.2.2. General scaling 283
      4.2.3. Ohmic case 283
5. Conclusion 285

FOREWORD

Scale Invariance and Beyond

B. Dubrulle, F. Graner* and D. Sornette**,***

CNRS, CEA/DSM/DAPNIA/service d'Astrophysique,
L'Orme des Merisiers, 709, 91191 Gif-sur-Yvette, France
** Laboratoire de Spectrométrie Physique, B.P. 87,*
38402 Saint-Martin-d'Hères cedex, France
*** Laboratoire de Physique de la Matière Condensée,*
CNRS and Université de Nice - Sophia Antipolis,
Parc Valrose, 06108 Nice, France
**** Department of Earth and Space Sciences*
and Institute of Geophysics and Planetary Physics,
University of California, Los Angeles, California 90095-1567, U.S.A.

1. INTRODUCTION

In physics, any symmetry is a bargain. It provides tools to analyze the solutions of a problem without need to explicitly solve the problem, or it can also suggest methods to simplify the problem. In real life, boundary conditions or even physical processes themselves break the initial beautiful symmetry. One has to deal with "approximate symmetry" and means to obtain useful informations from this broken symmetry have to be built.

The topic of the school held in Les Houches in March 1997 is the illustration of these general ideas in the case of scale symmetry. Part of the school involved descriptions of tools or concepts used in scale invariant systems, and examples pertaining to them ("Scale Invariance..."); the second half was devoted to recent attempts to go beyond the invariance or symmetry breaking, discuss causes and consequences, and extract useful informations about the system ("...and beyond").

Scale invariance has became very fashionable since the work of Mandelbrot on fractals, and the successes of the renormalization group of Kadanoff-Wilson.

When we first thought about organizing a school on that topic, our goal was not to add one more conference piling up descriptive examples of self-similar systems. It was rather to make a synthesis of our present understanding of the underlying mechanisms and more generally the origins of scale invariance. The school thus brought together workers with various multidisciplinary view points, such as axiomatic generalizations, partial symmetry breaking, classification of physical mechanisms, statistical approach, carefully worked out examples in condensed matter physics and other fields. It was also felt useful to enlarge the scope by inviting experts of neighboring disciplines that could present the key problems in their field related to the subject of the school. This cross-disciplinary approach was reflected also in the interests of the participants, which ranged from plasma physics to finance, from turbulence to telecom signal analysis, from percolation theory to foundations of quantum mechanics.

As the result of such tremendous diversity and regarding the novelty of many of the selected approaches, it was not possible to find a limited number of lecturers to teach this school. To help the reader to get a somewhat more synthetic view about the topic, we therefore decided to start the book by an extended foreword discussing the foundations of scale invariance and its extensions, in order to replace each individual lecture in a broader context. Contributions have been specifically edited towards more coherence and cross-references. Lecturers made an effort to provide their manuscript very quickly, so that this book will appear less than two years after we began to organize this school. We thus hope this book is an useful and timely tool for anybody interested in discovering that scale invariance is now evolving towards a mature field.

2. SCALE INVARIANCE

2.1. Concept

2.1.1. A definition of scale invariance

Speaking about a (material or mathematical) object, scale invariance refers to its invariance over changes of scales of observation. You sometimes hear that a properties "is the same whatever the scale you are looking at". In fact, you quickly notice that some people are talking about the invariance of their equations, some are talking about one physical quantity or another. Some people consider invariance in a statistical sense: geometrically, they are discussing fractal shapes; analytically, they are discussing the invariance of a probability distribution function or a correlation function, or sometimes only the shape of this function. So you first need to identify and clearly state what observable you are considering, and how you define this variable at each scale. Then, a nice mathematical way to express such symmetry for a scale dependent observable $\mathcal{O}_\ell$ is to require that when the scales is changed by a factor λ, the observable is changed by a factor $\mu(\lambda)$:

$$\mathcal{O}_\ell = \mu(\lambda)\mathcal{O}_{\lambda\ell}. \tag{1}$$

Three remarks can be made on such simple definition. First, the solution of (1) is a power law $\mathcal{O}_\ell = C\ell^\alpha$, with $\alpha = -\ln\mu/\ln\lambda$. Second, the definition of scale invariance (1) is even simpler when viewed in the log-space $(\log\mathcal{O}, \log\ell)$. It becomes then simply a translation invariance! This simple remark is the starting point of many recent approaches on scale invariance and is developed below. Third, we note that the relation (1) can be seen as a simple invariance with respect to changes of units on $\mathcal{O}$ (the μ factor) or ℓ (a $1/\lambda$ factor). This innocent remark suggests a strong link between *scale invariance* and *dimensional analysis*. These three remarks will be exploited below.

2.1.2. Open questions

Here are a few examples of questions that emerged during the school, especially at the round tables:

- (i) What is the definition of scale invariance, of scale covariance, what is the difference between them?
- (ii) Is scale symmetry an exact symmetry? If yes, why are there specific scales in physics, such as set by the Planck constant?
- (iii) What are the mechanisms responsible for power laws: change of variable, exponentiation, random walk, constrained multiplicative processes, SOC, disorder critical points...? Is there a generic mechanism, or a specific mechanism for each system ? And what about the mechanisms for breaking the scaling?
- (iv) How to exploit optimally scale symmetry (homogeneity, scaling hypothesis and reduction of variables, Barenblatt classification and beyond, conformal invariance)?

A coherent and unanimous answer to all of them has not always been found, and on some of them, it appears that a great deal of confusion is present. We will try to summarize the present state of the art as well as we can.

2.1.3. Analogy with translation invariance

The easiest of these questions can be tackled by noting that, as we have mentioned, the relation (1) is equivalently written as a translation invariance acting on log-variables $(X, T) = (\log\mathcal{O}, \log\ell)$. This means that scale and translation symmetries are formally analogous. Of course, we are more familiar with physical space, and thus translation, than with the (Fourier) scale space and its dilations. Of course again, scale symmetry applies to qualitatively different problems and objects: irregularity, multi-scale statistics, fractals, which do not necessarily have obvious analogs. But when it comes to quantitative description, the analogy is both exact and profound. So let us see how this analogy can make for the insufficiencies of our intuition.

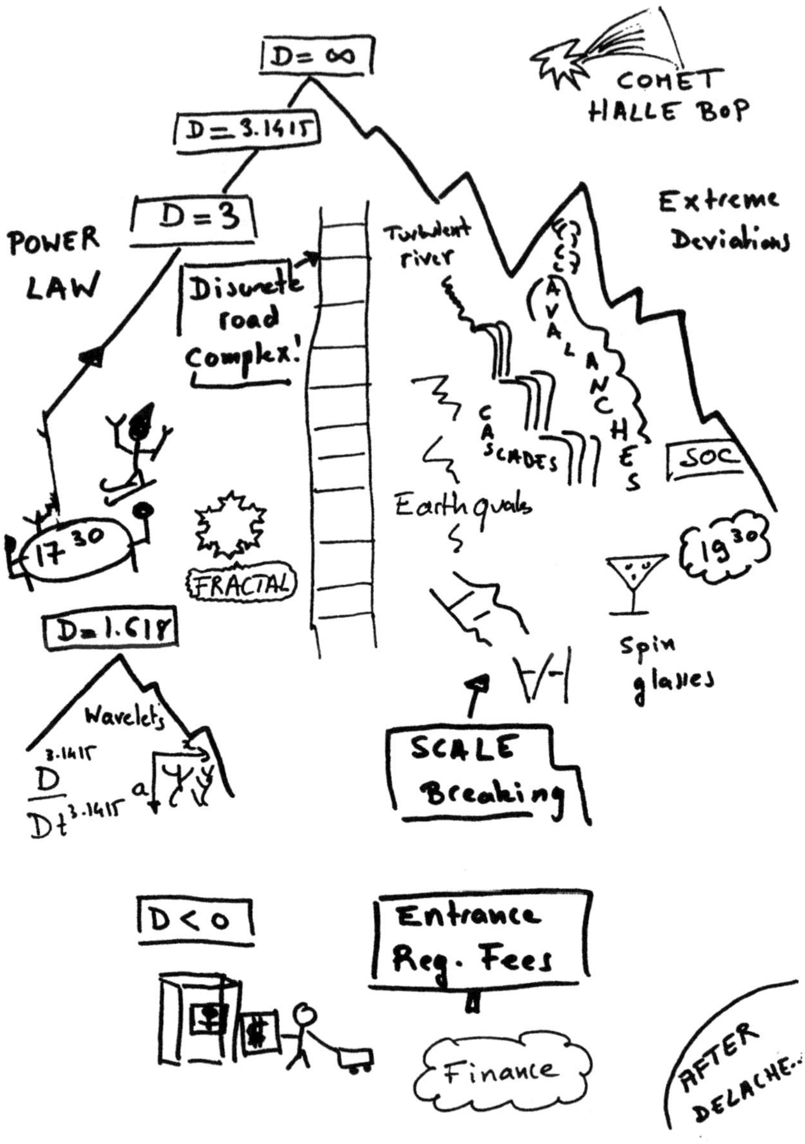

Fig. 1. — The many roads to scale invariance: to reach the highest altitude peak, at $D = \infty$, one can take a number of paths. The direct road, via power law, is the more direct one, but it is dangerous and slippery. One can try an easiest road, via discrete scales, but it is a complex path, and can involve dangerous scale breaking! A third path, via SOC, or cascades involve extreme deviations and the ascent is very slow. Before trying these difficult roads, one can practice on the school-rock, via wavelets, fractional derivatives, dimensional analysis or stable laws. However this small peak only leads you up to $D = 1.618$. This cartoon, used in the welcome talk, was inspired from a similar cartoon drawn about turbulence by Philippe Delache.

First, all concepts developed in translation invariance, and the corresponding vocabulary, can be transposed into scale invariance, and thus leads to a wealth of new ideas. Homogeneity of space translates into dimensional analysis. Symmetry is described by similar operators, and even Lorentz boosts find their scale counterparts, explored e.g. in the contributions of Nottale and Dubrulle. The notions of exact symmetry versus symmetry breaking also apply. A minimum and a maximum cut-off provide boundary conditions, of course in the sense of the scale space. As Sornette explains, continuous and discrete symmetries have similar meanings in the physical and in the scale space.

These statements help to solve many questions and paradoxes, often deeply hidden in our way of thinking. For instance, question (i) led to an heated debate in a round table, until somebody remarked that it was equally difficult to answer it for translation invariance. In fact, there are many operational definitions, adapted to each problem, as is the case for any symmetry in physics. See for instance Nottale for some axioms, Henriksen for mathematical formulations, Pocheau for the discussion of covariance...

Question (ii) also benefits from this analogy. Physical theories postulate that physical laws and equations are invariant under translation. We are living at a specific position in the universe, but now that the geocentric cosmology is out fashioned, we do not necessarily think of our house as the center of the world or as a particularly privileged position. Even the Universe as a whole can be of a finite size. You can displace an isolated system without modifying its physics, which means that fundamental, internal interactions must be invariant by translation; but external, effective forces are not necessarily so. Let us now propose an analogy with scale space. Postulate that fundamental laws that couple different scales do not explicitly depend separately on the particular scales considered. It does not matter whether our physical world involves constants with such or such value, nor whether the universe has finite upper and lower cut-offs. If you can, in your mind, isolate some scales from others, you can change their characteristic size without altering the physics. For instance, macroscopic physics in the XIXth century was scale invariant; if atoms ever existed, their scale was not relevant. In the XXth century, macroscopic physics is not scale invariant anymore; it is invariant only under a global rescaling which also affects the Planck constant and other micro-scales.

As for question (iii), the analogy will tell us that the answer is not simple! In fact, is there a unique physical mechanism responsible for the fact that a given system is, or is not, invariant by translation? As for scale, Nottale points out that scale symmetry is related to the (possibly non-differentiable) structure of space-time, thus suggesting a general frame for physical laws; but when coming to a given physical object, identifying the mechanism coupling different scales necessitates a case-by-case analysis, what Frisch called a branching cascade model. The vocabulary can be derived using the analogy, as in Dubrulle and Graner theory: "renormalization trajectories, Lagrangian..."; but we are often short when trying to physically understand what actually happens. However, there is still hope, and Castaing and Sornette provide examples of mechanisms

2.1.4. *Link with dimensional analysis*

A basic postulate in physics is the independence of the laws with respect to changes of units: the physics should be the same for a French experimentator recording in meters and kilograms, or for his English colleague thinking in terms of inches and pounds. As we have just seen, the invariance by changes of units necessarily involves power laws. As a result, any physical quantity can then be expressed in terms of monomials of a few "basic units" such as meters, second or kilograms. The exponents appearing in these monomials describe how the quantity varies with respect to unit changes. They define the "dimension" of the quantity. A quantity which remains invariant under changes of units is called "dimensionless".

Laws of physics must be such as to be expressed using only dimensionless variables. This is a powerful statement, mathematically described by the famous Π theorem [1]. It helps reducing the number of physically pertinent parameters, and evidences the equivalence between similar systems of different sizes. Fluid mechanics is notorious for its extensive use of similitude ratios, such as the Reynolds number. A famous example occurred during the cold war: the Americans got very worried because each time they were doing an atmospheric atomic bomb explosion, the Russians seemed to know the intensity of the corresponding explosion. Was there a spy in the heart of their most secured secret? In fact, the Russians had bright physicists and good ways to get pictures of the explosion at different times. The physicists knew about dimensional analysis, and they found that the speed of propagation of the explosion front only depended on the initial intensity and on time. By looking at profiles at different times, they could then reconstruct the profile at initial time, and hence, the strength of the explosion!

Another powerful prediction is due to Kolmogorov in 1941. He proposed that, in a turbulent fluid, there exist a range of scales called "inertial range" where the velocity u_ℓ of an eddy of size ℓ is only determined by the energy transfer rate $\epsilon = du^2/dt$. Dimensional analysis then tells you that $u_\ell = C(\epsilon\ell)^{1/3}$. This simple reasoning predicts that the turbulent spectral energy spectrum $E(k) = u(k)u(-k)$ scales like $k^{-5/3}$ in the inertial range. This prediction has been checked in a variety of turbulent flows and is extremely well satisfied [2].

Scale invariant quantities are power laws, and power laws are naturally generated by dimensional analysis. Does it mean that any scale invariant system can be studied by dimensional analysis? Alas, no. This is strictly true when you have only one scale, i.e. when you are dealing with an isotropic problem, or with a time scale, because power laws are the only homogeneous functions of one variable. As soon as you have more than one variable, you must examine the homogeneous function describing your problem more carefully. There are conspiracies, first discussed by Barenblatt [1], which can drive a system away from a "dimensional" scaling (one then speaks about "anomalous" scaling). These conspiracies involve additional non dimensional variables, which couple to the initial one. To understand this, let us take again the case of turbulence, and imagine that in fact the largest scale of the system also influence the eddy velocity. This means that we have now two independent dimensionless quantities, $\Pi_1 = u_\ell/(\epsilon\ell)^{1/3}$ and $\Pi_2 = \ell/L$, and by the Π theorem, we can write $u_\ell = (\epsilon\ell)^{1/3}\Phi(\ell/L)$. If Φ is arbitrary, u_ℓ is not a power law, and this solution is not scale invariant anymore. For u_ℓ to be scale invariant, Φ must be a power law $\Phi(x) \sim x^\alpha$, and the power law of u_ℓ is now $1/3 + \alpha$, different from the dimensional one. This new form of scaling was named "similarity of second kind" and is studied in great details in [1]. Another non-trivial example of similarity of second kind is given in the lecture by Castaing, on flows near a wall. In critical phenomena, this anomalous scaling is understood as the consequence of fluctuations of the order parameter at all scales between the microscopic cut-off up to the macroscopic scale. The renormalization group has provided the framework to calculate systematically the anomalous corrections to the mean field exponents. The renormalization group has even recently been shown [3] to explain the "similarity of second kind" classified by Barenblatt and found in nonlinear PDE's.

To sum up, we just learned that scale invariance is not simply dimensional analysis, and that it leads to power laws. Reciprocally, is any power law the signature of an underlying scale invariance? For many physicists, the answer seems to be "yes", but some caveat should be indicated: as underlined by Graner, almost any log-log plot displays a straight line over a small range of parameters, and this can be "accidental" with no physical mechanism! Graner makes his point using the Titius-Bode law describing a remarkable regularity in the distribution of planetary distances to the sun. He presents a mathematical formulation similar to that of Henriksen, which focuses on the scale symmetry of the underlying equations. He shows that all models that have been proposed to explain the Titius- Bode law share the common ingredient of scale symmetry. Assuming a discrete symmetry breaking in the rotation invariance, he thus shows that any such mechanism is compatible with the Titius-Bode law. As a consequence, this law cannot a priori be used to constraint the mechanism of planet formation and their organization around the sun. What is however not understood is the physical mechanism, if any, at the basis of the breakdown of continuous to discrete scale invariance embodied in the Titius-Bode law.

2.2. Tools to study scale invariance

So far, we spoke a lot about scale invariance, but not really about "tools" to study it. The simplest tool used by anybody is the log-log plot: you take the log of your data, you plot it versus the log of the scale, and if you see a straight line, you compute its slope (the scaling exponent) and the intercept at the origin (the prefactor). That's it! There are however more refined methods to get power laws, and also powerful tools to exploit better the scale invariance. Some of them were described in the school.

2.2.1. Wavelets

A sometimes ill-posed problem is to define a "quantity at scale ℓ". The wavelet approach is a powerful tool to answer this problem. It often enables a cleaner determination of scaling exponents via mathematically motivated simple treatments. This topic was addressed by Arnéodo, using turbulence and DNA as illustrative examples. Arnéodo summarizes the unified multifractal description of singular distributions, including measures and functions, based on the continuous wavelet transform. This approach relies upon the definition of partition functions from the wavelet transform modulus maxima (WTMM). He demonstrates that, very much like thermodynamic potentials, the generalized fractal dimensions and the spectrum of singularities can be readily determined from the scaling behavior of these partitions functions. This method provides a natural generalization of the classical box-counting and structure function techniques. Applications of the WTMM method include fully-developed turbulence data, DNA sequences, financial time-series, satellite images of fractal clouds and fractal growth phenomena. The method is particularly good at detecting anomalous scaling as well as some breaking of scale invariance. The WTMM skeleton allows one to identify the nature of the underlying statistics at a given scale and the nature of the cascade process from a given scale to a finer scale (if there is any such process).

2.2.2. Fractional derivative

Hilfer provides a brief but clear pedagogical introduction into the mathematics of fractional derivatives and integrals, and then proceeds to their recent emergence and applications in physics. In particular, it is shown how fractional derivatives arise naturally in the theory of phase transitions in thermodynamics. Next, it is demonstrated that fractional time derivatives emerge from some novel long time limit in ergodic theory as the generators of time evolution on macroscopic time scales. In general, fractional derivation is an ideal tool for describing long-range memory processes and allows often for exact solutions in arbitrary dimensions. Several exactly solved examples are presented which illustrate that many known solvable models exhibit new features of scale invariance in a rigorous fractional treatment.

2.2.3. Dimensional similarity

As discussed earlier, the dimensional analysis (the Π theorem) leads to the reduction of the number of independent variables by one. When applied to PDE, this sometimes enables to go from an PDE to an ODE. More generally, this considerably simplifies the solving of the PDE problem. The systematic formulation of this reduction in complexity was presented by Henriksen. He used a systematic mathematical formulation that Carter and himself have introduced to combine the intuitive and physical approaches towards self-similar geometry. His goal is reconcile the intuitive method of reasoning that allows one to find simple, physical solutions to non-linear equations with the mathematical perspective of invariance under Lie transport. This formalism reveals clearly that the effect of self-similar symmetry in general is not merely to reduce a set of partial differential equations to ordinary equations, but rather to reduce the number of independent variables by one. In a multi-dimensional problem there is in general a complete set of variables, termed 'self-similar', which are the natural variables in that they are invariant under the self-similar 'motion'. The use of these variables in an appropriate way renders any problem in which a self-similar symmetry is present effectively 'stationary'. The applications of this method include the study of phase mixing and violent relaxation in the Vlasov-Poisson coarse-grained and mean-field gravitational problem, as well as a search for hierarchical turbulence in the Burgers' equation.

The similarity analysis is also greatly used in statistical physics, to predict the shape of the quantity away from the critical point, and to compute the scaling exponents. This is called "Widom scaling", and it is used by Sethna.

2.2.4. *Stable laws*

Levy laws occur when discussing scale invariance from a statistical point of view: if one impose a pdf (probability density function) of a quantity to be scale invariant, one expects a power law distribution $P(x) \sim Cx^{-(1+\alpha)}$ for the variable x. The most famous distributions with power law tails are called *Levy* or *stable* laws, with $0 \leq \alpha \leq 2$, because they are stable by convolution. Some properties of these stable laws are discussed by Cont et al, and by Zajdenweber. A peculiar (and annoying) property of these laws is that they have infinite variance. This led people to consider instead "truncated Levy law" -it is a Levy law regularized by an exponential term to insure convergence of any moments-. Of course, this is a purely descriptive attitude, in a case where extremely large fluctuations are too probable to allow for any realistic predictions. It is not by accumulating data on earthquakes or typhoons that you can predict when your house will be destroyed. This is the fundamental difference between insuring on natural risks and on life: life expectancy and its variance are finite and provide to a life insurer a relevant parameter for predicting his benefit and calculating premia. These points are discussed further in the section "Scale symmetry breaking".

Frisch warns against a common trap in which he saw even famous scientists fell. The central limit theorem has to be used with caution for products of random variables, even when transposed easily in log-variables. Take the product W_n of n independent, identically distributed random variables $m_i; i = 1, ..., n$. The law of large numbers tells you that the *normalized* sum $(1/\sqrt{n}) \sum \log(m_i/ < m_i >)$ tends towards a Gaussian. It would however be incorrect to infer that the *non normalized* sum $\sum \log(m_i)$ is itself converging towards a Gaussian distribution, even if they are many terms: $P(\sum \log m_i)$ has a maximum at $< \log m >$ and is thus of course approximately parabolic around this maximum. Replacing the sum by a Gaussian distribution thus amounts to keep only the first two terms in the Taylor expansion around the mean $< \log m >$. Deviations from this expansion can be very strong, as quantified by the *theory of large deviations*. This point is further discussed in Section 3.2.

What we need to remember, here, is that even if log-normal laws can be widespread in physics or in finance, one must resist the temptation to justify this ubiquity using the central limit theorem for the logarithm of the product of random variables. Alternative to such justification can be found in the lecture of Castaing, Pocheau or Dubrulle, using scale covariance arguments.

2.2.5. Computation of scaling exponents

It is great to get experimentally measured exponents, but it is sometimes even greater to know how to compute them! When they are dimensional, a simple power counting analysis is sufficient. The difficulty arises when they are anomalous. In fact, there is no general procedure to compute them and it is often necessary to invent one procedure per problem (often working in analogy with another already solved problem!). In the present school, two example were provided, by Derrida, and Duplantier. Their presentation are not in this volume. Derrida presented a review of the recently discovered so-called "first-passage" exponents in Ising, Potts and other well-known models. Until recently, it was believed that all the exponents describing the scaling properties of correlation functions at criticality were known. The recent discovery is that there is a novel non-trivial and apparently unrelated exponent describing the first flipping of a spin, or more generally the change of sign of a diffusing random field [4].

Duplantier presented the coulomb gas and conformal invariance techniques that have allowed him and other workers to get a wealth of novel exact results on the scaling properties of 2D manifolds, including a variety of situations involving self-avoiding walks and interacting crumpled manifolds. A few guiding references are given in [5]. The main lesson is that the difficulty in calculating multi-point correlation functions is to control correctly the "fusion" process, i.e. when two of the points in the multi-point correlation function become equal. For instance, in turbulence, when this occurs, this means that one crosses the Kolmogorov dissipation scale in which the viscosity plays an inherent part and must thus be tackled to get access to the correct statistical description of the inertial range.

2.3. Examples

As we have seen earlier, power laws are the hallmark of scale invariance and are the quantitative counterpart of fractal descriptive tools. All around us, physical system display power law, whether in physical properties (metallurgy) or in pdf (biology, economics, geophysics). In some cases, one does not have any sound explanation for them: see e.g. the lecture by Bréchet. In other cases, one has an explanation but it is not the right one: see again the lecture by Bréchet and also Graner). In a few cases, one has just a simplified model (see e.g. the lecture by Marsili or Chopard). In any case, the variety of examples, and their importance in the school (they represented about one half of the school!)

reflect the importance of the topics in all areas of physics, and the need for new tools to study them. We now briefly discuss each of them, with emphasis on physical mechanism generating scale invariance.

2.3.1. *Condensed matter systems*

Sethna presents an example of scale invariance from condensed matter physics, in the presence of quenched disorder and in an out-of-equilibrium situation. This is a situation that turns out to be understood in great depth. Hysteresis, the lag between the force and the response, is often associated with noisy, jerky motion which have recently been called "avalanches". The interesting question is why the avalanches come in such a variety of sizes : naively one would expect either all small events or one large one. Power–law distributions are often seen near transitions, or critical points. He studies the zero temperature random field Ising model as a model for noise and avalanches in hysteretic systems. Tuning the amount of disorder in the system, he finds an ordinary critical point with avalanches on all length scales. The study of this critical point in $6-\epsilon$ dimensions, and with simulations in 2, 3, 4, 5, 7, and 9 dimensions with systems as large as 1000^3 allows him to relate this avalanche phenomenology to the physics of standard critical points. For this result, one should expect that the power–law distributions should in principle only occur for a special value of the randomness (the critical point), but many decades of scaling occur quite far from this special value : three decades of scaling (a publishable experiment) are obtained a factor of two away from the critical value. An explanation of this is suggested. Perhaps many of the power laws observed in nature are just systems near their critical points?

In a similar vein, Bouchaud discusses the large scale phenomenology of disordered systems and the method of turbulent renormalization. The goal is to understand the large-scale effective potential for elastic objects in the presence of a random pinning potential. Different theoretical approaches suggest that the energy landscape at large scales is a succession of parabolic wells of random depth, matching on singular points where the effective force is discontinuous. These parabolas are themselves subdivided into smaller parabolas, corresponding to the motion of smaller length scales, in a hierarchical manner. The energy depth of these local minima is found to be exponentially distributed, which leads to a power-law distribution for the residence times, and in turn to 'aging' at low temperatures. The analogy with Burgers' turbulence is emphasized. In particular, the natural renormalization group description of these problems is a forced Burgers equation. This work thus exhibits an unexpected relation between the physics of quenched disordered systems and turbulence.

The following presentations deal with even more complicated systems, in which scale invariance is obtained but the underlying physical mechanism is not easy to pinpoint as many ingredients and effects compete in general.

Bréchet teaches us that many empirical laws relating either processing conditions to microstructures, or microstructures to macroscopic properties have been identified over the years in metallurgy. These empirical laws have sometimes been derived from theories describing the behavior of the elementary events, but very often their range of experimental validity is wider than the theoretical explanation would predict. Conversely, some observations seem to be extremely robust, irrespectively of the underlying microscopic mechanisms. He suggests that, in many cases, these scaling laws are probably consequences of collective behavior rather than the individual properties of dislocations, grain boundaries... This introductory lecture illustrates, on examples of observed scaling behavior (dislocation structures, mechanical instabilities, fracture mechanisms...), the problem of universality versus specificity in material science. This will also be a theme to keep in mind for the other domains discussed below.

Naively, a paper on liquid water would not seem to be relevant here. However, this ubiquitous fluid, which is far from simple and from being understood, exhibit highly anomalous thermodynamic properties which may probably be attributed to the existence of a percolation network on one hand and of a novel critical point on the other hand. Stanley presents indications, both theoretical, numerical and experimental for the latter. Indeed, liquid water is believed to be a transient polymer gel, with the connectivity lasting only a picosecond. The metastable extension of the phase diagram of liquid water exhibits rich features that manifest themselves in the equilibrium properties of water. For example, the maximum in the density at 4C and the minimum in the isothermal compressibility at 46C are likely manifestations of singularities in the behavior of thermodynamic quantities occurring in the supercooled region. The hypothesis presented here is that the observed supercooling anomalies are caused by a critical point above which two metastable amorphous phases of ice, previously shown to be separated by a line of first order transitions, become indistinguishable. This model, which incorporated the two amorphous ice phases into our understanding of the liquid state of water, provides a more complete picture of the metastable and stable behavior of water.

2.3.2. From cosmology to earthquakes

The presentations that follow discuss systems from natural sciences to social sciences. Such a broad range of applications is a source of new problems and may enrich the same question in different contexts.

Starting from the cosmic scale, Lachièze-Ray presents a brief synthesis on the present evidence of scaling in the distribution of matter in the Universe. Here, we realize that, despite many observational surveys of galaxies and other cosmic objects using many tools (correlation functions, void and counting probabilities, multifractal dimensions) used to characterize this distribution and explore its properties, the distribution of matter in the Universe is still not well described, let alone understood. The various arguments suggesting a possible scale invariance are critically analyzed and their relevance and implications, especially in relation with the various aspects of dynamics and gravitational instability, are questioned. In sum, we do not know yet if the distribution of matter in the universe obeys some sort of scaling or self-similarity. Due to the limitations of the data, this is a domain where progress can only be done through the interplay between observations, their careful analysis and theory.

Peltier brings us back to earth, in fact below earth, by presenting a theoretical model of mantle convection, exhibiting avalanches that come in a large variety of sizes, quantified by a power law distribution. Because of the large temperature variation that is impressed across the mantle (approximately 3500 K) by virtue of its contact with the molten iron outer core, and because the rheology is essentially viscous even though the region is in the solid state, the mantle is convectively unstable. In fact, the Rayleigh number for the system is so high, approximately 10^4 times the critical value for the onset of convection, that the region is vigorously mixed on geological time scales. In axisymmetric spherical geometry and at the appropriate Rayleigh number for the "real" earth, the results of direct numerical simulations are such that the circulation must be characterized as being intensely intermittent. Internal boundary layer instability in such flows leads to the development of "avalanches" of material from the transition zone into the lower mantle and to transitions from the layered state of flow into a "whole mantle" state. It is through the continuous reactivation and suppression of this internal boundary layer instability that the intermittency of the circulation is engendered. This mechanism explains the intrinsic time scale of the so-called "Wilson Cycle" of supercontinent accumulation and dispersal. The origin of the power law distribution of the "avalanches" is not understood but could be related to the spatio-temporal coupling of first-order phase transitions, recently proposed [6].

Earthquakes come also in a large variety of sizes, with a power law distribution known as the Gutenberg-Richter law. The engine driving earthquakes is mantle convection described above, but the physics of earthquakes has more to do with the intermittent rupture of the upper brittle crust rather than with the mantle dynamics. The essential message of Knopoff, similar in philosophy to that of Pocheau, is that one needs to understand the full law including the cross-over rather than just the power law regime. In the case of earthquakes, the Gutenberg-Richter law cannot extrapolate to arbitrarily large values as the exponent is less than one and thus a mechanism (probably different that just the finite size effect) to limit the size of the large earthquakes must be present. There is furthermore the addition problem that most faults on which strong earthquakes occur in Southern California do not display small earthquake activity. Therefore we are obliged to study the self-organization of large earthquakes on time and distance scales for which power-law distributions are verifiably inappropriate. The time scales appropriate to large earthquakes are very long compared with either the time intervals over which earthquake catalog data are available, or with the repetition times of the greatest earthquakes. Knopoff proposes that the physical limits to the sizes of the largest earthquakes are to be found in the geometrical deviations from planarity of the faults. These barriers to large ruptures can store great amounts of deformational stress energy which must ultimately themselves relax on at least three time scales : the time scale of the earthquake rupture itself, the time scale of the rheology involved in causing the precursory clustering of earthquakes and aftershocks, and a time scale of relaxation of stress that is involved in the development of a complex network of secondary faults, which can be estimated to be of the order of tens of thousands of years. Fluctuations in periodicity of strong earthquakes and extended episodes of localized lacunarity, associated with earthquakes on a 2-D network of faults, suggest non-stationarity of the statistics of the largest earthquakes on the decamillenial scale. Fluctuations on the decadal scale may be useful in earthquake prediction studies.

2.3.3. *From biology to finance*

Extinctions of biological species over hundreds of million of years also seem to exhibit intense intermittency, with maybe a power law distribution of size of extinctions per million year. Marsili presents a generalized Lotka-Volterra model to study the co-evolution of a large number of interacting species. In contrast with the Bak-Sneppen extremal model, Marsili's approach is more flexible in not only allowing for a varying number of species but also for an evolving network of interactions between the species. In his work, the predator-prey relationship is not permanent, but rather forms a flexible network. When extinction happens, a species is wiped out as well as its links to its neighbors. A new species is introduced to occupy the niche with new links to randomly chosen neighbors. This model allows him to study the evolution history, especially the rare events when massive extinction happens. Its relevance for an evolving macro-economy is also presented.

The last three presentations focus on the nature of fluctuations in economy and finance. The text of Cont et al. presents a careful study of the probability distribution of returns on financial assets (stocks, currencies, bonds and commodities), where analysis of financial time series is made using techniques such as linear autocorrelation, moments, rank histogram and Hurst exponent. This text complements Zajdenweber's presentation of the evidence for the presence of stable non Gaussian laws, i.e. Lévy laws, in economics. This emphasis on Lévy laws is due to their stability under convolution, i.e. the same law describes the variations over different time scales for instance. This emphasis, which has a long history in finance and economy, is still debated as witnessed in the discussion of Cont et al. Zajdenweber presents some evidence that Lévy laws appear with two symmetrical or nearly symmetrical tails in finance (stock market prices, exchanges rates, futures and options prices) and in spatial economics (gravitational field of a market place). They appear with one tail only, that is with a completely asymmetric density, in many rank-ordering size distributions: firms, incomes (the famous Pareto laws), profits, sales, property losses (either insured or not), business interruptions, etc. Empirical evidences in economics show that the Gaussian law, the sole Lévy law with two moments, is not so frequently encountered. Lévy laws appear with different characteristic (or stability) exponents α. Nearly all the values corresponding to non-Gaussian range $[0-2[$ are found in economics: between 1.5 and 1.7 in most time series of speculative prices and in the size distributions of income, around 1 in spatial economics and in business interruption insurance, less than 1 in fire-hazards and in natural or man-made catastrophes. Those values mean that these economic phenomena are scale invariant without second moments, and sometimes, without first moment (i.e. without mathematical expectation). For the economists and the managers, those values mean that extreme values are really extreme and that they do appear with a relatively high frequency.

Zajdenweber proposes two sources for the Lévy laws in economics: exogenous and endogenous. Exogenous: many economic phenomena are linked to (or triggered by) natural phenomena with Lévy distributions and more generally with fractal behaviors: size distribution of natural resources (ore, oil, etc) and of natural hazards (earthquakes, hurricanes, droughts and floods, river discharges, etc.). Endogenous: in the theory of finance, the value of a firm is the ratio of two random variables, that may be independent or negatively correlated. Even in a non-stochastic universe, the theoretical financial value of a firm is determined by a divisor that is empirically small and worse, that could be zero! Zajdenweber then outlines some consequences, such as the problem of risk control and derivative instrument pricing.

Chopard and Chatigny present an example of a bottom-up approach to finance modeling, a market microsimulation, in the spirit of the lattice models commonly used in physics. They focus their attention on foreign exchange markets, which is one of the most active. They first review the most important technical properties that should be incorporated in their model, with the aim of reproducing the behavior of each actor (traders, banks). The still preliminary results are encouraging as several qualititative features are correctly captured by the model. However, the fluctuations of the exchange rates are too predictable and not rough enough, with a scaling different from the empirical one. This is probably due to the lack of sufficient "noise" in the dynamics of traders (to mimic the effect of external news) and to the absence of imitation rules, such as the "crowd" effect, that are known to lead to large intermittent price fluctuations. An important property that is missing in this "mechanical" approach is that agents are humans and can learn. Thus their rules may evolve with time as a function of the past history. Furthermore, the effect of "self-realization" is not incorporated, i.e. the fact that the belief of agents may lead them to irrational behaviors that finally realize this belief. This is thus a barely touched field with the exciting problem of unraveling the origin of the observed intermittent approximately self-similar price variations.

Incidentally, during a round table Zajdenweber elaborated on the political usage of scale invariance (laughs in the audience). Neo-economists who advocate against the necessity of a state argue that economics should be a purely scale-invariant, contradiction-free hierarchy with equal rules applying from the peanuts vendor to the largest company; on the opposite, keynesians ask for more state policy in order to compensate for the scale-symmetry breaking between a private individual and a large trust. In finance, it is similarly discussed whether an equal treatment should or should not be applied to well-informed day-to-day traders and uninformed slowly reacting investors, which were supposed to be radically different until Mandelbrot's stable Levy laws incorporated all time scales into an unique frame.

3. BEYOND SCALE INVARIANCE

3.1. From scale invariance to scale covariance

So far, we only studied the influence of scale symmetry on objects, and we stressed the importance of power laws. When applying the concept of scale symmetry to physical laws, one needs to proceed from the concept of scale *invariance* to that of scale *covariance*. This new concept applies to laws and require the independence of the *shape* of the laws with respect to changes of scales. Note that this notion is more general, and actually encompass the notion of scale invariance: solutions of scale covariant laws can sometimes be scale *invariant*, i.e. power laws. In general, however, boundary conditions, forcing or dissipation spoil this invariance, and the solution are not power law anymore. The requirement of scale *covariance* is then very useful to study the breaking of scale invariance, as is discussed in the next subsection.

Historically, one of the first use of the notion of scale covariance was made in the framework of critical systems, via the Renormalization Group. This is simple to understand within the "spin-block" construction first introduced by Migdal and Kadanoff, and then systemized by Wilson in a powerful theory. The idea is the following: consider an array of spin $\{S_1(x)\}$ interacting e.g. via the Ising Hamiltonian. At the critical or near the critical point, the correlation length of the system is very large, and the spins are correlated over large distances. A block of spins thus should behave as a single spin. The procedure is then to construct a spin-block $S_\lambda(x)$, obtained by adding λ^d single spins (in d dimensions): $S_\lambda(x) = \sum_1^{\lambda^d} S_1(x)$. This new spin can be seen as a coarse graining of the initial spin: it is a "spin at scale λ". Now, scale symmetry requires that this new spin, obtained by a change of scale, should follow the same kind of laws than the initial group of spin, i.e. up to a possible rescaling of the coupling constants the new spin S_λ has the same Hamiltonian than the initial spins. This rescaling makes all the difference between scale *invariance* and scale *covariance*: the Hamiltonian actually changes (its coupling constants are different), but its shape remains the same (up to irrelevant terms)! The physical solutions at the critical points are then build via this construction as the "fixed points" of the rescaling procedure. They correspond to fully symmetric states under global scale dilations. A similar situation occurs for more general field theories. In the last two decades, the status of the process of "renormalization", allowing to get rid of infinities in the quantum theory of fields, has been clarified: it is now understood that quantum field theories are only effective theories (in contrast to the belief of their founders, including Feynman, who thought of them as fundamental theories) that are essentially critical and the removal of infinities is deeply associated to a scale (or energy) invariance of the theory, only broken at very high and/or very low energies.

More recently, the issue of *scale covariance* has been re-investigated by Pocheau in the domain of turbulence and front flame, to constraint the nature of the laws. He shows an impressive application to the description of complex flame fronts propagating in turbulent gas. He focuses on two stochastic systems satisfying scale-invariance in a statistical sense, namely turbulent propagating interfaces and turbulent flows. His goal consists in exploiting this symmetry to select the physical laws governing these systems. This is achieved by requiring the invariance of laws by scale changes, i.e. their scale covariance. The relevant changes include not only unit changes preserving scale ratio but also scale changes modifying them. In the first system, the turbulent propagating interfaces, the property of scale invariance applies not to the geometry of interfaces nor to the velocity fluctuations of turbulent flows but to the wrinkling process of the former by the latter. The relevance of the selected laws to practical situations is then shown experimentally and the scale covariance of the wrinkling process is directly evidenced. This analysis reveals a deterministic transition from Euclidean to fractal forms which points out the much larger universality of scale covariance compared to its geometrical analog, the scale similarity of fractal forms.

A last example of scale covariance in the domain of turbulence can be found in the second part of the lecture of Castaing. This scale covariance describes how the probability distribution function (pdf) of the logarithm of the velocity difference δu at one scale ℓ is related to the same pdf at another scale ℓ' via convolution by a "propagator", formally equivalent to the "renormalization group" operator, which describes the changes in pdf occurring with scale. The statistical properties of this operator, encoding all the scale information, are studied in the lecture of Arnéodo by wavelet analysis. The description of Castaing in term of propagator is interesting for two reasons. First, it enables a simple and elegant formulation of the pdf in terms of solutions of a Fokker-Planck equation. Second, it provides a general framework for the analysis of "scale symmetry" breaking, occurring when the propagator does not depends on the ratio ℓ/ℓ' but on a more complicated function of ℓ and ℓ', taking into account the presence of cut-offs. This brings us to the second important aspect of "Beyond scale invariance", which we discuss now.

3.2. Scale symmetry breaking

Like in any other symmetry, scale symmetry breaking can occur for a variety of reasons and can take different forms. Two specific aspects were discussed in the course of the school: the appearance of complex exponents generated by the breaking of the continuous scale symmetry into the discrete scale symmetry; the modifications of the laws induced by taking into account the presence of cut offs, or finite size effects.

The first aspect is discussed in details by Sornette. He centers his presentation on a brief classification of the known mechanisms leading to discrete scale invariance, the property of systems to be invariant under powers of specific scaling ratios. He shows how discrete scale invariance leads to complex critical exponents (or dimensions), i.e. to the log-periodic corrections to scaling. After their initial suggestion as formal solutions of renormalization group equations in the seventies, complex exponents have been studied in the eighties in relation to various problems of physics embedded in hierarchical systems. Only recently has it been realized that discrete scale invariance and its associated complex exponents may appear "spontaneously" in Euclidean systems, i.e. without the need for a pre-existing hierarchy. Examples are diffusion-limited-aggregation clusters, rupture in heterogeneous systems, earthquakes, animals (a generalization of percolation) among many other systems. He provides an extensive list of situations where complex exponents have been found. The main motivation for studying discrete scale invariance and its signatures is that it provides new insights in the underlying mechanisms of scale invariance. It may also be very interesting for prediction purposes.

The second aspect was discussed by Nottale, Dubrulle, Pocheau and Cont et al. The deterministic case was explored by Nottale, via a very original formulation which inspired the studies of statistical aspects by Pocheau, and Dubrulle and Graner. Nottale summarizes his past efforts to develop a general theory of scale relativity, encompassing scales in addition to the usual space-time coordinates, with consequences on particle physics, string theory, quantum mechanics, astrophysics and cosmology. He proposes the generalization that space-time has an internal fractal structure, with a small scale and large scale cut-offs (respectively the Planck scale and the size of the Universe). He then builds explicit dilation operators taking into account these cut-off and discusses how the couple *(length trajectory, exponent)* transforms under a change of resolution. His theory still awaits experimental confirmations. It is also still in a developing stage waiting for a fully five-dimensional covariant theory coupling space-time- scale transformations.

Dubrulle summarizes an axiomatic group theoretical approach to scale relativity for random statistical fields. This approach is very similar to that of Nottale, with however a different interpretation of the variables and a different choice of the required symmetries: the variables of interest are the set of exponents that characterize the various moments of a statistical distribution in turbulence, so that the role of the exponent and of the resolution is reversed in this approach compared to that of Nottale. Among other things, this approach exhibit new conserved quantities associated with the scale symmetry, and stress the possibility of apparition of new scaling laws, generalizing the power laws in the case where only a "weak form " of scale symmetry breaking is realized. This provides an elegant explanation of a recently empirically discovered scaling property found in turbulent flows and known as "extended self-similarity" (ESS). This is described by Biferale: in isotropic turbulence, the velocity structure functions present an extended scaling range when one plots one structure function as a function of the other. Another explanation of this extended scaling is provided by Biferale using an extension of the multifractal formalism. A hierarchical cascade model allows him to justify a "generalized ESS" in which the scaling exponents vary as a function of both orders of the structure functions. This provides another general classification in which the multifractal formalism appears as a special case.

Cont et al include finite size effects directly in the pdf of their Levy laws by truncating their range of application, so as to ensure a finite variance. Due to the truncation, the second moment exists and leads, in absence of severe correlations, to the slow convergence to the Gaussian law for large time intervals (months). This slow cross over from (truncated) Lévy laws to a Gaussian law as the time scale increases reconcile the antagonist claims found in the literature, and allows a better fit of the stock market price variations, even at small time scales (5 minutes).

Finally, Frisch discussed in his oral presentation the role of large and of extreme fluctuations in breaking down the statistical scale covariance described by the Gaussian pdf and the central limit theorem of the sum or the product of a large number of random variables. This presentation is not in this volume but can be found in [7]. The concept of "large deviations" is well-studied and is understood as the study of the exponentially-low-probability events where the sum of n independent equally distributed random variables, divided by n, instead of being within $1/\sqrt{n}$ of the mean (as governed by the central limit theorem) is within $O(1)$ of the mean. Extreme deviations are even larger. The motivation to study this regime comes from the observation of stretched exponential probability density functions (pdf), having the form of the exponential minus a fractional power of the argument, commonly found in turbulence and other areas. Frisch showed that they can arise because of an underlying random multiplicative process. For this, a theory of *extreme* deviations has been developed, devoted to the far tail of the pdf of the sum X of a finite number n of independent random variables, with a common pdf. The tail behavior of the sum comes then mostly from individual variables in the sum close to X/n and the tail of the pdf can be explicited. This theory can then be applied to products of independent random variables, such that their logarithms are in the above class, yielding usually stretched exponential tails. Applications to turbulence, power-law distributions of sizes in fragmentation and anomalous relaxation of glasses have been discussed.

4. CONCLUSION

We hoped that the school would help classify and clarify the status on the mechanisms for scale invariance and/or covariance as well as their breaking. This volume does not present a genuine synthesis of the physical mechanisms, even if many physical mechanisms are discussed, and this is probably due to the large variety of origins of scale invariance and the still limited understanding of many systems. The two cases we understand best are dimensional analysis and critical phenomena. Beyond this, we have understanding, sometimes quite detailed, only on specific cases. There is a consensus that a given problem can display simultaneously both a universal and a specific mechanism, ... and there is almost a consensus that both deserve to be studied. Even the general concept of self-organized criticality discussed by Bak in his oral presentation, where he applied it to "explain" stock market price fluctuations, is now seen as a composite of mechanisms. At least, we think that one of the goals of the school has been attained, and that the different approaches can be related to a global and self-consistent theory of scale symmetry, as we tried to suggest in the present foreword.

"Fractals in physics: where is the physics?", asked Procaccia in his synthesis of the school, following the famous title of the Physics Today letter by Kadanoff. As he pointed out, a few years ago the name of Mandelbrot was quoted in all talks on scale invariance, while it was almost absent from this school. Procaccia sees this as an indication that the field is getting mature; in view of the presentations, it seems fair to say that the field has now evolved from descriptive to the beginning of understanding.

References

[1] G.I. Barenblatt, Dimensional analysis (translated from the Russian by Paul Makinen), New York : Gordon and Breach Science Publishers, 1987; Scaling, self-similarity, and intermediate asymptotics, New York : Cambridge University Press, 1996.
[2] U. Frisch, *Turbulence*, The legacy of A.N. Kolmogorov (Cambridge University Press, Cambridge, 1995).
[3] N. Goldenfeld, Lectures on phase transitions and the renormalization group (Addison-Wesley, Advanced Book Program, Reading, Mass., 1992).
[4] B. Derrida, V. Hakim and V. Pasquier, J. Stat. Phys. 85, 763-797, 1996; B. Derrida, V. Hakim and R. Zeitak, Phys. Rev. Lett. 77, 2871-2874, 1996; B. Derrida, V. Hakim and V. Pasquier, Phys. Rev. Lett. 75, 751-754, 1995; B. Derrida, A.J. Bray and C. Godrèche, J. Phys. A 27, L357-L361, 1994.
[5] F. David, B. Duplantier and E. Guitter, Phys. Rev. Lett. 72, 311-315, 1994; B. Duplantier, Physica A163, 158-182, 1990; Phys. Reports 184, 229-257, 1989.
[6] L. Gil and D. Sornette, Phys. Rev.Lett. 76, 3991-3994, 1996.
[7] U. Frisch and D. Sornette, Extreme deviations and applications, J. Phys. I France, in press (1997) (http://xxx.lanl.gov/abs/cond-mat/9705132)

Acknowledgments

This school was organized in the frame of Formation Permanente du CNRS, Direction Régionale Alpes. A large number of students could be financially supported thanks to our sponsors and partners, whom the troika of organizers would like to thank warmly:

- le Ministère français des Affaires Etrangères
- CNRS, via le Groupement de Recherches
- CEA
- Integrated Seismic System International (South Africa)
- le département "Produits dérivés de taux" de la Société Générale, et son directeur Monsieur Sébastien Cahen.

SCALE INVARIANCE

Concept, Tools, Examples

LECTURE 1

Scale Invariance Without Mechanism ?

F. Graner

UMR 5588 du CNRS, Laboratoire de Spectrométrie Physique,
Université Grenoble I - Joseph Fourier, B.P. 87,
38402 Saint-Martin-d'Hères cedex, France

1. INTRODUCTION

A log-log plot of experimental data has the wonderful property of displaying, more often than not, a portion of a straight line. However, there is a widely spread tendency to look for simple, ideal situations; one often interprets a straight line as a scaling law reflecting an important information. We want to stress, on a curious example, how deeply misleading this quest for simple interpretation can be.

Our plan is as follow. We briefly review past attempts at observing, then explaining, a regularity in the distribution of planetary distances to the sun. We try to understand how such question, easy to enounce, can suscitate hundreds of different and sometimes incompatible theoretical explanations. We then show how an argument based on scale invariance helps finding a way in this jungle, identifies the relevent ingredients and guards against pure numerology. Moreover, a judicious utilization of scale symmetry enables to simplify the mathematical treatment of equations, in an elegant way similar to what Henriksen presents in his lecture. We conclude with renewed warnings.

2. MAGIC IN THE SOLAR SYSTEM ?

Quest for regularity and simplicity in the position of planets is not recent, see references in [1, 4]. For instance, within a geocentric cosmology, intervals powers of 3 by Plutarque ($\sim$ 120 AD), as regular musical intervals by Cassius Dio ($\sim$ 200 AD), and as alternating powers of 2 and 3 by Hippolytus around ($\sim$ 230 AD). Within a heliocentric cosmology Kepler assumed in 1596 each planet orbit was a circle circumscribed to one of the five regular polyhedra: cube, tetra-, dodeca-, icosa- and octahedron, until he worked with Tycho Brahe and was convinced that Mars' orbit is actually an ellipse.

The search for an "order" in the solar system was revived in 1766 by Titius, who noticed that planet orbits would follow a geometric progression provided a "missing" planet was inserted between Mars and Jupiter. Reformulation by Bode in 1772, then Wurm in 1787, lead to an expression of the type $r_n = 0.4 + 0.3 \times 2^n$ where the distance r_n of the nth planet to the sun is expressed in astronomical units: $n = -\infty$ for Mercury, $n = 0$ for Venus, $n = 1$ for the Earth, and so forth. A good example indeed of how one can arbitrarily assign an integer number to a planet ! Amidst a never ceasing polemics, interest in this expression was raised by the discovery of Uranus in 1781 and of the first asteroid Ceres in 1801, both close to the locations expected. But Pallas, a second asteroid, in 1802, Neptune in 1846 and Pluto in 1930 were not consistent with this progression of ratio 2.

2.1. A "log versus rank" plot

If one forgets about this magical ratio 2, one can obtain a much better fit of the observed planetary distances, with the additional simplicity of using only positive integer values of n and no more additive constant. See for instance Fig. 1, where the best fits finally seem to be a true geometrical progression:

$$r_n = r_0\, K^n, \tag{1}$$

where r_0 is a normalizing distance and the progression ratio K is constant: for the solar system $K \sim 1.7$, $n = 1$ for Mercury, $n = 2$ for Venus, and so forth. Note that r_0 has no special significance: a shift by any factor K^m, m integer, just results in the shift in the numbering of planets. By the way, similar laws were also searched in the orbits of satellites of the four major planets: K is respectively 1.6, 1.5 and 1.4 for the Jupiter, Saturn and Uranus system [2].

Fig. 1 is not a log-log plot! It is a "log versus integer" plot. This is reminiscent of rank-ordering diagrams (see e.g. the presentation by Potters in this book) but with a major difference. By construction, rank-ordering diagrams are built using cumulated histograms, robust against noise and errors on data sampling. On the opposite, the present Titius-Bode-like diagram is very sensitive to subjective choices. The value $K \sim 1.7$ presented here changes when planets are added or suppressed. Additions consist for instance in including rocks with eccentric orbits, like Chiro; or in splitting the asteroid belt into two subcomponents placed at the maxima of density, instead of one component at the average position (see Fig. 2). Conversely, some authors obtain different fits by removing planets presenting suspicious characteristics (eccentricity, in-

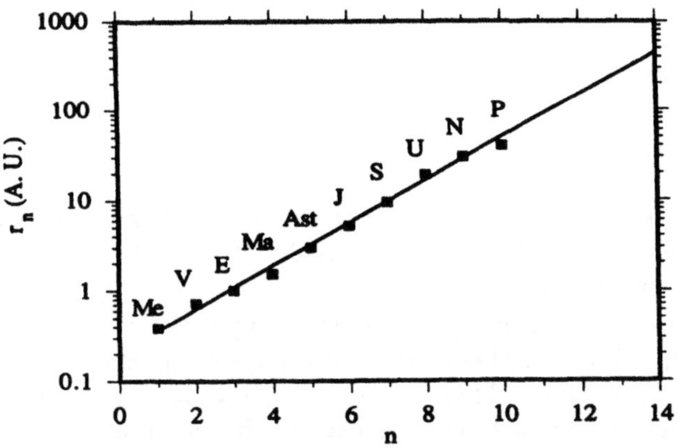

Fig. 1. — A log-versus-rank plot: actual positions of planets in the solar system expressed in Astronomical Units (squares) and best fit by a geometrical law (line).

clination,...) which could be interpreted as evidences for capture or collision events.

Some authors prefer to divide the solar system into two subsystems (inner and outer), with a different normalizing radius r_0. This extra degree of freedom allows for instance Nottale to present in this school an interesting non-geometric progression: as a consequence of long term behavior of chaotic systems, possible orbits for planets are quantified as $r_n = r_0\, n(n+1)$.

2.2. Explanations of this geometrical progression

The difficulty to find a mechanism common to both solar and planetary systems, which are a priori rather different in nature (temperature, density, etc..), does not seem to be a major limitation to the imagination of the theoreticians. There are hundreds of articles published on that subject: see [3, 4] for critical reviews.

Many models (i) start from a flat protoplanetary disk in rotation around a massive sun; (ii) invoke a certain physical mechanism of instability in the disk; (iii) obtain the geometrical progression (1) for mass density maxima; (iv) assume that present planets are still placed at the positions of those maxima; (v) thus use the present observation (1) to set constraints on the physical characteristics of the initial disk.

These models mainly assume one of four physical mechanisms to take place in the protoplanetary disk : planetesimal accretion; competition between gravity and electromagnetic forces; self-gravitational instability; and hydro-dynamic instabilities. Models sometimes contradict each other; some invoke resonances, other anti-resonances between planet orbits.

Facing this tremendous diversity, skeptical authors used an excluded distance procedure (adjacent planets cannot be "too close to each other") to generate random numbers: e.g. Hénon [5] obtained resonances relations, and Lecar [6] a geometrical progression, as good as those observed in the Solar System.

2.3. Comments

In our opinion, this easiness to produce a geometric progression is even more puzzling than the reality of the observation itself. It suggests there is a "hidden" order behind all the explanations.

In fact, within such a tremendous diversity, the above models share a common methodology. Each of them assumes a physical phenomenon to be the origin of planet formation. Then, coming to quantitative predictions, each model needs to add an hypothesis to constrain the unknown physical properties of the primordial system. The most natural assumption, as long as one is totally ignorant, avoids introducing unnecessary parameters. Thus the first reflex is to propose a model with no supplementary length scale, other than the radial length scale r itself. Only Prentice [7] seems to explicitly mention his hypothesis, and apparently no author points out its relevance. Such hypotheses include: homologous collapse, constant eccentricity, disk height H or vortex size $\propto r$, r^{-2} force fields, turbulent viscosity $\nu_t \propto r^{1/2}$, surface density $\sigma \propto r^{-2}$, or instability growth rate $s \propto \sigma\nu_t$. In this list, we must not forget the hypothesis of excluded distance in log-variables used by Hénon and Lecar to discredit the other models!

This method seems reasonable; it is very popular, but simultaneously not innocent, for a very precise reason. Indeed, it amounts to an (implicit !) assumption of scale symmetry, in addition to the explicit assumption of rotation symmetry. This never fails to produce a geometric progression in orbit sizes, even if not desired, and whatever the underlying physical model.

3. HAND-WAVING SYMMETRY ARGUMENT

We are interested in "explaining the explanations" and do not comment the physical relevance of one model or the other. We want to point out that geometrical progressions arise in a whole class of physical problems (see Sornette's lecture); sufficient properties, by no mean necessary, are to be (i) invariant by rotation around an axis Oz, (ii) invariant by scale dilatation in the plane perpendicular to z, and (iii) unstable under a scale-invariant mechanism.

3.1. Generic equations

Consider a given physical quantity g. To use rotational invariance, it is natural to use the Fourier serie decomposition on the basis of azimuthal modes $g(r,\theta) = Re(\Sigma_m a_m(r) \exp[i(m\theta + \theta_m)])$, where the symbol Re stands for real part, m is integer and θ_m real. Analytical functions of g which are invariant by rotation develop as series of the rotational invariant terms $(a_m)^p.(\bar{a}_n)^q$ where the bar denotes the complex conjugate, $(a_m)^p$ denotes the pth power of the mth azimuthal mode, and $mp = nq$.

Analogously, analytical functions of g which are invariant by scale transformations develop over the various scale invariant terms such as g/r^γ or $r(\partial_r g)/g$, as well as terms built with functions of these, e.g. $r^{\gamma-1}\partial_r g$, or higher derivatives, e.g. $r^2(\partial_{r^2} g)/g$.

Quantities invariant by both rotation and scale dilatation are $r(\partial_r a_m)/a_m$, $a_m \bar{a}_m / r^{2\gamma}$, and so on. Any physical equation governing an instability of the system can be written using such invariants [8] or, even better, using only the scale invariant variables:

$$\begin{aligned} b_m &= \frac{a_m}{r^\gamma}, \\ x &= \ln r/r_0, \end{aligned} \tag{2}$$

where r_0 is a normalizing radius. If we assume time independence, the simple example of such generic equation takes the form :

$$\partial_x b_m = b_m G_m(|b_m|^2). \tag{3}$$

where G_m is a regular function.

3.2. Generic solutions

As long as the amplitude of the perturbation is small, the function G_m can be developed in Taylor serie. Neglecting non-linear terms in G_m, straightforward integration yields:

$$b_m = B_0 \left(\frac{r}{r_0}\right)^\mu e^{ik\ln(r/r_0)} = B_0 \exp[(\mu + ik)x]. \tag{4}$$

As expected, the exponent is complex when the continous scale invariance is broken into a discrete one. These symmetry arguments provide no indication regarding the underlying physical mechanism of this symmetry breaking, see section (5.1) and Sornette's lecture.

If r_n and r_{n+1} are the radial location of two consecutive equal phase cylinders, they obey the relation (1):

$$\begin{aligned} x_{n+1} &= x_n + 2\pi/k = x_0 + 2n\pi/k, \\ r_{n+1} &= r_n \exp[2\pi/k] = r_0 K^n. \end{aligned} \tag{5}$$

where the constant $K \equiv \exp[2\pi/k]$ depends on the physics involved in the problem.

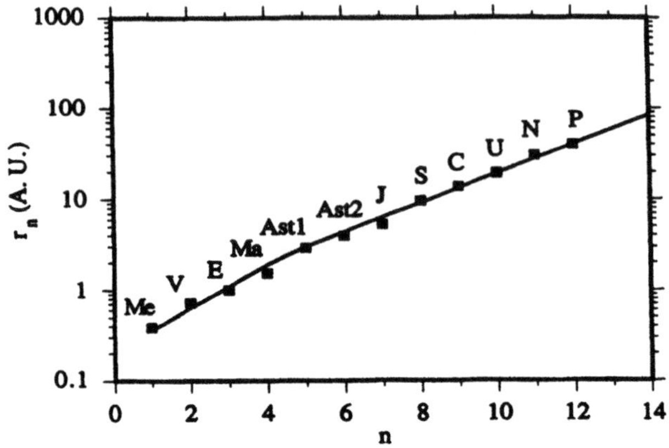

Fig. 2. — Arbitrariness in the definition of planets results in extra degrees of freedom for both fits and models. This figure is similar to Fig. 1, with a more enlarged definition of planets which includes Chiro and two asteroid belts. A solution of (3), including non-linear corrections (continuous line), is fitted by hand to observed positions (squares).

3.3. Comments

This geometric progression is lurking in any rotation & scale invariant system. Therefore, if one assumes that planets are forming at the maxima (or minima) of one of the dynamical variable (the most popular being the density), he will automatically find a close Titius-Bode law within a linear development.

Planet formation is fundamentally a non-linear phenomenon, since it corresponds to a saturation (in the density amplitude). The present symmetry considerations offer no arguments in the debate whether planet positions were determined by linear or non-linear instabilities. For the Titius-Bode hunter, however, nonlinear corrections to G_m would alter the spacing of surfaces of equal phase, i.e. introduce additional parameters. As an illustration, we show in Fig. 2 a flexible fit to the positions of the "total solar system", which is nothing more than numerology: since this list of the planets is of course arbitrary, we attach no physical significance to it.

4. EXAMPLE: THE FLAT, COLD DISK

Up to now we used general symmetry arguments. Let us now turn to practical applications of scale invariance and use it to simplify the resolution of tedious equations.

As an example, we simply chose to treat the cold, flat disk, in which the only physical processes are the self-gravitation competing with the stabilizing Keplerian central gravity field. Two reasons motivate us: first, is is the denominator common to all models reviewed above; second, it is often treated, especially in galactic dynamics, so that we will validate our method against more brutal calculations. But we do *not* believe it is physically pertinent.

4.1. Symmetry of the equations

The most general equations describing such a cold primordial solar nebula are :

$$\begin{aligned} \partial_t \rho + \nabla(\rho \mathbf{v}) &= 0, \\ \partial_t \mathbf{v} + (\mathbf{v}\nabla)\mathbf{v} &= -\nabla\phi - \frac{GM_C}{r^2}\mathbf{e}_r, \\ \nabla^2\phi &= 4\pi G\rho. \end{aligned} \qquad (6)$$

ρ, $\mathbf{v}$, and ϕ are the density, velocity and self-gravitation potential; $\mathbf{e}_r$ is the unit vector in the radial direction, M_C the mass of the central object and G the gravitational constant.

Apart from its cylindrical symmetry, the set (6) exhibits an interesting symmetry. Since it exhibits no other length scale than r itself, it is invariant under the scale transformation:

$$\begin{aligned} S_\Lambda : \quad & r \to \Lambda r, & t \to \Lambda^{3/2} t, \\ & \rho \to \Lambda^{-3}\rho, & v \to \Lambda^{-1/2} v, \\ & \phi \to \Lambda^{-1}\phi, & \lambda \to \Lambda^{-1}\lambda. \end{aligned} \qquad (7)$$

where Λ is an arbitrary real number, and where we have defined a "linear mass density" $\lambda = \int 2\pi r \rho(r,z) dz$. A system invariant through all S_Λ, Λ real, is called a scale invariant system; if a physical solution of (6) is not scale invariant, then at least all its transformed through the S_Λ are also solutions of (6).

In (6) we set $\rho(r,z,t) = \lambda(r,t)\delta(z)/2\pi r$, where δ is the Dirac function; so that the vertical velocity along z vanishes everywhere. Without radial accretion v_r vanishes as well.

4.2. Linear stability analysis

The stability of a self-gravitating disk mostly depends on the ratio between the mass of the central object, M_C, and the disk mass M_D. When $M_D/M_C \ll 1$, the disk is keplerian and stable; when $M_D/M_C \gg 1$, the disk is gravitationally unstable; the threshold in M_D/M_C is of the order of unity.

To study the stability of the disk, we now linearize (6) around its simplest scale-invariant equilibrium state. The calculation is straightforward, using scale-invariant notations(1)

(1) The first attempt at explaining Titius-Bode law that way [9] used a two-dimensional conjugated-space treatment of the Poisson equation which dissimulates

• the variables $x = \ln(r/r_0)$, where r_0 is a still undefined normalizing radius; and a pseudo-time $\tau = r^{-3/2}t$, not homogeneous to a time.
• the functions $\tilde{u} = r^{1/2}v_r$, $\tilde{v} = r^{1/2}v_\theta$, $\tilde{\phi} = r\phi$, $\tilde{\lambda} = r\lambda$.
• the keplerian angular velocity $\Omega_e^2(x) = G(M_C + M_{eg}(x))$. Here $M_{eg} = (\partial_x - 1)\phi_0$ is the effective gravitational mass of the disk felt at the position x, where ϕ_0 is the gravitational potential defined using the equilibrium linear density $\tilde{\lambda} = L_0(x)$. It differs from the real disk mass $M_D = \int_{x_{min}}^{x_{max}} L_0(x)dx$.

Perturbations around the equilibrium state, labeled with an index 1 and with the tilde omitted, obey the linearized equations:

$$\begin{aligned}
\partial_\tau u_1 - 2\Omega_e v_1 &= (1 - \partial_x + \frac{3}{2}\tau\partial_\tau)\phi_1 \\
\partial_\tau v_1 + \frac{1}{2}(\Omega_e + \partial_x\Omega_e)u_1 &= 0 \\
\partial_\tau\lambda_1 - \frac{3}{2}L_0u_1 + (\partial_x - \frac{3}{2}\tau\partial_\tau)(L_0u_1) &= 0.
\end{aligned} \tag{8}$$

4.3. Unstable modes

The resolution of such problem of course depends on boundary conditions, which introduce a coupling between modes and must be treated with due care [10]. The most natural are $\lambda_1(x_{min}) = \lambda_1(x_{max}) = 0$.

Scale invariance translates into a translation invariance for the variable x. A natural choice of eigenfunctions is therefore the set of Fourier modes

$$\{\Psi_k(x) = \exp[ikx];\ k = nk_0;\ |n| = 0, 1, 2...\},$$

where k_0 is the fundamental wavenumber:

$$k_0 = \frac{\pi}{x_{max} - x_{min}} = \frac{\pi}{\ln(r_{max}/r_{min})}. \tag{9}$$

The boundary conditions are then satisfied provided either $r_0 = r_{min}$ or $r_0 = r_{max}$. The solutions of (8) will then be expressed as linear combinations of $\sin(nk_0x)$.

Thanks to scale invariant variables, the temporal stability analysis can be completely calculated. The evolution of a given k component obeys linear equations, where the pseudo-time still appears explicitly. In the early stage, the system is unstable for any k. The thick curve drawn in Fig. 3 delimitates the low wavenumber modes, which are slowly growing and are restabilized at later stages of the evolution, and the higher wavenumber modes, always quickly growing. The transition between the two regimes is characterized by a "bend" in the curve $\omega(k)$. WKB approximation, which holds only for massive disks, small-scale, and at early-stage, validates the present analysis [10].

---

actual exponential divergences along the z axis. Here, using the scale-invariance symmetry enables us to avoid such flaw. We evaluate ϕ only in the $z = 0$ plane, and formally treat only two-dimensional equations in the real space [10].

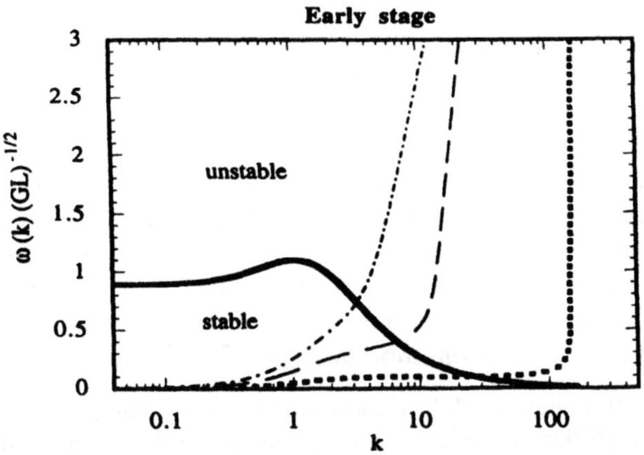

Fig. 3. — Solutions for the flat disk problem obtained using scale symmetry, at the early stage of the evolution. Growthrate ω of instability, expressed in units of $(GL)^{1/2}$, is plotted versus the wavenumber k for various reasonable values of the ratio M_{eg}/M_C: $M_{eg}/M_C = 8.4 \times 10^{-2}$ (dashed dot line); $M_{eg}/M_C = 2.5 \times 10^{-2}$ (dashed line); $M_{eg}/M_C = 2.5 \times 10^{-3}$ (dotted line). Solid line marks the critical growthrate ω_c below which the late stage is algebraically stable.

5. DISCUSSION

5.1. From continuous to discrete scale invariance

Didier Sornette, in his lecture, makes a point of explicitly identifying the underlying physical mechanism wherever he can. He states that the symmetry breaking CSI $\to$ DSI involves a non-trivial mechanism, i.e. a strong constraint on the underlying physics. Hence his question: in the case discussed here, what is the physical mechanism breaking the continous scale invariance (CSI) into a discrete one (DSI) ?

Let us first recall that we are not dealing here with full invariance of dilatation of the scale ℓ, but only with invariance under dilation *of the distance r to an origin*. Now, if one introduces CSI in a model by hand, solutions will obey either CSI or DSI. Observing a complex exponent could constitute a trace of DSI, not a proof of the validity of the physical model. And that is the point of the present work. Since most authors of models missed this point and were happy enough to predict a geometrical progression, they did not question in detail why one mode, and only one, should be selected.

We now split Didier's question into four parts: (i) what is the unstability mechanism in the solar system? (ii) and what selected a particular mode? (iii) and (iv), same questions in our simplistic model? Since we do not claim our model to be physically realistic, answers can sensibly differ.

In the solar system, answer to (i) is unknown. Gravitation certainly played a role! And accretion too, probably. Obviously, at small scales, both mechanisms lead to strong non-linear instabilities, i.e. an existing density fluctuation is drastically enhanced. Thus imaginary parts of exponents (fluctuations) develop. The ultimate non-linear restabilization can come only very late in the history of planet formation: from excluded volume constraints, since present planets are not massive enough for gravitation to make them collapse against atomic repulsion). As for (ii), we will elude this question... since we do not know whether a particular mode is selected at all ! DSI implies eq. (1), which in turns implies an observation similar to fig. 1; but there is no reverse implication, and fig.1 is *not* an unambiguous proof of DSI.

In our cold flat disk model, answer to (iii) is evident, since gravitation provides both a stabilization and a destabilization, and symmetry-breaking occurs when their relative importance is inverted. Answer to (iv) is not evident. Boundary conditions (r_{min}, r_{max}) select a fundamental mode k_0; but the actual selection should be dynamical, and does not occur unless other ingredients are added to the model, as we now discuss.

5.2. Warnings

This model of idealized flat disk does not select any preferred wave length, but only a marginal wavevector k_c. In any real disk, non-linearities introduce characteristic length scales and the resulting k is a priori very far from k_c. Even within the linear regime, more realistic boundary conditions would select a different mode. Moreover, the actual selection of the instability wavelength probably arises from the competition with another physical mechanism, which results in a maximum in the dispersion relation $\omega(k)$, i.e. a mode growing faster than the others. A typical example is the viscous stabilization of large k.

Such processes can be included in the type of model described here without breaking the Titius-Bode law, provided they respect the scale invariance (7). Explicit recipes to cook up a more complex Titius-Bode law include a vertical scale height $H \propto r$, a temperature profile $T \propto r^{-1}$, a turbulent viscosity $\nu_t = \alpha c_s^2/\Omega$ with α constant and sound velocity $c_s \sim r^{-1/2}$, r^{-2} force fields, a vortex size $\propto r$, a state equation $P \propto \rho^{4/3}$, a constant Mach number, etc... Therefore, the possibilities of cooking up Titius-Bode laws are endless and only restricted by our own imagination.

But this kind of models is hardly reconciled with astrophysical constraints. For instance, the surface density predicted can be compared with the "actual" surface density of the solar system protoplanetary disk, of course unknown, but crudely interpolated from masses of present planets and estimated evaporation rates. Both surface densities differ sensibly, indicating lack of physical significance of this kind of models.

6. SUMMARY

We have shown the following results. (i) In any two-dimensional systems characterized by both rotational and scale invariance, the most natural Fourier basis vectors are $\exp[ik \ln(r/r_0)]$. (ii) Any model of planet formation which respects both symmetries will predict planet orbit sizes in a geometric progression of the form $r_n = r_0 K^n$. This is independent of its physical content and does not constitute a proof of its validity. (iii) Observed value of K is as at worst numerology, at best very sensitive to subjective choices. Predicted value of K is extremely sensitive to various parameters such as disk mass, radial and vertical density distribution, but also boundary conditions, viscosity, temperature, or chemical composition. Comparison between observed and predicted values does not set any relevant constraint on cosmogonic models of solar system formation. (iv) In addition, we showed how scale invariant variables simplify an explicit calculation, here on the academic example of a self-gravitating disk.

Acknowledgments

The author wishes to thank B. Dubrulle who is co-author of the work presented here.

References

[1] Jaki, S. *Amer. J. Phys.* **40**, (1972) 93
[2] Neuhäuser, R. and Feitzinger, J. *Astron. Astrophys.* **170**, (1986) 174
[3] Nieto, M. *The Titius-Bode law of planetary distances: its history and theory*, Pergamon Press, Oxford, 1972
[4] Graner, F. and Dubrulle, B., *Astron. Astrophys.* **282**, (1994) 269
[5] Hénon, M. *Icarus* **11**, (1969) 93
[6] Lecar, M. *Nature* **242**, (1973) 318
[7] Prentice, A., in S. Dermott (ed.), *The Origin of the Solar System*, Wiley, New York, 1977
[8] Gückenheimer, J. and Holmes, P., *Non linear oscillations, dynamical systems and bifurcations of vector fields*, Springer, Berlin, 1983
[9] Fridman, A. and Polyachenko, V., *Physics of gravitating systems*, Springer, 1984
[10] Dubrulle, B. and Graner, F., *Astron. Astrophys.* **282**, (1994) 277

LECTURE 2

Scale Invariance and Beyond: What Can We Learn from Wavelet Analysis ?

A. Arneodo, B. Audit, E. Bacry*, S. Manneville, J.F. Muzy and S.G. Roux

Centre de Recherche Paul Pascal, avenue Schweitzer, 33600 Pessac, France
** Centre de Mathématiques Appliquées, École Polytechnique, 91128 Palaiseau, France*

1. FROM GLOBAL TO LOCAL CHARACTERIZATION OF THE REGULARITY OF FRACTAL FUNCTIONS

In many situations in physics as well as in some applied sciences, one is faced to the problem of characterizing very irregular functions [1–8]. The examples range from plots of various kind of random walks, *e.g.* Brownian signals [9], to financial time-series [1], to geological shapes [1,6], to medical time-series [5], to interfaces developing in far from equilibrium growth processes [3,4,8], to turbulent velocity signals [7,10] and to "DNA walks" coding nucleotide sequences [11,12]. These functions can be qualified as *fractal functions* [1,2,9] whenever their graphs are fractal sets in $\mathbb{R}^2$ (for our purpose here we will only consider functions from $\mathbb{R}$ to $\mathbb{R}$). They are commonly called *self-affine* functions since their graphs are similar to themselves when transformed by anisotropic dilations: $\forall x_0 \in \mathbb{R}, \exists H \in \mathbb{R}$ such that for any $\lambda > 0$, one has

$$f(x_0 + \lambda x) - f(x_0) \simeq \lambda^H (f(x_0 + x) - f(x_0)). \tag{1}$$

If f is a stochastic process, this identity holds in law for fixed λ and x_0. The exponent H is called the *roughness* or Hurst exponent [1,2,4]. The graph of f is *self-similar* if it is invariant under some isotropic dilations, *i.e.*, when the Hurst exponent $H = 1$. Let us note that if $H < 1$, then f is not differentiable and the smaller the exponent H, the more singular f. Thus the Hurst exponent provides indication of how globally irregular the function f is.

Different methods [4] (*e.g.*, height-height correlation function, variance and power spectral methods) have been used to estimate the roughness exponent which is supposed to be related to the fractal dimension $D_F = 2-H$ of the graph of the considered function. However, D_F as well as H are global quantities that do not account for possible fluctuations (from point to point) in the local regularity of f. To describe these ***nonhomogeneous*** functions, one thus needs to change slightly the definition of the Hurst regularity so that it becomes a local quantity [13–16]. This "local roughness exponent" $h(x_0)$ corresponds to the mathematical notion of ***Hölder exponent***. It is defined as the largest exponent such that there exists a polynomial $P_n(x - x_0)$ of order n satisfying [17–20]

$$|f(x) - P_n(x - x_0)| \leq C|x - x_0|^h, \tag{2}$$

for x in the neighborhood of x_0. If $h(x_0) \in]n, n+1[$, one can easily prove that f is n times, but not $n+1$ times differentiable at the point x_0. Thus the higher the exponent $h(x_0)$, the more regular the function f at the point x_0.

2. STATISTICAL ANALYSIS OF THE REGULARITY OF FRACTAL FUNCTIONS: THE MULTIFRACTAL FORMALISM

The *multifractal* formalism [21–24] has been originally established to account for the statistical scaling properties of singular measures arising in various situations in physics, chemistry, geology or biology. Notable examples include the invariant probability distribution on a strange attractor [21–24], the distribution of voltage drops across a random resistor network [2], the distribution of growth probabilities on the boundary of a diffusion-limited aggregate [2,3] and the spatial distribution of dissipative regions in a turbulent flow [10,23,25,26]. This formalism lies upon the determination of the so-called $f(\alpha)$ ***singularity spectrum*** [21] which characterizes the relative statistical contribution of each singularity of the measure.

In the past few years, the multifractal approach has proved particularly fruitful in many contexts. However, as pointed out previously, fractals appear not only as singular measures but also as singular functions. There have been several attempts to extend the concept of multifractal to singular functions [13,14]. In the context of fully developed turbulence [10], the intermittent character of turbulent velocity signals was investigated by calculating the moments of the probability density function (pdf) of (longitudinal) velocity increments $\delta v_l(x) = v(x + l) - v(x)$, over inertial separation [10,13,27]:

$$S_p(l) =< \delta v_l^p > \sim l^{\zeta_p}. \tag{3}$$

By Legendre transforming the scaling exponents ζ_p of these ***structure functions*** (SF's) of order p, one gets the Hausdorff dimension $D(h)$ of the subset of $I\!R$

for which velocity increments behave as $\delta v_l \sim l^h$ [13]:

$$D(h) = \min_p (ph - \zeta_p + 1). \tag{4}$$

In a more general context, $D(h)$ will be the spectrum of Hölder exponents for the singular signal under study and thus will have a similar status than the $f(\alpha)$ singularity spectrum for singular measures. However, as discussed in the next section, there are some fundamental limitations to the SF approach which intrinsically fails to fully characterize the $D(h)$ singularity spectrum [28].

3. A THERMODYNAMICS OF FRACTAL SIGNALS BASED ON WAVELET ANALYSIS

The *wavelet transform* (WT) is a mathematical technique introduced in signal analysis in the early eighties [29]. Since then, it has been the subject of considerable theoretical developments and practical applications in a wide variety of fields [30–36]. The WT has been recently emphasized as a mathematical microscope that is well adapted to reveal the hierarchy that governs the spatial distribution of the singularities of multifractal measures. What makes the WT of fundamental use in the present study is that its singularity scanning ability equally applies to singular functions than to singular measures [15,16,36–38].

3.1. Singularity detection and processing with wavelets

The WT is a space-scale analysis which consists in expanding signals in terms of *wavelets* which are constructed from a single function, the ***analyzing wavelet*** ψ, by means of translations and dilations. The continuous WT of a real-valued function f is defined as [29]

$$T_\psi[f](x_0, a) = \frac{1}{a} \int_{-\infty}^{+\infty} f(x)\, \psi\left(\frac{x - x_0}{a}\right) \mathrm{d}x, \tag{5}$$

where x_0 is the space parameter and $a(> 0)$ the scale parameter. ψ is generally chosen to be well localized in both space and frequency. Usually, ψ is required to be of zero mean for the WT to be invertible. The main advantage of using the WT for analyzing the regularity of a function, is its ability to be blind to smooth (polynomial) behavior by an appropriate choice of the analyzing wavelet ψ. Indeed, let us assume that according to (2), f has a local scaling exponent $h(x_0)$ at the point x_0; then one can easily prove that the local behavior of f is mirrored by the WT which locally behaves like [17–20,37]:

$$T_\psi[f](x_0, a) \sim a^{h(x_0)}, \quad a \to 0^+, \tag{6}$$

provided $n_\psi > h(x_0)$, where n_ψ is the number of vanishing moments of ψ. Thus, one can extract $h(x_0)$ from a log-log plot of the WT amplitude versus the scale a. Throughout this paper, we will mainly use the class of analyzing wavelets defined by the successive derivatives of the Gaussian function [15,16].

3.2. The wavelet transform modulus maxima method

A natural way of performing a multifractal analysis of fractal functions consists in generalizing the "classical" box-counting techniques using wavelets instead of boxes[15,16]. By taking advantage of the freedom in the choice of the "generalized oscillating boxes" that are the wavelets, one can hope to get rid of possible smooth behavior that could mask singularities or perturb the estimation of their strength h. But the major difficulty consists in defining a covering of the support of the singular part of the function with our set of wavelets of different sizes. The wavelet transform modulus maxima (WTMM) method [15,16,20,28,39] consists in building a partition function from the modulus maxima of the WT. The maxima are defined, at each scale a, as the local maxima of $|T_\psi[f](x,a)|$ considered as a function of x. As illustrated in Figure 1(c), these WTMM are disposed on connected curves called *maxima lines*. Let us define $\mathcal{L}(a_0)$ as the set of all the maxima lines that exist at the scale a_0 and which contain maxima at any scale $a \leq a_0$. An important feature of these maxima lines is that, each time the analyzed signal has a local Hölder exponent $h(x_0) < n_\psi$, there is at least one maxima line pointing towards x_0 along which (6) holds [19,20]. In the case of fractal signals, we thus expect that the number of maxima lines will diverge in the limit $a \to 0^+$. In fact, as emphasized in Refs. [15,16,39], the branching structure of the WTMM *skeleton* in the (x, a) half-plane enlightens the hierarchical organization of the singularities. The WTMM method consists in taking advantage of the space-scale partitioning given by this skeleton to define the following partition function:

$$Z(q,a) = \sum_{l \in \mathcal{L}(a)} \left(\sup_{\substack{(x,a') \in l \\ a' \leq a}} |T_\psi[f](x,a')| \right)^q , \quad \forall q \in I\!R. \tag{7}$$

From the scaling behavior of $Z(q,a)$ one defines the scaling exponents $\tau(q)$ as:

$$Z(q,a) \sim a^{\tau(q)}, \quad a \to 0^+. \tag{8}$$

Then, by using both the WT scaling behavior (6) along the maxima lines and the definition of the singularity spectrum $D(h) = \dim_H\{x | h(x) = h\}$, where $\dim_H$ stands for the Hausdorff dimension, one can show that $D(h)$ can be determined by Legendre transforming $\tau(q)$ [20,40]:

$$D(h) = \min_q (qh - \tau(q)). \tag{9}$$

From the properties of the Legendre transform, it is easy to convince oneself that *homogeneous* fractal functions that involve singularities of unique Hölder exponent $h = \partial\tau/\partial q$, are characterized by a *linear* $\tau(q)$ spectrum. On the contrary, a *nonlinear* $\tau(q)$ curve is a signature of nonhomogeneous functions that display *multifractal* properties, in the sense that the Hölder exponent $h(x)$ is a fluctuating quantity that depends upon the spatial position.

3.3. The structure function approach versus the WTMM method

It is tempting to relate the exponents $\tau(q)$, defined from the WT partition function (7,8), to the scaling exponents ζ_p of the structure functions (3). A simple comparison of (4) and (9) gives immediately:

$$\tau(q) = \zeta_q - 1. \tag{10}$$

But this relationship does not hold for all values of q [28]. As pointed out in previous studies [28,41], the local increment of a function can be seen as its WT computed with the "poor man's wavelet" $\Delta(x) = \delta(x-1) - \delta(x)$. In this spirit, the SF's (3) are analogous to the naive partition functions:

$$S_q(l) = \int |\delta f_l(x)|^q dx. \tag{11}$$

There are two main differences between (7) and (11) from which result the insufficiencies of the SF method [28].

(i) *The continuous integral used to define* $S_q(l)$. Indeed, there is no reason, *a priori*, that the increment pdf vanishes around $\delta f_l = 0$. Thus $S_q(l)$ generically diverges for $q \leq -1$. The first crucial advantage of the WTMM method is that the partition function (7) is computed using a discrete summation over the WT skeleton where the WT coefficients do not vanish (by definition); this keeps the calculation of $Z(q,a)$ clear of divergences for $q < 0$.

(ii) *The poorness of the analyzing wavelet* $\Delta(x)$. The second main disadvantage of the SF method is that $\Delta(x)$ does not satisfy the criteria to be an efficient analyzing wavelet. First $\Delta(x)$ is only orthogonal to constants but not to higher order polynomials ($n_\psi = 1$). This precludes the detection of singularities with $h \geq 1$. Moreover, $\Delta(x)$ is a singular analyzing wavelet which is not adapted to study irregular distributions which involve singularities with $h \leq 0$. Because one has the freedom to select an analyzing wavelet ψ which is smooth and which has enough vanishing moments, the WTMM method does not possess any of these drawbacks [28].

4. NUMERICAL AND EXPERIMENTAL APPLICATIONS OF THE WTMM METHOD

The WTMM method has been tested on pedagogical examples, *e.g.*, generalized devil's staircases, synthetic multifractal signals and fractional Brownian motions [15,16,20,28,39]. It has already been successfully applied to numerical and experimental data from various domains [15,16,36]. Here we will report on recent results obtained in two unrelated fields, namely fully developped turbulence and DNA sequences.

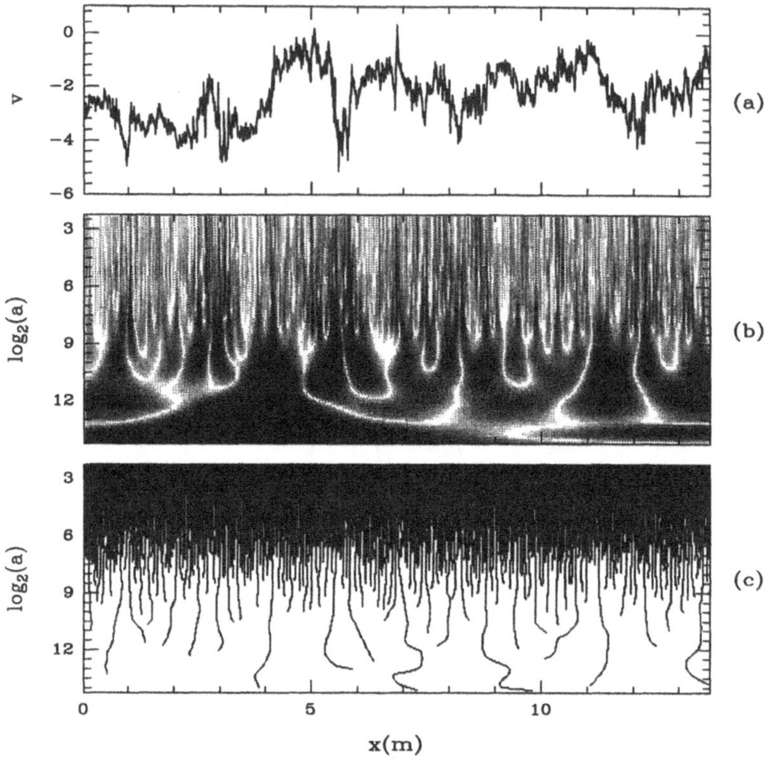

Fig. 1. — Continuous WT of fully developed turbulence from wind tunnel data. (a) The turbulent velocity signal over about two integral scales. (b) WT of the turbulent signal; the amplitude is coded, independently at each scale a, using 32 grey levels from white ($|T_\psi[v](x,a)| = 0$) to black ($\max_x |T_\psi[v](x,a)|$). (c) WT skeleton defined by the set of all the maxima lines. In (b) and (c), the small scales are at the top. The analyzing wavelet is a first-order ($n_\psi = 1$) compactly supported wavelet [47].

4.1. Fully developed turbulence

It is now well-accepted [10,27] that in the fully developed regime, a turbulent flow is likely to be in a universal state that can be experimentally characterized by statistical quantities such as the scaling exponents ζ_p of the longitudinal velocity structure functions or the shape of the velocity increment pdf's. For more than thirty years, one of the main features recognized experimentally, is the intermittency of small scales [7,10,26,27]; this phenomenon manifests in a nonlinear behavior of ζ_p that does not seem to depend on the Reynolds number [42–44]. This observation reveals some departure of the experimental data from the prediction $\zeta_p = p/3$ of Kolmogorov (K41) [45] based on the assumption

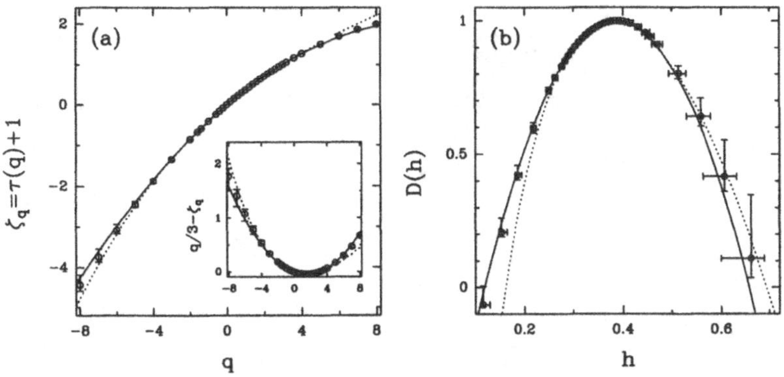

Fig. 2. — WTMM measurement of the $\zeta_q = \tau(q)+1$ and $D(h)$ spectra of the Modane turbulent velocity signal. (a) ζ_q vs q. (b) $D(h)$ vs h. The solid lines correspond to a fit of the data with the theoretical quadratic spectra for log-normal random cascade processes. The dotted lines correspond to the predictions of the She and Leveque's model [51] when assuming log-Poisson statistics for the energy dissipation (see text).

that, at each point of the fluid, the velocity field has the same scaling behavior $\delta v_l(x) \sim l^{1/3}$, which yields the well known $E(k) \sim k^{-5/3}$ energy spectrum. This nonlinear ζ_p spectrum actually characterizes some evolution of the velocity increment pdf from a Gaussian shape at large scales to stretched exponentials towards smaller scales in the inertial range [43,46]. Here we report on the use of the WTMM method to achieve a multifractal analysis of single point velocity data from high Reynolds three-dimensional turbulence [15,16,47].

The data were obtained by Gagne and collaborators in the large wind tunnel S1 of ONERA at Modane. The Taylor scale based Reynolds number is $R_\lambda \simeq 2000$ and the extent of the inertial range following approximately the Kolmogorov $k^{-5/3}$ law is about four decades . Figure 1 illustrates the WT of a sample of the (longitudinal) velocity signal of length of about two integral scales, when using the Taylor hypothesis [10,43]. In Figure 2(a) are reported the results of the estimate of the $\zeta_q = \tau(q) + 1$ scaling exponent of the partition function $Z(q, a)$ (7), as computed with the WTMM method for an overall statistical sample of about 28000 integral scales. As advocated by Benzi *et al.* [48], we have used the extended self-similarity (ESS) concept to improve scaling: $Z(q,a)/Z(0,a) \sim (Z(3,a)/Z(0,a))^{\zeta_q}$ (fixing $\zeta_3 = 1$). Over a range of values of q that extends from -8 to $+8$, for which statistical convergence is achieved, the data remarkably fall on a nonlinear quadratic curve $\zeta_q = mq - \sigma^2 q^2/2$ ($m = 0.388, \sigma^2 = 0.036$), as predicted for log-normal random cascade processes [47,49,50]. In the insert of Figure 2(a), the deviation of the ζ_q's from the K41 linear spectrum $q/3$, is shown to be significant and quite compatible with a quadratic correction. In Figure 2(a) is also shown for comparison the ζ_q spec-

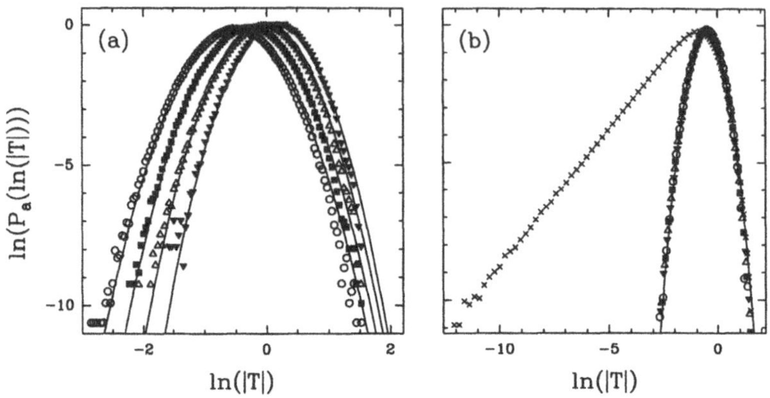

Fig. 3. — Probability density functions of the WTMM of the Modane turbulent velocity signal. (a) $\ln(P_a(\ln(|T|)))$ vs $\ln|T|$ as computed at different scales $a = 385\eta$ (○), 770η (■), 1540η (△), and 3080η (▼). (b) The same pdf's after being transformed according to (12) with a Gaussian kernel $G_{aa'}(x)$ with $s(a,a') = (a'^{-\beta} - a^{-\beta})/\beta$ where $\beta = 0.095$. The (×) represent the corresponding velocity increment pdf. The solid lines correspond to the Gaussian approximations of these histograms.

trum predicted by She and Leveque [51], $\zeta_q = q/9 + 2[1 - (2/3)^{q/3}]$, when postulating log-Poisson statistics for the local nondimensional energy dissipation [52,53] and assuming the validity of the local Kolmogorov scaling relation (K62) [49]. This model provides also a rather good fit of the data for ζ_q, even though some systematic deviation becomes perceptible for $q \geq 5$. The $D(h)$ singularity spectrum obtained by Legendre transforming the experimental $\tau(q)$ spectrum is shown in Figure 2(b). Its characteristic single humped shape over a finite range of Hölder exponents, $h \in [0.12, 0.68]$, is a clear signature of the multifractal nature of the turbulent signal. Its parabolic shape strengthens the possible relevance of log-normal random cascade processes. For $q = 0$, the largest dimension is attained for singularities of exponent $h = 0.39 \pm 0.01$, *i.e.*, a value which is slightly higher than the K41 prediction $h = 1/3$. Moreover, the corresponding maximum of the $D(h)$ curve, $D(h(q=0)) = -\tau(0) = 0.999 \pm 0.002$, suggests that the turbulent velocity signal is singular everywhere. These observations are corroborated by the robustness of the $D(h)$ data with respect to changes in the choice of the analyzing wavelet.

In order to go further in the investigation of the velocity fluctuations, we have computed the pdf's of WT coefficients at different scales in the inertial range [47]. Very much like the velocity increment pdf's, these distributions are nearly centered with a shape that goes from Gaussian at large scales to more intermittent profiles with stretched exponential-like tails at smaller scales. As emphasized in Refs. [15,16,19,47], a way to break free from the redundancy of the continuous WT is to use the WTMM representation. In Figure 3(a) are reported the results of the computation of the WTMM pdf's when restricting our analysis to the WT skeleton (Fig. 1(c)). Since by definition the WTMM are different from zero, the so-obtained pdf's decrease exponentially fast to zero at zero, which makes the estimate of the exponents ζ_q tractable for $q < 0$ (7). When plotting $\ln P_a(|T|)$ vs $\ln |T|$, one can see that for any scale in the inertial range, all the data points fall on a parabola; this is a strong indication that the statistics of the logarithm of the WT coefficients is Gaussian with a mean and a variance that depend upon the scale parameters [47]. This implies that, along the line of Castaing *et al* ansatz [46,54], the WTMM pdf at a given scale a can be expressed in terms of the pdf at a coarser scale a' as:

$$P_a(T) = \int G_{aa'}(\ln r) \frac{1}{r} P_{a'}\left(\frac{T}{r}\right) \mathrm{d}\ln r, \tag{12}$$

where $G_{aa'}$ is a Gaussian kernel that depends upon a and a' only through a function $s(a, a') = s(a) - s(a')$ that accounts for the number of steps of the cascade from scale a' to scale a [54]. Using Novikov's definition [55], scale invariance amounts to write $s(a, a') = \ln(a/a')$, *i.e.*, $s(a) = \ln a$. In Figure 3(b), we have succeeded in collapsing all the WTMM pdf's, computed at different scales, onto a single parabolic curve when using (12) and $s(a) = (1 - a^{-\beta})/\beta$ with $\beta = 0.095$. This is some unambiguous evidence for scale invariance breaking [47,54,56]. We have reproduced our analysis for other turbulent flows with a Reynolds number ranging from $R_\lambda \simeq 200$ to 3000; when increasing R_λ, the exponent β is found to decrease to zero, which suggests that scale invariance is only reached asymptotically in the limit $R_\lambda \to +\infty$ (in the limit $\beta \to 0$, $s(a) = (1 - a^{-\beta})/\beta$ reduces to $\ln a$). Let us note that according to these findings, one would expect to observe some departure from scaling of the partition function $Z(q, a)$ at finite Reynolds number. Hopefully this breaking of scale invariance does not invalidate the ESS hypothesis [47]. This means that the WTMM estimates of the multifractal spectra ζ_q and $D(h)$ in Figure 2 are likely to be valid in the infinite Reynolds number asymptotic limit.

4.2. DNA sequences

The possible relevance of scale invariance and fractal concepts to the structural complexity of genomic sequences is the subject of considerable increasing interest [12]. During the past few years, there has been intense discussion about the existence, the nature and the origin of *long-range correlations* in DNA sequences. Different techniques including mutual information functions [57], autocorrelation functions [58], power spectra [59], "DNA walk" representation [11,12], Zipf analysis [60] were used for statistical analysis of DNA sequences. But despite the effort spent, there is still some continuing debate on rather struggling questions. In that respect, it is of fundamental importance to corroborate the fact that the reported long-range correlations are not just an artefact of the compositional heterogeneity of the genome organization [61–63].

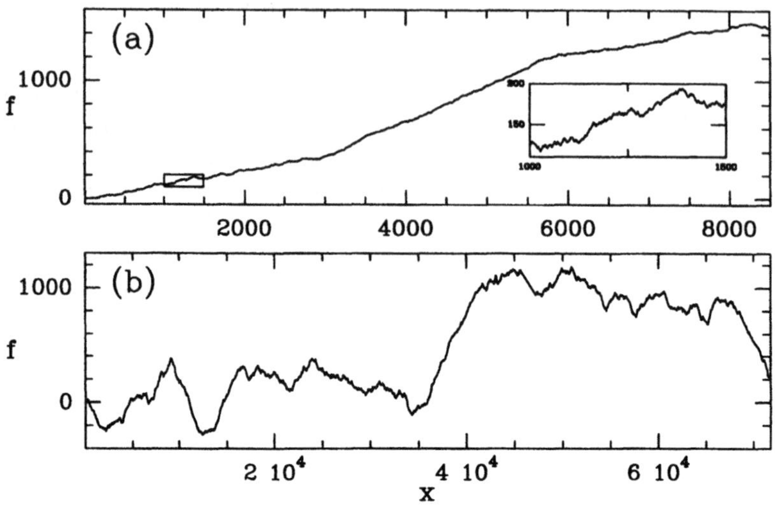

Fig. 4. — The "DNA walk" representation of genomic sequences based on the purine-pyrimidine distinction. (a) The human desmoplakin I CDS ($L = 8499$). (b) The largest intron ($L = 71718$) in the human retinoblastoma susceptibility gene

Furthermore, it is still an open question whether the long-range correlation properties are different for protein-coding (exonic) and noncoding (intronic, intergenic) sequences [11,12,57–60,64].

One of the main obstacles to long-range correlation analysis is the *mosaic structure* of DNA sequences which are well known to be formed of "patches" ("strand bias") of different underlying compositions [65]. As illustrated in Figure 4, these patches appear as trends in the "DNA walk" landscapes and are likely to introduce some breaking of scale invariance [61–63]. Most of the technique used so far for characterizing the presence of long-range correlations are not well adapted to study patchy sequences. In a previous work [66], we have emphasized the wavelet transform as a very powerful technique for fractal analysis of DNA sequences. By taking advantage of the freedom in the choice of the analyzing wavelet, one can use wavelets that oscillate enough in order to make the WT microscope blind to low frequency trends. Here we report on recent results obtained by applying the WTMM method to various genomic sequences selected in the human genome [67,68]. To graphically portray these sequences, we follow the so-called "DNA walk" analysis [11] which requires first to convert the four letter (A,C,G,T) text into a binary sequence. This can be done, for example, on the basis of purine (A,G) vs pyrimidine (C,T) distinction, by defining the incremental variable that associates to position i the value $\chi(i) = 1$ or -1, depending on whether the ith nucleotide of the sequence is a purine or a pyrimidine. (We refer the reader to Refs. [67,68] for similar analysis with the two complementary pair-base identifications). In Figure 4 are illustrated the "DNA walk" graphs corresponding to both a protein coding and a non coding sequences. The patchiness of the coding sequence in Figure 4(a) is patent; one clearly recognizes three regions of different bias. When applying the WTMM method to this coding "DNA walk" landscape, one gets a $\tau(q)$ spectrum (8) which remarkably falls on a straight line (the hallmark of homogeneous fractal signals) of slope $h = \partial\tau/\partial q = 0.50 \pm 0.01$, provided the analyzing wavelet has at least two vanishing moments ($n_\psi = 2$) in order to be orthogonal to the observed linear trends [67]. A similar result is obtained for the noncoding sequence in Figure 4(b), except that the slope of the $\tau(q)$ spectrum $h = 0.60 \pm 0.02$, is now significantly larger than 1/2. These results are quite representative of the results obtained for our statistical of 47 coding and 74 noncoding sequences [67,68]. When averaging the partition functions over these two statistical samples, we get $\bar{Z}_{\mathrm{C}}(q,a)$ and $\bar{Z}_{\mathrm{NC}}(q,a)$ which both scale with the exponent predicted for homogeneous fractional Brownian motions, *i.e.*, $\tau(q) = qH - 1$ [15,16,67]. Within that prospect, we have studied the probability distribution function of wavelet coefficient values $P(T_{\psi^{(2)}}(.,a))$, as computed at a fixed scale a in the fractal scaling range, and found that it perfectly fits a Gaussian distribution for both the coding and noncoding DNA sequences [67]. The main difference which allows us to distinguish coding from noncoding sequences is the presence of long range correlations in the latter: $\bar{H}_{\mathrm{NC}} = 0.59 \pm 0.02$, while the former look like uncorrelated random walks: $\bar{H}_{\mathrm{C}} = 0.51 \pm 0.02$ [67,68].

In Figure 5 are reported the results of a similar statistical WTMM analysis when classifying these DNA sequences into categories that correspond to a given GC content [67,68]. When investigating the scaling behavior of $a^{1/2}Z(1,a) \sim a^{H-1/2}$ for our set of introns, one notices some significant tendancy of the curves to become steeper when continuously increasing the GC content. The corresponding values of the roughness exponent H are reported in Figure 5(a). H clearly increases from values close to 1/2 at low GC content (~30%) up to values significantly larger than 0.6 at high GC content (>60%). In Figure 5(a) are also shown the estimates of H for the coding sequences. Whether the CDS be poor or rich in GC, it does not seem to possess strong long-range correlations as indicated by an exponent H close to 1/2. Because of the "period three" codon structure of coding DNA, it is natural to investigate separately the three subsequences relative to the position (1, 2 or 3) of the bases within their codons. We have built up these subsequences from our 35 largest CDS sequences and we have repeated the WTMM analysis [68]. As shown in Figure 5(b), for the first and the second subsequences, one gets results quite consistent with the estimates obtained with the overall CDS: whatever the GC content, H does not significantly depart from the value 1/2. For the third subsequence, H is found to increase up to values close to 0.60 at high GC content, which brings the clue that this subsequence exhibits [GC]-dependent long-range correlations very much like those observed in Figure 5(a) for introns.

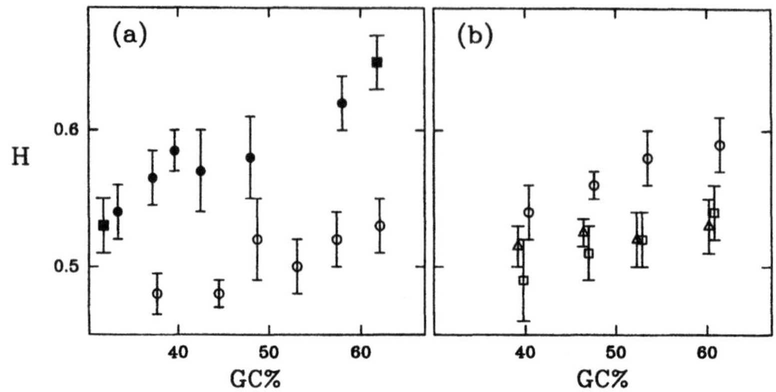

Fig. 5. — Roughness exponent H vs the GC content of the DNA sequence. (a) Introns: (•) $L \simeq 50000$, (■) $L \simeq 15000$; CDSs: (o) $L \simeq 50000$. (b) Coding subsequences relative to position 1(△), 2(□) and 3(o) of the bases within the codons: $L \simeq 20000$. L is the number of nucleotides involved in our WTMM statistical analysis.

The results of our WTMM analysis clearly show that the GC content is likely to be relevant to the long-range correlation properties observed in both intronic and exonic DNA sequences. This observation is likely to be of fundamental importance since it suggests that the long-range correlations are related to the mechanisms involved in the "isochore structure" formation of the human genome [65,69]. The human genome is well known to be compartimentalized into wide ($\sim 10^5$ nucleotides) specific domains with uniform GC content, called *isochores*; appreciable scatter of the isochore GC content is actually observed when comparing different domains. Thanks to the functional constraints acting on coding sequences embedded in the GC rich regions, these non-equilibrium processes should be less active on exons, with a concomittant lack of long-range correlations as compared to the surrounding introns. Since these constraints are less stringent on the third base of the codons (most of the synonymous in the genetic code are due to the change of the third base in the codon), this would explain the correlations observed between these nucleotides in high GC containing exons. This observation favors the hypothesis that the exonic correlations are produced by the same mechanisms which lead to intronic correlations.

5. FURTHER DEVELOPMENTS AND PERSPECTIVES

As we have just seen in the previous section, the application of wavelet analysis has already produced very promising results in various contexts. Let us emphasize that it has also been successfully applied to many other fields including fractal analysis of functions involving oscillating singularities [70,71], multifractal image analysis [72], two-point space-scale analysis of turbulent signals and financial time-series [73], iterated-function system theory with application to diffusion-limited aggregation [74]. We are convinced that futher application of wavelet analysis will lead to other major breakthroughs in various areas where multi-scale phenomena are ubiquitous.

Acknowledgments

We would like to thank Y. d'Aubenton-Carafa, P.V. Graves and C. Thermes for their collaboration to the DNA part of the work reported in this manuscript. We are very grateful to Y. Gagne and Y. Malecot for the permission to use their experimental turbulent signals. This work was supported by "Direction des Recherches, Etudes et Techniques" under contract (DRET n°95/111), the GIP-GREG (project "Motifs dans les Séquences) and by the Ministère de l'Education Nationale, de l'Enseignement Supérieur, de la Recherche et de l'Insertion Professionnelle ACC-SV (project "Génétique et Environnement").

References

[1] Mandelbrot B.B., Fractals: Form, Chance and Dimension (Freeman, San Francisco, 1977); The Fractal Geometry of Nature (Freeman, San Franscisco, 1982).
[2] Feder J., Fractals (Pergamon, New York, 1988).
[3] Vicsek T., Fractal Growth Phenomena (World Scientific, Singapore, 1989).
[4] Family F. and Viscek T. Eds., Dynamics of Fractal Surfaces (World Scientific, Singapore, 1991).
[5] West B.J. and Deering W., *Phys. Rep.* **254** (1994) 1.
[6] Wilkinson G.G., Kanellopoulos I. and Mégier J. Eds., Fractals in Geoscience and Remote Sensing, Image Understanding Research Series, Vol. 1 (ECSC-EC-EAEC, Brussels, Luxemburg, 1995).
[7] Tabeling P. and Cardoso O. Eds., Turbulence: A Tentative Dictionary (Plenum, New York, 1995).
[8] Barabási A.-L. and Stanley H.E., Fractal Concepts in Surface Growth (Cambridge Univ. Press, Cambridge, 1995).
[9] Peitgen H.O. and Saupe D. Eds., The Science of Fractal Images (Springer, Berlin, 1987).
[10] Frisch U., Turbulence (Cambridge Univ. Press, Cambridge, 1995).
[11] Peng C.K., Buldyrev S.V., Goldberger A.L., Havlin S., Sciortino F., Simons M. and Stanley H.E., *Nature* **356** (1992) 168.
[12] Stanley H.E., Buldyrev S.V., Goldberger A.L., Havlin S., Ossadnik S.M., Peng C.K. and Simons M., *Fractals* **1** (1993) 283.

[13] Frisch U. and Parisi G., Turbulence and Predictability in Geophysical Fluid Dynamics and Climate Dynamics, M. Ghil, R. Benzi and G. Parisi Eds. (North-Holland, Amsterdam, 1985) pp. 84.
[14] Barabási A.L. and Vicsek T., *Phys. Rev. A* **44** (1991) 2730.
[15] Muzy J.F., Bacry E. and Arneodo A., *Int. J. Bifurcation & Chaos* **4** (1994) 245.
[16] Arneodo A., Bacry E. and Muzy J.F., *Physica A* **213** (1995) 232.
[17] Jaffard S., *C.R. Acad. Sci. Paris* **308**, serie I (1989) 79; serie I (1992) 19.
[18] Holschneider M. and Tchamitchian P., in Ref. [32], pp. 102.
[19] Mallat S. and Hwang W.L., *IEEE Trans. Inform. Theory* **38** (1992) 617.
[20] Bacry E., Muzy J.F. and Arneodo A., *J. Stat. Phys.* **70** (1993) 635.
[21] Halsey T.C., Jensen M.H., Kadanoff L.P., Procaccia I. and Shraiman B.I., *Phys. Rev. A* **33** (1986) 1141.
[22] Collet P., Lebowitz J. and Porzio A., *J. Stat. Phys.* **47** (1987) 609.
[23] Paladin G. and Vulpiani A., *Phys. Rep.* **156** (1987) 148.
[24] Rand D., *Ergod Th. and Dyn. Sys.* **9** (1989) 527.
[25] Mandelbrot B.B., *J. Fluid. Mech.* **62** (1974) 331.
[26] Meneveau C. and Sreenivasan K.R., *J. Fluid Mech.* **224** (1991) 429.
[27] Monin A.S. and Yaglom A.M., Statistical Fluid Mechanics, Vol. 2 (MIT Press, Cambridge, MA, 1975).
[28] Muzy J.F., Bacry E. and Arneodo A., *Phys. Rev. E* **47** (1993) 875.
[29] Grossmann A. and Morlet J., *SIAM J. Math. Anal. Appl.* **15** (1984) 723; in Mathematics and Physics, Lecture on Recent Results, L. Steit Ed. (World Scientific, Singapore, 1985) pp. 135.
[30] Combes J.M., Grossmann A. and Tchamitchian P. Eds., Wavelets (Springer, Berlin, 1989).
[31] Meyer Y., Ondelettes (Hermann, Paris, 1990).
[32] Lemarié P.G. Ed., Les Ondelettes en 1989 (Springer, Berlin, 1990).
[33] Meyer Y. Ed., Wavelets and Applications (Springer, Berlin, 1992).
[34] Daubechies I., Ten Lectures on Wavelets (SIAM, Philadelphia, 1992).
[35] Meyer Y. and Roques S. Eds., Progress in Wavelet Analysis and Applications (Editions Frontières, Gif sur Yvette, 1993).
[36] Arneodo A., Argoul F., Bacry E., Elezgaray J. and Muzy J.F., Ondelettes, Multifractales et Turbulences: de l'ADN aux croissances cristallines (Diderot Editeur, Arts et Sciences, Paris, 1995).
[37] Holschneider M., *J. Stat. Phys.* **50** (1988) 963; Thesis, University of Aix-Marseille II (1988).
[38] Arneodo A., Grasseau G. and Holschneider M., *Phys. Rev. Lett.* **61** (1988) 2281; in Ref. [30], pp. 182.
[39] Muzy J.F., Bacry E. and Arneodo A., *Phys. Rev. Lett.* **67** (1991) 3515.
[40] Jaffard S., "Multifractal formalism for functions: Part I and Part II", to appear in *SIAM Journal of Mathematical Analysis.*
[41] Vergassola M. and Frisch U., *Physica D* **54** (1991) 58.
[42] Anselmet F., Gagne Y., Hopfinger E.J. and Antonia R.A., *J. Fluid Mech.* **140** (1984) 63.

[43] Gagne Y., Thesis, University of Grenoble (1987).
[44] Arneodo A. *et al.*, *Europhys. Lett.* **34** (1996) 411.
[45] Kolmogorov A.N., *C. R. Acad. Sci. USSR* **30** (1941) 301.
[46] Castaing B., Gagne Y. and Hopfinger E.J., *Physica D* **46** (1990) 177.
[47] Arneodo A., Muzy J.F. and Roux S.G., *J. Phys. II France* **7** (1997) 363.
[48] Benzi R., Ciliberto S., Tripiccione R., Baudet C., Massaioli F. and Succi S., *Phys. Rev. E* **48** (1993) R29; *Phys. Rev. E* **51** (1995) 2672.
[49] Kolmogorov A.N., *J. Fluid Mech.* **13** (1962) 82.
[50] Obukhov A.M., *J. Fluid Mech.* **13** (1962) 77.
[51] She Z.S. and Leveque E., *Phys. Rev. Lett.* **72** (1994) 336.
[52] She Z.S. and Waymire E.C., *Phys. Rev. Lett.* **74** (1995) 262.
[53] Dubrulle B., *Phys. Rev. Lett.* **73** (1994) 959; *J. Phys. II France* **6** (1996) 1825.
[54] Castaing B., in this volume.
[55] Novikov E.A., *Phys. Fluids A* **2** (1990) 814; *Phys. Rev. E* **50** (1995) 3303.
[56] Dubrulle B., in this volume.
[57] Li W., *Int. J. Bifurcation & Chaos* **2** (1993) 137.
[58] Herzel H. and Große I., *Physica A* **216** (1995) 518.
[59] Voss R.F., *Phys. Rev. Lett.* **68** (1992) 3805; *Fractals* **2** (1994) 1.
[60] Mantegna R.N., Buldyrev S.V., Goldberger A.L., Havlin S., Peng C.K., Simons M. and Stanley H.E., *Phys. Rev. E* **52** (1995) 2939.
[61] Nee S., *Nature* **357** (1992) 450.
[62] Chazidimitriou-Dreismann C.A. and Larhammar L., *Nature* **361** (1993) 212.
[63] Karlin S. and Brendel V., *Science* **259** (1993) 677.
[64] Buldyrev S.V., Goldberger A.L., Havlin S., Mantegna R.N., Matsa M.E., Peng C.K., Simons M. and Stanley H.E., *Phys. Rev. E* **51** (1995) 5084.
[65] Bernardi G., Olofsson B., Filipski J., Zerial M., Salinas J., Cuny F., Meunier-Rotival M. and Rodier F., *Science* **228** (1985) 953.
[66] Arneodo A., Bacry E., Graves P.V. and Muzy J.F., *Phys. Rev. Lett.* **74** (1995) 3293.
[67] Arneodo A., D'Aubenton-Carafa Y., Bacry E., Graves P.V., Muzy J.F. and Thermes C., *Physica D* **96** (1996) 291.
[68] Arneodo A., D'Aubenton-Carafa Y., Audit B., Bacry E., Muzy J.F. and Thermes C., "Nucleotide composition effects on the long-range correlations in human genes", preprint (1997).
[69] Bernardi G., *Annu. Rev. Genet.* **23** (1989) 637.
[70] Arneodo A., Bacry E. and Muzy J.F., *Phys. Rev. Lett.* **74** (1995) 4823.
[71] Arneodo A., Bacry E., Jaffard S. and Muzy J.F., *J. Stat. Phys.* **87** (1997) 179.
[72] Arrault J., Arneodo A., Davis A. and Marshak A., Phys. Rev. Lett. (1997), to appear.
[73] Arneodo A., Muzy J.F. and Sornette D., "Causal cascade in the stock market from the infrared to the ultraviolet", preprint (1997).
[74] Arneodo A., Argoul F., Bacry E., Muzy J.F. and Tabard M., *Fractals* **1** (1993) 629.

LECTURE 3

Fractional Derivatives in Static and Dynamic Scaling

R. Hilfer

ICA-1, Universität Stuttgart, 70569 Stuttgart, Germany
and
Institut für Physik, Universität Mainz, 55099 Mainz, Germany

1. INTRODUCTION

A derivative or integral of fractional order is usually defined by analytically continuing a suitable definition of the derivative or integral of integer order $n \in \mathbb{N}$ to real or complex values of n.

My purpose here is to give some of the definitions and then to review briefly recent applications of fractional derivatives and integrals in physics with emphasis on static and dynamic scaling.

Derivatives of fractional order have recently emerged in physics as generators of time evolutions in ergodic theory [1, 2, 3, 4, 5, 6], and as a tool for classifying phase transitions in thermodynamics by generalizing the classification scheme of Ehrenfest [1, 7, 8, 9, 10].

Given the fact that a fractional integral is, loosely speaking, a convolution operator with a power law kernel it is perhaps not too surprising that fractional integrals and derivatives are useful tools in scaling theory.

2. FRACTIONAL DERIVATIVES

Let $f : [a,b] \to \mathbb{R}$ be a real function and recall that its n-th order integral is given by

$$
\begin{aligned}
(I^n_{a+}f)(x) &= \int_a^x \int_a^{y1} \dots \int_a^{y_{n-1}} f(y_n)\, dy_n \dots dy_1 \qquad (x > a) \\
&= \frac{1}{(n-1)!} \int_a^x (x-y)^{n-1} f(y)\, dy \qquad (1)
\end{aligned}
$$

as is readily proven by induction. Generalizing equation (1) to noninteger n defines the ***Riemann-Liouville fractional integral of order*** $\alpha > 0$ as [11, 12]

$$(I^\alpha_{a+}f)(x) = \frac{1}{\Gamma(\alpha)} \int_a^x (x-y)^{\alpha-1} f(y)\, dy \qquad (2)$$

for $x > a$ and $f \in L_1([a,b])$ to ensure that the integral will be finite. A dual form is the right-sided integral

$$(I^\alpha_{b-}f)(x) = \frac{1}{\Gamma(\alpha)} \int_x^b (x-y)^{\alpha-1} f(y)\, dy \qquad (3)$$

for $x < b$. The Riemann-Liouville integral (2) is perhaps the most commonly employed definition [13, 14]. For $0 < a < 1$ these definitions are extended to the whole real axis $\mathbb{R}$ as

$$
\begin{aligned}
(I^\alpha_{+}f)(x) &= \frac{1}{\Gamma(\alpha)} \int_{-\infty}^x (x-y)^{\alpha-1} f(y)\, dy \qquad (4) \\
(I^\alpha_{-}f)(x) &= \frac{1}{\Gamma(\alpha)} \int_x^{\infty} (x-y)^{\alpha-1} f(y)\, dy \qquad (5)
\end{aligned}
$$

for $f \in L_p(\mathbb{R})$ with $1 \leq p < 1/\alpha$. Formula (5) is often referred to as the ***Weyl fractional integral of order*** α [15].

The definition of fractional derivatives is frequently based on that of fractional integrals through

$$(\partial^\alpha_i f)(x) = \frac{d^{n+1}}{dx^{n+1}} (I^{n+1-\alpha}_i f)(x) \qquad (6)$$

where $i \in \{a+, b-, +, -\}$ and $n = [\alpha]$ is the largest integer less than or equal to α. An alternative would be to interchange the orders of differentiation and fractional integration in (6).

Another approach to define a fractional derivative tries to replace α with $-\alpha$ directly in the definitions (2)–(5). However the resulting integral is divergent and needs to be regularized. The traditional regularization uses Hadamards finite part. Starting from (6) and (4) and proceeding formally one arrives at [15, 16]

$$\begin{aligned}
(\partial_+^\alpha f)(x) &= \frac{d}{dx}(I_+^{1-\alpha} f)(x) \\
&= \frac{1}{\Gamma(1-\alpha)} \frac{d}{dx} \int_{-\infty}^{x} (x-y)^{-\alpha} f(y)\, dy \\
&= \frac{1}{\Gamma(1-\alpha)} \frac{d}{dx} \int_{0}^{\infty} t^{-\alpha} f(x-t)\, dt \\
&= \frac{\alpha}{\Gamma(1-\alpha)} \int_{0}^{\infty} f'(x-t) \int_{t}^{\infty} \frac{1}{z^{1+\alpha}}\, dz dt \\
&= \frac{\alpha}{\Gamma(1-\alpha)} \int_{0}^{\infty} \frac{f(x)-f(x-t)}{t^{1+\alpha}}\, dt
\end{aligned} \tag{7}$$

where the last equality serves to define $(\partial_+^\alpha f)$ for $0 < \alpha < 1$. Here f' denotes the first derivative of f with respect to its argument. Similar definitions can be given for $\alpha > 1$. The main difference between the definitions (7) and (6) is that (7) allows more freedom for the behaviour of f at infinity. Thus (7) is defined for $f(x) = const$ while (6) is not.

The preceding definitions define fractional differentiation as an inverse operation to fractional integration, i.e. as integration of order $-\alpha$. A more direct approach arises from the fact that derivatives are limits of difference quotients. Let T^h denote the translation

$$(T^h f)(x) = f(x-h) \tag{8}$$

by h. The finite difference of order α is defined as

$$(\Delta_h^\alpha f)(x) = (\mathbf{1} - T^h)^\alpha f(x) = \sum_{k=0}^{\infty} (-1)^k \binom{\alpha}{k} f(x-kh) \tag{9}$$

with $\mathbf{1} = T^0$ the identity, and

$$\binom{\alpha}{k} = \frac{(-1)^{k-1} \alpha \Gamma(k-\alpha)}{\Gamma(1-\alpha)\Gamma(k+1)} \tag{10}$$

It reduces to the familiar finite difference of integer order for $\alpha \in \mathbb{N}$. The *Grünwald fractional derivative of order* α is then defined as [17, 18]

$$(\partial_\pm^\alpha f)(x) = \lim_{h \to 0+} h^{-\alpha} (\Delta_{\pm h}^\alpha f)(x) \tag{11}$$

as the limit of a fractional finite difference quotient. There are several possibilities to define this limit, e.g. pointwise, almost everywhere, or in the norm of a Banach space. The choice depends upon the question at hand. It is possible to show that, in a suitable sense, the definitions (11) and (7) coincide.

As equation (9) suggests it is also possible to define fractional derivatives as fractional powers of the differentiation operator, i.e. $\partial_+^\alpha = (d/dx)^\alpha$ [19, 18]. More generally one considers fractional powers of the infinitesimal generators of strongly continuous semigroups. A strongly continuous semigroup on a Banach space X is a one parameter family of maps $T(t) : X \to X$ fulfilling

$$1. \quad T(t)T(s) = T(t+s) \tag{12}$$

$$2. \quad T(0) = \mathbf{1} \tag{13}$$

$$3. \quad \lim_{t\to s} ||T(t)f - T(s)f|| = 0 \qquad \text{for all} \quad f \in X \tag{14}$$

whose infinitesimal generator A is defined by

$$Af = \lim_{t\to 0+} \frac{1}{t}(T(t) - \mathbf{1})f \tag{15}$$

The fractional power $(-A)^\alpha$ can be defined analogous to eq. (11) or analogous to eq. (7) through

$$(-A)^\alpha f = \frac{1}{\Gamma(-\alpha)} \int_0^\infty t^{-\alpha-1}(T(t) - \mathbf{1})f \, dt \tag{16}$$

for $0 < \alpha < 1$ and f within the domain of A. Equation (16) is known as *Balakrishnan's formula*. Obviously, (16) generalizes (7) and (11).

Finally, fractional derivatives may also be defined through analytic continuation in Fourier space. If

$$\widehat{f}(k) = \int_{-\infty}^{\infty} e^{ikx} f(x) \, dx \tag{17}$$

denotes the Fourier transform of f then a fractional derivative of order α can be defined as the operator

$$(\partial_\pm^\alpha f)(x) = \frac{1}{2\pi} \int_{-\infty}^{\infty} (\mp ik)^\alpha \widehat{f}(k) e^{-ikx} \, dk \tag{18}$$

for functions f such that the integrals exist (e.g. smooth with compact support).

3. SCALING IN THERMODYNAMICS

Fractional derivatives have been employed in thermodynamics to generalize the Ehrenfest classification scheme for phase transitions [1, 7, 8, 9]. It is found that this generalization gives rise to a generalized form of static scaling at the transition point.

To be concrete consider the pressure $p(T, \mu)$ as a function of temperature T and chemical potential μ. Let $p(T, \mu)$ have a phase transition (i.e. a singular point) at (T_c, μ_c). Let furthermore $\mathcal{C} : \sigma \to (T(\sigma), \mu(\sigma))$ be a thermodynamic process parametrized by σ such that $\sigma = 0$ corresponds to the critical point, i.e. $(T(0), \mu(0)) = (T_c, \mu_c)$. Let p_s denote the singular part of p, i.e. $p = p_r + p_s$ where p_r is the regular part. Let

$$F(\mathcal{C}, q; \sigma) = \begin{cases} (\partial^q_{0+} p_s)(\sigma) & : \text{ for } \sigma > 0 \\ (\partial^q_{0-} p_s)(\sigma) & : \text{ for } \sigma < 0 \end{cases} \tag{19}$$

denote the q-th order derivative of the function $p_s(\sigma)$.

Ehrenfest [20] suggested to classify phase transitions according to their order. The phase transition at $\sigma = 0$ is defined to have Ehrenfest order $n \in \mathbb{N}$ if and only if $F(\mathcal{C}, n; \sigma)$ has a jump discontinuity at $\sigma = 0$, i.e. iff

$$\lim_{\sigma \to 0+} F(\mathcal{C}, n; \sigma) \neq \lim_{\sigma \to 0-} F(\mathcal{C}, n; \sigma) \tag{20}$$

In [1, 7, 8, 9] Ehrenfests classification was generalized to allow noninteger orders and slowly varying confluent behaviour. The phase transition at $\sigma = 0$ is defined to be of order $\lambda^{\pm} > 0$ if and only if [1, 7, 8, 9]

$$\lim_{\sigma \to 0\pm} \frac{F(\mathcal{C}, \lambda^{\pm}; b\sigma)}{F(\mathcal{C}, \lambda^{\pm}; \sigma)} = 1 \tag{21}$$

for all $b > 0$, i.e. iff $F(\mathcal{C}, \lambda^{\pm}; \sigma)$ is a slowly varying function in the sense of Karamata [21].

It follows readily from this definition that $p_s(\sigma)$ is regularly varying of order $\lambda^{\pm}(\mathcal{C})$. Note that the order depends on the choice of the curve $\mathcal{C}$. Hence

$$p_s(T(\sigma), \mu(\sigma)) = \begin{cases} \Lambda^+(\mathcal{C}; \sigma)\sigma^{\lambda^+(\mathcal{C})} & : \text{ for } \sigma > 0 \\ \Lambda^-(\mathcal{C}; \sigma)\sigma^{\lambda^-(\mathcal{C})} & : \text{ for } \sigma < 0 \end{cases} \tag{22}$$

where the functions $\Lambda^{\pm}(\mathcal{C}; \sigma)$ are slowly varying functions at $\sigma = 0$.

These results show that phase transitions with noninteger Ehrenfest order exhibit algebraic scaling behaviour with an exponent that is in general path dependent.

What is the interest of this new classification ? First of all classifying phenomena is an important task of science. An exhaustive classification of fixed points in renormalization group theory of phase transitions does not exist [22]. An exhaustive classification allows to predict that certain cases cannot occur or that new cases beyond known ones must be expected. Secondly the generalized Ehrenfest scheme has led to a wealth of new predictions ranging from new classes of phase transitions [1] to the prediction of scaling functions and universal amplitude ratios for finite size scaling [23, 24]. Finally the same ideas giving rise to a classification of static macroscopic phenomena are also useful to classify macroscopic dynamics as discussed next.

4. SCALING IN MACROSCOPIC DYNAMICS

First order time derivatives play a fundamental role as the infinitesimal generators of time evolution for a dynamical system. A growing number of publications uses fractional time derivatives to describe the time evolution of physical systems [6, 25, 26, 27, 28, 29, 30, 31, 32, 33, 34, 35, 36, 37, 38]. This raises the question whether fractional time derivatives are also infinitesimal generators of time evolution. Recent work has given a positive answer to this question [3, 4, 5]. The purpose of this section is to present this result.

Let Γ be the microscopic phase or state space of a dynamical system in the sense of ergodic theory [39, 40], with σ-algebra $\mathcal{G}$, measure μ, and time evolution $T(t) = T^t$ with either $t \in \mathbb{R}$ or $t \geq 0$, i.e.

$$T(t)\mu(G, t_0) = \mu(G, t_0 - t) \tag{23}$$

for all $G \in \mathcal{G}$. Let $\Gamma^* \subset \Gamma$ be the microscopic phase or state space associated with a subsystem. If $\mu(\Gamma^*) > 0$ then the time evolution $T(t)$ on Γ canonically induces also a time evolution on Γ^* [39, 40]. In situations of physical interest, however, one often has $\mu(\Gamma^*) = 0$. An example would be the solid phase in a system of interacting classical particles exhibiting macroscopic solid-fluid coexistence at low temperatures.

The physically interesting case with $\mu(\Gamma^*) = 0$ was first discussed in [3, 4, 5] in the context of ergodic theory with the following result: If the subset Γ^* is suitably chosen, then the limit of macroscopic times (ultralong time limit) of the induced time evolution $T^*(t^*)$ on Γ^* is given by a one-parameter family of semigroups parametrized by $0 < \alpha \leq 1$

$$\begin{aligned} T^*_\alpha(t^*)\mu^*(B, t^*_0) &= \int_0^\infty \mu^*(B, t^*_0 - t) h_\alpha\left(\frac{t}{t^*}\right) \frac{dt}{t^*} \\ &= \frac{1}{t^*}\int_0^\infty h_\alpha(t/t^*)T(t)\mu^*(B, t^*_0)dt \end{aligned} \tag{24}$$

with $t^* \geq 0$. Here $B \subset \Gamma^*$, μ^* is a measure on Γ^*, and

$$h_\alpha(x) = \frac{1}{x\alpha} H^{10}_{11}\left(\frac{1}{x} \middle| \begin{matrix} (0,1) \\ (0,1/\alpha) \end{matrix}\right) \tag{25}$$

is a one-parameter familiy of functions with H^{10}_{11} defined through its Mellin transform [41]

$$\int_0^\infty H^{10}_{11}\left(x \middle| \begin{matrix} (0,1) \\ (0,1/\alpha) \end{matrix}\right) x^{s-1}\, dx = \frac{\Gamma(s/\alpha)}{\Gamma(s)} \tag{26}$$

or through the Laplace transform

$$\int_0^\infty \frac{e^{-ux}}{x\alpha} H^{10}_{11}\left(\frac{1}{x} \middle| \begin{matrix} (0,1) \\ (0,1/\alpha) \end{matrix}\right) dx = e^{-u^\alpha}. \tag{27}$$

Obviously

$$h_1(x) = \lim_{\alpha \to 1^-} h_\alpha(x) = \delta(x-1) \tag{28}$$

and hence from (24)

$$T_1^*(t^*)\mu^*(B,t_0^*) = \mu^*(B,t_0^* - t^*) \tag{29}$$

is again a translation as in the microscopic evolution (23) except that here the macroscopic time parameter obeys always $t^* \geq 0$ while the microscopic time parameter can also obey $t \in \mathbb{R}$. This means that the macroscopic time evolution is always irreversible while the microscopic evolution may be either reversible or irreversible.

Fractional derivatives appear when the infinitesimal generator, (15), of the induced time evolution $T^*(t^*)$, given in (24), is determined as

$$A_\alpha^* = \lim_{t^* \to 0+} \frac{1}{t^*}(T_\alpha^*(t^*) - \mathbf{1}) = c^+ \int_0^\infty t^{-\alpha-1}(T(t) - T(0))\, dt \tag{30}$$

where $c^+ > 0$ is a constant. Comparison with Balakrishnans formula (16) shows that

$$A_\alpha^* = c\partial_+^\alpha \tag{31}$$

for $0 < \alpha < 1$ with c a constant, and

$$A_1^* = \frac{d}{dt^*} \tag{32}$$

for $\alpha = 1$ are the infinitesimal generators of the induced time evolution $T_\alpha^*(t^*)$.

An immediate consequence of (31) was proposed in [1]. It identifies A_α^* with a generalized Liouvillian, and suggests to generalize the Liouville equation for the time evolution of probability measures in phase space into a *fractional Liouville equation*

$$\partial_+^\alpha \mu^*(B,t^*) = A_\alpha^* \mu^*(B,t^*) \tag{33}$$

where A_α^* is interpreted as a Liouville operator. Explicit examples are given by anomalous fractional diffusion where the right hans side is the Laplacian [33] or fractional relaxation [1, 31, 32].

A second consequence emerges from the condition of invariance (or stationarity) of a measure under the induced macrotime evolution $T^*(t^*)$ which reads

$$T_\alpha^*(t^*)\mu^*(B,t_0^*) = \mu^*(t_0^*). \tag{34}$$

This becomes infinitesimally a fractional differential equation

$$\partial_+^\alpha \mu^*(B,t^*) = 0 \tag{35}$$

whose solution is

$$\mu^*(B,t^*) = C_0 t^{*^{\alpha-1}}. \tag{36}$$

Thus algebraic decay in time can be a sign of stationarity for induced dynamics on subsets of measure zero. The algebraic time decay shows that order α of the derivative plays the role of a dynamic scaling exponent.

References

[1] R. Hilfer, "Classification theory for anequilibrium phase transitions," *Phys. Rev. E*, vol. 48, p. 2466, 1993.

[2] R. Hilfer, "On a new class of phase transitions," in *Random Magnetism and High-Temperature Superconductivity* (W. Beyermann, N. Huang-Liu, and D. MacLaughlin, eds.), (Singapore,), p. 85, World Scientific Publ. Co., 1994.

[3] R. Hilfer, "Fractional dynamics, irreversibility and ergodicity breaking," *Chaos, Solitons & Fractals*, vol. 5, p. 1475, 1995.

[4] R. Hilfer, "Foundations of fractional dynamics," *Fractals*, vol. 3, p. 549, 1995.

[5] R. Hilfer, "An extension of the dynamical foundation for the statistical equilibrium concept," *Physica A*, vol. 221, p. 89, 1995.

[6] R. Hilfer and L. Anton, "Fractional master equations and fractal time random walks," *Phys.Rev.E, Rapid Commun.*, vol. 51, p. 848, 1995.

[7] R. Hilfer, "Thermodynamic scaling derived via analytic continuation from the classification of Ehrenfest," *Physica Scripta*, vol. 44, p. 321, 1991.

[8] R. Hilfer, "Multiscaling and the classification of continuous phase transitions," *Phys. Rev. Lett.*, vol. 68, p. 190, 1992.

[9] R. Hilfer, "Scaling theory and the classification of phase transitions," *Mod. Phys. Lett. B*, vol. 6, p. 773, 1992.

[10] R. Hilfer, "Classification theory for phase transitions," *Int.J.Mod.Phys.B*, vol. 7, p. 4371, 1993.

[11] J. Liouville, "Mémoire sur quelques questions de geometrie et de mecanique, et sur un nouveau genre de calcul pour resoudre ces questions," *J. l'Ecole Roy. Polytechn.*, vol. 13, p. 1, 1832.

[12] B. Riemann, "Versuch einer allgemeinen Auffassung der Integration und Differentiation, (Januar 1847)," in *Bernhard Riemann's gesammelte mathematische Werke und wissenschaftlicher Nachlass* (H. Weber, ed.), p. 353, Dover Publications, 1953.

[13] A. Erdelyi (et al.), *Tables of Integral Transforms*, vol. II. New York: Mc Graw Hill Book Co., 1954.

[14] K. Oldham and J. Spanier, *The Fractional Calculus.* New York: Academic Press, 1974.

[15] H. Weyl, "Bemerkungen zum Begriff des Differentialquotienten gebrochener Ordnung," *Vierteljahresschrift der Naturforsch. Ges. Zürich*, vol. 62, p. 296, 1917.

[16] A. Marchaud, "Sur les derivess et sur les differences des fonctions de variables reelles," *J.math. pures et appl.*, vol. 6, p. 337, 1927.

[17] A. K. Grünwald, "Über "begrenzte" Derivationen und deren Anwendung," *Z.f.angew.Math. und Phys.*, vol. 12, p. 441, 1867.

[18] U. Westphal, "An approach to fractional powers of operators via fractional differences," *Proc. London Math. Soc.*, vol. 29, p. 557, 1974.

[19] A. Balakrishnan, "Fractional powers of closed operators and the semigroups generated by them," *Pacific J. Math.*, vol. 10, p. 419, 1960.

[20] P. Ehrenfest, "Phasenumwandlungen im üblichen und erweitereten Sinn, classifiziert nach den entsprechenden Singularitäten des thermodynamischen Potentiales," *Suppl. Mitteilungen aus dem Kamerlingh-Onnes Institut, Leiden*, vol. 75b, p. 153, 1933.

[21] E. Seneta, *Regularly Varying Functions*. Berlin: Springer Verlag, 1976.

[22] R. Fernandez, J. Fröhlich, and A. Sokal, *Random Walks, Critical Phenomena, and Triviality in Quantum Field Theory*. Berlin: Springer Verlag, 1992.

[23] R. Hilfer, "Absence of hyperscaling violations for phase transitions with positive specific heat exponent," *Z. Physik B*, vol. 96, p. 63, 1994.

[24] R. Hilfer and N.B.Wilding, "Are critical finite size scaling functions calculable from knowledge of an appropriate critical exponent ?," *J. Phys. A: Math. Gen.*, vol. 28, p. L281, 1995.

[25] B. Mandelbrot and J. van Ness, "Fractional brownian motions, fractional noises and applications," *SIAM Review*, vol. 10, p. 422, 1968.

[26] B. Mandelbrot, *The Fractal Geometry of Nature*. San Francisco: Freeman, 1982.

[27] R. Nigmatullin, "The realization of the generalized transfer equation in a medium with fractal geometry," *phys. stat. sol. b*, vol. 133, p. 425, 1986.

[28] R. Hilfer, "The continuum limit for selfsimilar Laplacians and the Greens function localization exponent," 1989. UCLA-Report 982051.

[29] T. Nonnenmacher, "Fractional integral and differential equations for a class of Levy-type probability densities," *J. Phys. A: Math. Gen.*, vol. 23, p. L697, 1990.

[30] C. Friederich, "Relaxation functions of rheological constitutive equations with fractional derivatives: Thermodynamical constraints," in *Rheological Modeling: Thermodynamic and Statistical Approaches* (J. Casas-Vasquez and D. Jou, eds.), (Berlin), p. 309, Springer, 1991.

[31] T. Nonnenmacher and W. Glöckle, "A fractional model for mechanical stress relaxation," *Phil. Mag. Lett.*, vol. 64, p. 89, 1991.

[32] H. Schiessel and A. Blumen, "Hierarchical analogues to fractional relaxation equations," *J. Phys. A: Math. Gen.*, vol. 26, p. 5057, 1993.

[33] R. Hilfer, "Exact solutions for a class of fractal time random walks," *Fractals*, vol. 3(1), p. 211, 1995.

[34] A. Gemant, "A method of analyzing experimental results obtained from elastoviscous bodies," *Physics*, vol. 7, p. 311, 1936.

[35] M. Riesz, "L'integrale de Riemann-Liouville et le probleme de Cauchy," *Acta mathematica*, vol. 81, p. 1, 1949.

[36] G. Scott-Blair and J. Caffyn, "An application of the theory of quasi-properties to the treatment of anomalous stress-strain relations," *Phil. Mag.*, vol. 40, p. 80, 1949.

[37] K. Oldham and J. Spanier, "The replacement of Fick's law by a formulation involving semidifferentiation," *J. Electroanal. Chem. Interfacial Electrochem.*, vol. 26, p. 331, 1970.

[38] S. Westlund, "Dead matter has memory !," *Physica Scripta*, vol. 43, p. 174, 1991.

[39] I. Cornfeld, S. Fomin, and Y. Sinai, *Ergodic Theory*. Berlin: Springer, 1982.
[40] K. Petersen, *Ergodic Theory*. Cambridge: Cambridge University Press, 1983.
[41] C. Fox, "The G and H functions as symmetrical Fourier kernels," *Trans. Am. Math. Soc.*, vol. 98, p. 395, 1961.

LECTURE 4

Multi-Dimensional Self-Similarity, and Self-Gravitating N-Body Systems

R.N. Henriksen

Department of Physics, Queen's University at Kingston, Ontario, K7L 3N6, Canada

1. INTRODUCTION

We consider a single system that is determinate in the sense that the governing equations are known. The property variously described as 'Scale Invariance', 'Rescaling Symmetry', 'Self-Similarity', or even 'Renormalizability' suffers evidently from a non-standard terminology but also from an *apparent* familiarity. The familiarity is only apparent because, as we have seen at this school, it is often rather unclear just what quality of the system is invariant and what quality is rescaling. The restriction to a single system excludes the relation of 'similarity' between systems, which exists when certain purely numerical quantities are invariant between them. Such quantities are the various 'numbers' (e.g. Reynolds, Froude, Strouhal, Rayleigh–) familiar from engineering modelling.

The basic notion stems from the idea of geometric similarity. We recall that in Euclidian plane geometry, polygons are said to be similar if their sides are rescaled in such a way as to leave their internal angles invariant. If a polygon were made of elastic material so that it could be uniformly stretched or compressed in time while preserving its internal angles, then it would exhibit self-similarity. Both the invariant and rescaling qualities of this 'system' are obvious.

The extention to general physical systems is traditionally by Dimensional (this word written with a capital letter shall mean physical properties that require units-without a capital letter it shall mean degrees of freedom of a manifold) analysis. The arbitrary units correspond to the rescalable sides of a polygon. One infers an obligatory rescaling of any physical property under a unit 'resizing', from the dependence of its Dimension on the chosen basis Dimensions (e.g. mass, length, time in mechanics). Due to the rationality of the relation between any Dimension and the basis dimensions ***this Dimensional rescaling is always a power law.*** In fact one infers in this naive way that it should always be a ***rational*** power law, but this is not necessarily the case as the Dimensions themselves may be scale invariants (see next section). The invariant quantities correspond to the essential or 'internal' physical properties of the system, just as the angles of a polygon describe its essence independent of its 'size'.

In determinate physical systems the essential internal properties are expressed in terms of general mathematical relations (i.e. equations) between the rescaling invariant quantities. These quantities include most generally the underlying space-time manifold in suitable Dimensionless form. The natural expression of the relation between various physical quantities in a determinate physical system is by way of partial differential equations.

The most familiar type of determinate Self-Similarity is encountered in the attempt to reduce the partial differential system of equations to an ordinary system. This is always possible by the Buckingham Π theorem (e.g. [10]) whenever the number of essential, invariant quantities is reducible to one so that the system has a manifest Rescaling Symmetry. However in general the system will have multiple invariants under a unit resizing and thus will be *MULTI-dIMENSIONAL*. Nevertheless a Rescaling Symmetry can still be imposed, and if imposed may still be used to simplify the partial differential system.

What makes the apparently arbitrary imposition of a Rescaling Symmetry in such multi-dimensional problems important is the observed dynamical emergence of such symmetry in many systems well away from their initial conditions or boundaries [1]. In this sense they are very well described as "intermediate asymptotes", implying that they have a universal character 'between' bondaries. Thus we seek to have a general, practical method of imposing a multi-dimensional self-similarity ***as a way of identifying the possible asymptotic behaviour.*** In [1] a distinction was drawn between those symmetries where the power laws appearing in the rescaling factors and in the definition of the invariants were rational or irrational. In fact this is not mathematically sound since the rational numbers and the irrationals are not isolated [2]. We show in the next section that a better classification is simply in terms of the direction of the rescaling vector in the space of physical properties.

In this paper we do not treat explicitly systems that exhibit a statistical Rescaling Symmetry even though the relevant mathematical relations may not be known (see these proceedings generally and the lecture by B. Dubrulle especially), such as economic systems, nor do we treat discrete symmetries such as described by Sornette [3]. However the original treatment by Carter and Henriksen (CH) in [2] was meant to be abstract enough to include such possibilities. In the next section we present a pragmatic description of the method in the most familiar context that is designed to be 'user friendly'. Finally we give some recent examples from our studies in N-body, collisionless, self-gravitating systems ***where the intermediate asymptotes may replace the conventional notion of maximum entropy equilibrium*** (compare also [4]).

2. CH Prescription for Multi-dimensional Self-Similarity

The terminology in this section will be chosen to be suggestive rather than rigorously precise. The prescription can be presented in point form:

1. ***For the physical system at hand, choose an appropriate Dimension basis, which implies a corresponding manifold of physical quantities.*** This should be complete in the sense that the Dimensions of all physical properties of the system can be represented by powers of the elements of this basis. However it does not have to consist of conventionally (due to the existence of fundamental physical constants we can formally collapse the Dimension space in many cases (Graner; private communication), but it is useful to retain the perceived distinctions here) independent Dimensions, for it is sometimes convenient to include time, length and velocity separately. In Newtonian mechanics a standard choice is $[time, length, mass]$. If our focus is on time-dependent phase space then we might prefer to choose $[time, length, velocity, mass]$ and if on a stationary phase space, then simply $[length, velocity, mass]$.

Now multiplicative rescaling of physical quantities due to change in the units constitutes an Abelian group of which any element may be represented as $\mathbf{A} \equiv (e^{\alpha}, e^{\delta}, e^{\nu}, e^{\mu}, ----)$, where we have written the time (e^{α}), length (e^{δ}), velocity (e^{ν}) and mass (e^{μ}) rescaling factors explicitly as comprising the Dimension space most commonly used (and which we use for example subsequently). We will normally use the logarithms of quantities so that the group is additive with the element $\mathbf{a} \equiv (\alpha, \delta, \nu, \mu, ----)$. These elements may be thought of as elements in a Lie Algebra and the effect of the rescaling on any physical quantity F, be it scalar or tensor component, can be specified locally ***when the rescaling group is continuous*** by giving the result of the Lie derivative $\mathbf{k}\mathcal{L}F$, where the vector $\mathbf{k}$ is the generator of the rescaling symmetry or self-similarity in the Lie Algebra. This must be equal to a constant times F itself to ensure an additive increment in its logarithm. The constant must be dependent on both the applied rescaling $\mathbf{a}$ and on the dimensionality of F, and we proceed to describe it in the next section. Discrete rescaling groups, which we do not discuss here, would require another appropriate local generator.

Normally (when all 'coordinates' refer to the logarithms of quantities) we can identify the generator $\mathbf{k}$ with the vector $\mathbf{a}$. In cases of fundamental tensor fields such as the metric of space-time, one can use the Lie definition of rescaling to solve for the most general possible generator. This has been done in [2], but we will not discuss this special development here.

2. *For each scalar or tensor field F in the system we define a 'Lie dimensionality covector' $\mathbf{d}_F$ in the Lie Algebra defined above (in particular in the same basis).* The components are simply positive or negative integers. Thus for example the usual mass density would have $\mathbf{d}_\rho = (0,-3,0,1)$, velocity would have $\mathbf{d}_v = (0,0,1,0)$ and a mass distribution function would have $\mathbf{d}_f = (0,-3,-3,1)$. Then the scalar $\mathbf{a}.\mathbf{d}_F \equiv \alpha d_{F1} + \delta d_{F2} + \nu d_{F3} + \mu d_{F4}$ gives the required increment in the logarithm of F. That is

$$\mathbf{k}\mathcal{L}F = (\mathbf{a}.\mathbf{d}_F)F. \tag{1}$$

Here we use the normal definition of the Lie derivative in terms of quantities rather than their logarithms so that

$$\mathbf{k} \equiv \alpha t\partial_t + \delta r\partial_r + \nu v_r\partial_{v_r} + \mu m\partial_m, \tag{2}$$

where we have used (t, r, v_r, m) as an example of appropriate 'coordinates' in the manifold of physical quantities spanned by our Dimension basis.

Equation (1) is a local, necessary and sufficient by definition, condition for $\mathbf{k}$ to be the generator of a continuous Rescaling Symmetry or Self-Similarity. Notice that the result of these first two steps in the procedure are equivalent to the usual process of Dimensional analysis and the $\mathbf{a}.\mathbf{d}$ factor gives the usual Dimensional rescaling. However with this mathematically precise definition of a Rescaling Symmetry we can proceed to formulate our physical relations in terms of rescaling invariant quantities.

3. *Choose coordinates in the physical manifold such that the Lie differentiation is easy.* To this end we first introduce a complete set of scalar quantities $\{X^i\}$ that are invariant under the rescaling . These are the customarily named "self-similar" variables in the conventional 'ad hoc' treatment, but normally a complete set is not found. They satisfy

$$k^j\partial_j X^i = 0, \tag{3}$$

and can be readily found by the method of characteristics using equation (2). The index i will normally run from 1 to $M-1$ where M is the dimension of the physical manifold associated with the Dimension basis. We choose the one additional variable that ensures completeness in a special way. The vector $\mathbf{k}$ lies along the tangent to the self-similar motion in the physical manifold, and we choose one variable to be measured along this direction. In a time dependent manifold it is convenient to choose this to be an 'oblique time' T such that it is parametrized by the true coordinate time t as $T(t)$. Then we normalize it so that $\mathbf{k}\mathcal{L}T = 1$ or

$$k^j\partial_j T \equiv \alpha t\frac{dT}{dt} = 1. \tag{4}$$

Thus T is a logarithmic time that satisfies $e^{\alpha T} = |\alpha t|$.

In the new coordinates for the manifold $\{T, X^i\}$ we have the happy result that $\mathbf{k}\mathcal{L}F \equiv \partial_T F$ (the X^i being invariant) , provided $F = F(T, X^i)$, be it scalar or tensor component. The coordinates $\{T, X^i\}$ can be defined in this way relative to *any* direction in the physical manifold and thus no Rescaling symmetry has as yet been imposed. All quantities are functions of the set $\{T, X^i\}$. There is no restriction until we require that $\mathbf{k}$ be such that (1) applies. Because of the simple form of the Lie derivative in the constructed coordinates, we can integrate this latter equation immediately to

$$F = \overline{F}(X^i)\, e^{(\mathbf{a}.\mathbf{d}_F)T}. \tag{5}$$

Thus under a Rescaling Symmetry every quantity can be written as a power law in time (the exponential factor) times an invariant function. The invariant function is to be found from the governing equations in a determinate problem. This exhibits very clearly the respective roles of invariance and renormalization in such a Rescaling Symmetry. The mathematical treatment parallels very closely the arguments from Dimensionality that were recalled in the introduction, although there is now no reason why the renormalization factor should be a rational power. In fact it becomes the basis for the classification of the self-similarity that was proposed in [2], but its physical origin lies normally in 'remembered' initial conditions as in our examples to follow.

When time is not present in the manifold of interest, some other physical quantity must be chosen to parameterize the Self-Similar motion. In our first example below this is taken to be the radius of a stationary, spherical system, but this choice is basically arbitrary among the Dimensional quantities present.

4. *Reduce the Dimension basis by requiring the global constants of the system to be scale invariant.* Such constants may either be fundamental constants or those imposed by initial or boundary conditions. Thus consider Newton's constant G for example. It is one of the physical quantities F in our manifold which must satisfy equation (1) in the form

$$\mathbf{k}\mathcal{L}G = (\mathbf{a}.\mathbf{d}_G)G = 0. \tag{6}$$

Now using our representative Dimension basis $\mathbf{d}_G = (-2, 3, 0, -1)$ or $\mathbf{d}_G = (0, 1, 2, -1)$. In either case the invariance constraint is $\mathbf{a}.\mathbf{d}_G = 0$. Thus we conclude, using the latter representation as one that is convenient for our phase space examples,

$$\mu = 2\nu + \delta. \tag{7}$$

This shows us that the mass rescaling is no longer independent of the rescalings in velocity and length so that we can use a reduced Dimension basis according to

$$\mathbf{d}_F = (n_\alpha, n_\delta, n_\nu) + n_\mu(0, 1, 2). \tag{8}$$

Here the integers 'n' give the dependence of the Dimension of F on the corresponding basis Dimensions, and the second term redistributes the dependence on μ over the remaining independent dimensions according to the constraint provided by the invariance of G.

This completes the formulation of Multi-dimensional Self-Similarity for determinate systems. One has now only to substitute the forms (5) into the governing equations using the reduced Dimensionality imposed by the invariant constants. The result will always be a set of partial differential equations for the invariant functions $\overline{F}(X^i)$, ***which have had the number of independent variables reduced by one*** since T will not appear, [2], [5].

3. Rescaling Symmetries in N-body Self-Gravitating Systems

3.1. Stationary Example

As an example we turn first to a stationary self-gravitating system governed by the collisionless Boltzmann equation (CBE) and the Poisson equation [7]. That is, it is a collisionless system wherein all particle interactions are with the collective mean field. We impose spatial spherical symmetry although we allow the orbits to be anisotropic. For such a system the CBE can be written [6] as

$$v_r \partial_r f + \left(\frac{j^2}{r^3} - \partial_r \Phi\right) \partial_{v_r} f = 0, \tag{9}$$

and Poisson's equation becomes

$$\partial_r(r^2 \partial_r \Phi) - 4\pi^2 G \int dj^2 \int dv_r f(r, v_r, j^2) = 0. \tag{10}$$

Here r, v_r, refer to the radial position and velocity; j^2 is the square of the transverse angular momentum, namely $r^2(v_\theta^2 + v_\phi^2)$ in spherical components; $\Phi(r)$ is the mean potential and $f(r, v_r, j^2)$ is the mass distribution function (DF). We choose the dimensional basis to be ***length, velocity, squared specific angular momentum, mass*** and denote the corresponding rescaling algebra vector as $\mathbf{a} = (\delta, \nu, \lambda, \mu)$. However we wish G to be invariant in this problem and representing its Dimension in terms of the length, velocity and mass components of the basis leads as in the previous section to $\mu = 2\nu + \delta$. This leads to the reduced basis $\mathbf{d}_F = (n_\delta, n_\nu, n_\lambda) + n_\mu(1, 2, 0)$. We may now deduce that $\mathbf{d}_f = (-1, -1, -1, 1) \to (0, 1, -1)$ and $\mathbf{d}_\Phi = (0, 2, 0)$ in the reduced basis.

The Lie rescaling motion $\mathbf{k}$ can be taken equal to $\mathbf{a}$ in logarithmic coordinates, so that

$$k^j \partial_j \equiv \delta\ r\partial_r + \nu\ v_r \partial_{v_r} + \lambda\ j^2 \partial_{j^2}. \tag{11}$$

We use this to define the invariant coordinates X^i as independent solutions to $k^j \partial_j X^i = 0$. Using the method of characteristics on this latter 'quasi-linear' [5] equation, these may be taken as

$$\begin{aligned} X^{(1)} \equiv X &= \frac{v_r}{(|\delta| r)^{1/\delta}}, \\ X^{(2)} \equiv Y &= \frac{j^2}{(|\delta| r)^{\lambda/\delta}}. \end{aligned} \tag{12}$$

The coordinate measured along the rescaling motion **k** will be taken to be $R(r)$ in this stationary case. Hence $k^j\partial_j R = 1$, which is simply $\delta r \partial_r R = 1$ or

$$|\delta| r = e^{\delta R}. \tag{13}$$

Our Self-Similar motion coordinates are now (R, X, Y). We impose the symmetry on the DF and the potential by writing them in the form of equation (5), namely, remembering the Dimension covectors given in the reduced basis above,

$$\begin{aligned} f &= \overline{f}(X,Y)\; e^{(\nu-\delta)}R, \\ \Phi &= \overline{\Phi}(X,Y)\; e^{(2\nu R)}. \end{aligned} \tag{14}$$

These forms may now be substituted in equations (9) and (10) , which must be written in terms of the variables (R, X, Y), to obtain the equations governing our system under the assumed Self-Similar motion as

$$-(2\delta+1)\overline{f} - X\partial_X\overline{f} - 2(1+\delta)Y\partial_Y\overline{f} + \frac{1}{X^2}(\delta^3 Y - 2\overline{\Phi})X\partial_X\overline{f} = 0, \tag{15}$$

$$2\overline{\Phi}\left(\frac{1}{\delta} + \frac{2}{\delta^2}\right) - 4\pi^2 G \int dY \int dX\overline{f} = 0. \tag{16}$$

One finds such equations, from which R has been removed, iff $\nu = 1$ and $\lambda = 2(1+\delta)$. These conditions restrict the 'direction' of the Rescaling Symmetry in the manifold and reduce the choice to a one parameter set, labelled say by δ. ***Any solution to this pair of equations for any acceptable symmetry δ represents a multi-dimensional self-similar system*** in the sense of the theme of this paper. In fact as discussed in [2], the only reasonable classification of the Rescaling Symmetries is simply the value of δ. In time dependent problems normally both α and δ are free (corresponding physically to different initial conditions), but only the ratio affects the solution. This ratio then becomes the appropriate classification.

We will not discuss the physical import of these results here since this has been done in [7], where the general solution is given. Suffice it to say that it turns out that this class of rescaling symmetric, stationary, self-gravitating, collisionless systems contains most of the known simple analytic solutions. Their general form satisfies Jeans' theorem as it must, but it is now much more restricted by the Rescaling Symmetry.

3.2. Asymptotic Self-Similarity: Speculations

For the deeper question of the place of Self-Similarity in the asymptotic development of such systems (a question of importance for the formation of galaxies and of clusters of galaxies, particularly for the 'dark matter' component) we should treat a time-dependent system. A model system has been identified previously in astrophysics and has been treated recently in [8] using the present methods. It is about the simplest non-trivial case imaginable consisting of material at rest at $t = 0$ in a power law density distribution, which is then free to collapse under its own self-gravity. Only radial orbits are permitted for simplicity, but the physics is expected to be similar for eccentric orbits. Orbits that approach the circular state pose a different physical problem more related to that of the Solar System (e.g. Graner, these proceedings). Moreover the matter is collisionless, so spherical shells of particles pass through one another as they mix.

The formulation is in terms of a canonical 1d DF F defined by

$$f = \frac{F}{4\pi^2}\,\delta(j^2) = \frac{F}{4\pi r^2}\,\delta(v_\theta)\delta(v_\phi), \tag{17}$$

and the CBE and Poisson equations are respectively in the usual notation

$$\partial_t F + v_r\partial_r F - \partial_r\Phi\partial_{v_r}F = 0, \tag{18}$$

$$\partial_r(r^2\partial_r\Phi) = G\int F\,dv_r. \tag{19}$$

We proceed to introduce the self-similar variables (T, X, Y) following the method presented above as

$$X = re^{-\delta T}, \tag{20}$$

$$Y = v_r e^{-\nu T}, \tag{21}$$

$$e^{\alpha T} = \alpha t; \tag{22}$$

where the Dimension basis, again reduced by the invariance of G, is *time, length, velocity* so that $\mathbf{a} = (\alpha, \delta, \nu)$.

Now to impose Rescaling Symmetry we know that we proceed by substituting the forms (5) into the governing equations. However since we want to study the *approach* to Self-Similarity in this system, let us transform to the (T, X, Y) variables without imposing the Self-Similarity through the strict use of (5). We will still use (5) as an ansätz, but now the overlined functions are in general functions of T as well as X and Y. Any function may be written in this form without loss of generality [2].

Substitution of this Ansatz into the governing equations with the corresponding change in variables yields respectively

$$\partial_T\overline{F} + \nu\overline{F} + (Y - \delta X)\partial_X\overline{F} - (\nu Y + \frac{d\overline{\Phi}}{dX})\partial_Y\overline{F} = 0, \tag{23}$$

$$\frac{d}{dX}\left(X^2\frac{d\overline{\Phi}}{dX}\right) - G\int \overline{F}dY = 0, \tag{24}$$

provided that the consistency relation $\nu = \delta - \alpha$ holds. In fact we may take $\alpha = 1$ without loss of generality. These equations describe the system in complete generality without an imposed Self-Similarity. However they have

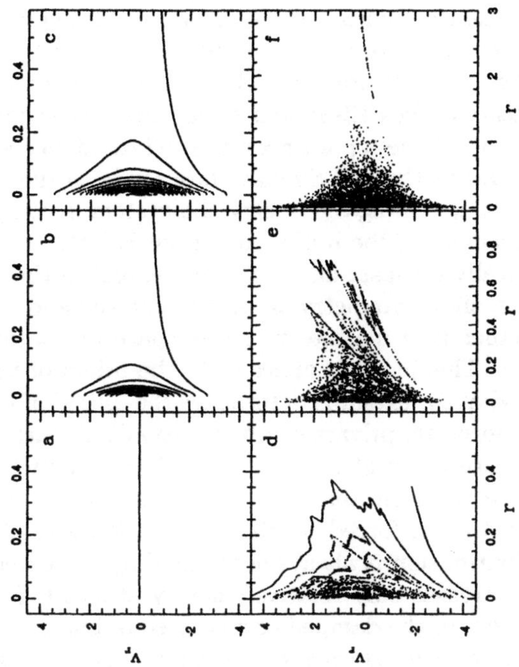

Fig. 1. — Numerical development from an initial density profile $\propto r^{-3/2}$ as reported in [8]. In units where $G =$ *total mass*$= 1$ the times corresponding to the ordered panels are $0.0, 0.4, 0.6, 1.0, 1.2, 5.6$ respectively. There is a discrete Rescaling Symmetry until panel d after which it relaxes to a continuous symmetry.

the merit of allowing a dynamical approach to such a symmetry to be readily recognized as the state with $\partial_T\overline{F} = 0$ (and hence $\partial_T\overline{\Phi} = 0$). The equations are exactly those of a multi-dimensional time dependent Self- Similarity when this constraint is applied [8].

Now the system may be easily simulated numerically [8] and the time development from the initial state is illustrated in figure 1.

We observe that there is an early stage of 'phase mixing' followed by a phase of 'violent relaxation' shown in panel d that is driven by a radial instability. ***In both phases we find Rescaling Symmetry to be present although it is discrete in the mixing phase.*** The Rescaling symmetry is proved to exist essentially by showing numerically that the system is stationary in the variables (T,X,Y). Moreover [8] certain dependences on the initial power law may be understood in terms of critical points in the family of characteristics of the self-similar equations (23) and (24). Specifically the initial power law index ϵ (defined so that $\rho_o(r) \propto r^{-\epsilon}$) is found to be related to δ through $\epsilon = 2/\delta$. Thus in this case the similarity 'class' is simply determined by the initial density distribution. Looking at the system of characteristic equations of equation (23) one finds that when $\delta < 1$ there is a separatrix or saddle type critical point in the system of trajectories. This seems to trap the relaxed or chaotic region since one finds that the mean density in the Self-Similar state varies once again with the power law index ϵ. When $\delta > 1$ the separatrix moves off to infinity and one finds that the mean density law in the Self-Similar state tends to the universal value 2. Thus in this class of Rescaling Symmetry the system is completely chaotic in that all detailed memory of the initial conditions is lost.

Thus asymptotically (at least until the finite extent of the initial system becomes evident) Rescaling Symmetry arises dynamically and it appears to make no difference whether it is discrete in phase space or continuous (after the violent relaxation). The question arises as to the origin of the discreteness in the early phase. What for example determines the 'pitch' of the spiral in phase space in figure 1 (Sornette, private communication)? A partial answer may be found by looking once again at the system of characteristics of equation (23). One can write these in a form that shows $X(T)$ to satisfy a damped harmonic oscillator equation that is forced by the term $-d\overline{\Phi}/dX(T,X)$. When $\delta > 1$ the initial mass distribution is always dominated by the exterior particles and the forcing term increases only relatively slowly at small X. In this case the homogeneous solution to the damped oscillator equation is a damped harmonic function *where the frequency is equal to the square root of the damping constant*, namely $\delta - 1$. To the extent then that we may ignore the forcing term in the early transient phases, this predicts a decay factor in X of $exp{-}(2\pi\sqrt{\delta-1})$ per period at least initially. By equation (20), we obtain the factor in r by multiplying by $exp\ (2\pi\delta/\sqrt{\delta-1})$ to obtain finally the *expansion* factor in r as $exp\ (2\pi/\sqrt{\delta-1})$. Unfortunately our present simulations do not have the resolution to test this simple argument in the early stages at small r (and it cannot be numerically correct later), but it suggests at least why the discreteness first appears.

Rather than relate all of the details of these results here I refer the reader to [8], and prefer to raise the question as to whether Self-Similarity in these systems is asymptotically inevitable and so represents perhaps a kind of generalized stationarity [4]. In the space remaining I have only room to note that the method of the Lyapounov functional (e.g. [9]) may be useful in connection with the formulation of equations (23) and (24).

Thus I find using equation (23) that the functional

$$H = -\int_{Y-\delta X\geq 0} h(X,Y,\overline{F})\, dX dY, \tag{25}$$

with $h \equiv \frac{\overline{F}(Y-\delta X)^2}{2}$ is strictly increasing with T for $\delta > 1$ provided that

$$\overline{F}(X=0;Y,T) = \mathcal{F}(T)\delta(Y), \tag{26}$$

$$\overline{F}(X=X_b;Y,T) = 0. \tag{27}$$

The first condition says that any particle found at the centre of the system is at rest, while the second condition requires the system either to be infinite, or to be compact in the invariant variable X with boundary X_b. Now evidently $\frac{dH}{dT} = 0$ in the Self-Similar state and H is bounded above by zero so that indeed the inevitable appearance of the Rescaling Symmetry is suggested.

For $2/3 < \delta < 1$ a similar conclusion holds with $h \equiv \frac{(1-\delta)\delta}{2}\overline{F}\ X^2$ and the double integral taken over the region $Y - \delta X \leq 0$. The condition at $X = 0$ is not strictly required in this case.

Such results are preliminary, but they suggest that the Rescaling Symmetry or Self-Similarity in these systems does arise dynamically from arbitrary initial conditions ***provided there is sufficient time during wich the system is not perturbed.*** However if the system is subsequently changed in character, say by cutting off the flow of particles, then the Self-Similar state is not likely to be the final state of the system which must rather find a finite stationary state. Thus Self-Similarity retains its intermediate character as expected [1].

Acknowledgments

I wish to acknowledge valuable discussions with my colleague L. Widrow, and to thank the organizers of this school for the opportunity to expound these ideas. It was good of them to arrange the Comet as entertainment for those of us too injured to ski!

References

[1] Barenblatt, G.E., Zel'dovich, Ia. B., *Ann.Rev. Fluid Mech.* **4** (1972) 285.
[2] Carter, B., Henriksen, R.N., *J. Math. Phys.* **32** (1991) 2580.
[3] Sornette, D., Ecole des Houches (Editions de Physique, Les Houches, 1997) p. 00.
[4] Hilfer, R.,*Physica A* **221** (1995) 89.
[5] Bluman, G.W., Kumei, S., Symmetries and Differential Equations (Springer, New York, 1989) p. 163.
[6] Fujiwara, T.,*PASJ* **35** (1983) 547.
[7] Henriksen, R.N., Widrow, L.,*Astrophys. J.* **276** (1995) 679.
[8] Henriksen, R.N., Widrow, L., *Phys. Rev. Let.* **78** (1997) 3426.
[9] Berryman, J.G., *J. Math. Phys.* **21** (1980) 1326.
[10] McCormack, F.D., Crane, L., Physical Fluid Dynamics (Academic Press, New York and London, 1973) p. 97.

LECTURE 5

Scaling in Stock Market Data: Stable Laws and Beyond

Rama Cont*,**,***, Marc Potters** and Jean-Philippe Bouchaud*,**

** Service de Physique de l'État Condensé, Centre d'Études de Saclay, 91191 Gif-sur-Yvette, France*
*** Science & Finance, 109-111 rue Victor Hugo, 92532 Levallois, France*
**** Laboratoire de Physique de la Matière Condensée, URA 190 du CNRS, Université de Nice, 06108 Nice cedex 2, France*

1. Introduction

The concepts of scale invariance and scaling behavior are now increasingly applied outside their traditional domains of application, the physical sciences. Their application to financial markets, initiated by Mandelbrot [1, 2] in the 1960s, has experienced a regain of interest in the recent years, partly due to the abundance of high-frequency data sets and availability of computers for analyzing their statistical properties.

This lecture is intended as an introduction and a brief review of current research in a field which is becoming increasingly popular in the theoretical physics community. We will try to show how the concepts of scale invariance and scaling behavior may be usefully applied in the framework of a statistical approach to the study of financial data, pointing out at the same time the limits of such an approach.

2. Statistical description of market data

Quantitative research mainly focuses on "liquid" financial markets i.e. organized markets where transactions are frequent and the number of actors is large. Typical examples are foreign exchange markets, organized futures markets and stock index markets and the market for large stocks. Prices are recorded several times a minute in such markets, creating mines of data to exploit. Such markets are complex systems with many degrees of freedom [3], where many internal and external factors interact at each instant in order to fix the transaction price of financial assets.Various factors such as public policy, interest rates and economic conditions doubtlessly influence market behavior. However, the precise nature of their influence is not well known and given the complex nature of the pricing mechanism, simple deterministic models are unable to reproduce the properties observed in financial time series. Furthermore, although the details of the price fixing mechanism- market microstructure - may be different from one market to the other, what is striking is the universality of some simple statistical properties of price fluctuations, prompting a unified approach to the study of different types of markets.

As in the case of other types of complex systems with universal characteristics, a stochastic approach has proved to be fruitful in this case. An investor buying a financial asset at time t and selling it at time $t+T$ is primarily interested in the (relative or absolute) variation of the price between t and $t+T$. The main object of study in this framework is therefore the probability density function (PDF) of the increments at a given time scale T i.e. the probability distribution P_T of $x(t+T) - x(t)$ where $x(t)$ is the price of the asset at time t. This approach was inaugurated by Louis Bachelier who first introduced the idea that stock market prices behave as a random walk [4], and who considered Brownian motion as a candidate for modeling price fluctuations.

Bachelier's model, applied to the logarithm of the prices (to ensure positivity of the price!), became very popular in the 1950s [5] and it is one of the main ingredients of the famous Black-Scholes option pricing formula [6]. This model implies that the increments of asset returns (or asset prices) are independent identically distributed (iid) Gaussian variables. Indeed if one considers each price change as a sum of many small and independent random contributions from various market factors, the Central Limit Theorem suggests the Gaussian as a natural candidate.

Figure 1. gives a typical example of the empirical distribution of the increments of asset prices, in this case the U.S. Dollar/ Deutschemark exchange rate, sampled every 5 minutes. The contrast with the Gaussian is striking: this 'heavy tailed' or 'leptokurtic' character of the distribution of price changes, has been repeatedely observed in various market data and may be quantitavely measure by the kurtosis of the distribution P_T defined by :

$$\kappa = \frac{\overline{(\Delta_T x - \overline{\Delta_T x})^4}}{\sigma(T)^4} - 3 \tag{1}$$

where $\sigma(T)^2$ is the variance of $\Delta_T x = x(t+T) - x(t)$. The kurtosis is defined such that $\kappa = 0$ for a Gaussian distribution, a positive value of κ indicating a slow asymptotic decay of the PDF. The kurtosis of the increments of asset prices is far from its Gaussian value: typical values for T=5 minutes are: $\kappa \simeq 74$ (US$/DM exchange rate futures), $\kappa \simeq 60$ (US$/Swiss Franc exchange rate futures), $\kappa \simeq 16$ (S&P500 index futures) [7, 8].

These observations imply that using a Gaussian PDF systematically underestimates the probability of large price fluctuations, an issue of utmost importance is financial risk management.

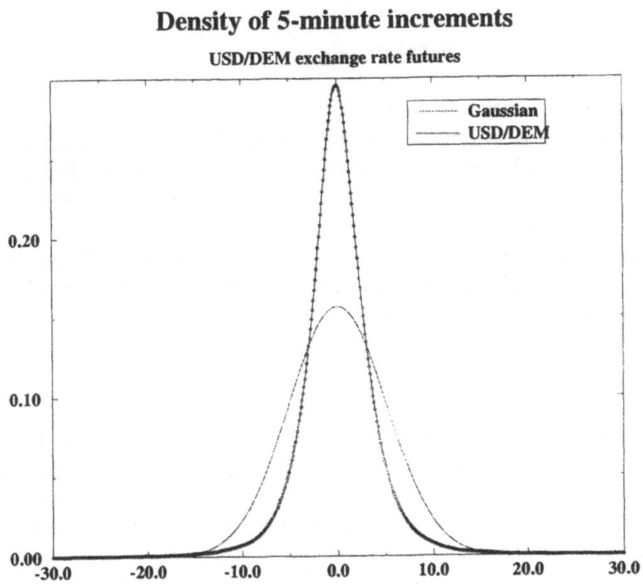

Fig. 1. — Probability density of 5 minute increments of USD/Deutschmark exchange rate futures. The lower curve is a gaussian with same mean and variance.

3. Scale invariance and stable laws

But why should stock price fluctuations have scale invariant properties in the first place? The answer lies in a generalized version of the Central Limit Theorem [9]: the distribution of the sum of a large number of independent identically distributed random variables belongs to a family of distributions known as *stable* or *stable Lévy* distributions, characterized by their Fourier transform (characteristic function) [9, 10]:

$$\phi_\mu(z) = \exp\left[-a_\mu |z|^\mu (1 + i\beta \tan(\mu\pi/2) \frac{z}{|z|})\right] \tag{2}$$

The parameter μ, called the stability index, belongs to the interval $]0,2]$, $\mu = 2$ corresponding to the Gaussian distribution. β is a skewness parameter, $\beta = 0$ for a symmetric distribution. Lévy distributions are characterized by the property of being stable under convolution: in other words, the sum of two *iid* Lévy-distributed variables is also Lévy distributed, with the same stability index μ. More precisely, if $X_i, i = 1..n$ are n *iid* μ-stable random variables with zero mean then the renormalized sum:

$$S_n = \frac{\sum_{i=1}^{n} X_i}{N^{1/\mu}}$$

has the same PDF as the X_i. In particular, the sum scales as $N^{1/\mu}$ and not as $\sqrt{N}$ which is the case in diffusive random walks. Lévy distributions, in addition to their stability under convolution, have other interesting properties: except for the Gaussian ($\mu = 2$) all μ-stable PDFs have power-law tails with exponent $1 + \mu$:

$$p_\mu(x) \simeq \frac{C}{x^{1+\mu}} \tag{3}$$

leading to an infinite variance and heavy tails.

These observations led Mandelbrot to propose stable Lévy distributions as candidates for the PDF of price changes of financial assets [1, 2]. Mandelbrot observed that stable distributions offer a heavy tailed alternative to Bachelier's model while still enabling a justification of the model in terms of the Central Limit Theorem (see above). Furthermore, their stability under convolution gives rise to the scale invariance of the process: if appropriately rescaled, the increment at scale $N\tau$ will have the same distribution as the increment at scale τ:

$$P_{N\tau}(x) = \frac{1}{\lambda} P_\tau\left(\frac{x}{\lambda}\right) \; ; \quad \lambda = N^{1/\mu}$$

The above relation means that the price process $x(t)$ is *self-similar* with a self-similarity exponent which is the inverse of the stability index μ. Self-similarity in market prices was first remarked by Mandelbrot in the 1960s in his seminal work on cotton prices [1]. Recent studies have confirmed the presence of self-similarity and scale invariant properties in various markets: the Milan stock exchange [11], the S&P500 index [12], the CAC40 (Paris stock market) index [13] and foreign exchange markets [14, 15] and individual french stocks [16]. The value of μ may be estimated for example by examining the scaling behavior of the probability of return to the origin $P_T(0)$, which scales as $\frac{1}{N^{1/\mu}}$; which is indeed the method used in [12, 13]. The value of μ found depends on the market considered but all values fall between 1.5 and 1.7, μ being lower for more volatile markets.

4. Beyond scale invariance: truncated Lévy flights

Although the representation of price increments as Lévy-stable *iid* variables accounts for their leptokurtic character as well as their self-similar properties, it has several drawbacks: careful comparison of the distributions at various time scales show that in fact it is *not* scale invariant [8, 15, 7]. The self-similarity properties described above do not hold at all time scales but only for short ones, typically less than a week. For example, if the price followed a Lévy walk, different methods of estimating the self-similarity parameter would give the same value, namely $1/\mu$. Yet in practice, data from different time resolutions give different values of the self-similarity exponent, between 0.5 (the value for a Brownian diffusion) and 0.6 ($= 1/\mu$ with $\mu = 1.7$).

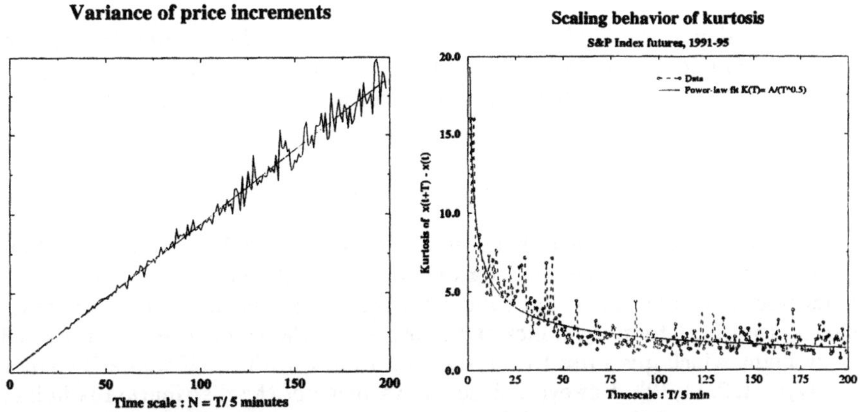

Fig. 2. — Left: Scaling behavior of the variance of price increments for S&P 500 index futures: the variance increases linearly with the time scale, a property which is consistent with the absence of significant linear correlation. Right: Scaling behavior of the kurtosis of price increments for S&P 500 index futures: the kurtosis decreases more slowly than in the case of iid increments where it decreases as $1/N$. The solid line represents a fit with $\kappa(T) = \kappa(1)/\sqrt{T}$.

Furthermore, while the best fit by a stable Lévy distribution [12] describes quite well the central region of the empirical distribution, it still overestimates the probability for extreme fluctuations: in other words, Lévy distributions have tails that are 'too fat' compared to fluctuations of real prices. Tails of empirical distributions may be studied more closely by rank-ordering techniques [17] (also called Zipf plots). They turn out to decay more slowly than a Gaussian but more quickly than any Lévy-stable distribution [8, 18]: their decay is better described by an exponential tail. Last but not least, real price changes turn out to have finite variance [8, 15]: the variance of a data set converges to a finite value when the size of the sample is increased rather than diverging to infinity as would be the case if the price changes where Lévy-stable *iid* variables.

These apparently contradictory aspects- self-similarity at short time scales, breakdown of scaling for longer time scales, truncated power law tails, finite variance- may be blended together into a consistent picture by a *truncated Lévy flight* description [19, 8, 20]. The idea is to use a distribution which coincides with a stable one around zero i.e. for small fluctuations, has heavy tails yet decreases to zero fast enough to assure a finite variance. This is done by truncating the stable distribution p_μ given above. The truncation may be done in various fashions: for example, $p(x)$ may be set to zero for x greater than some cut-off parameter as in [19]. We have chosen here a smooth exponential truncation as suggested in [21] which offers the advantage of resulting in an infinitely divisible distribution [9] and enabling an analytical expression for the characteristic function to be obtained:

$$\phi(z) = \exp\left(-a_\mu\left[\left((\alpha^2 + z^2)^{\mu/2}\frac{\cos\left(\mu\ arctan(|z|/\alpha)\right)}{\cos\left(\pi\mu/2\right)} - \alpha^\mu\right)\right]\right) \qquad (4)$$

The above expression differs from the characteristic function of a stable law by the presence of a cut-off parameter α. $\phi(z)$ is defined such that it has a regular

Taylor expansion at the origin, assuring the finiteness of all moments. In terms of the variable x, $p(x)$ decays exponentially for large values of x, which describes adequately the decay of the tails of the PDF of empirical price changes. The asymptotic behavior of ϕ is given by

$$\phi(z) \underset{z\to\infty}{\sim} \phi_\mu(z)$$

i.e. for small values of x, $p(x)$ behaves like a Lévy-stable law of index μ. We can now proceed to describe the behavior of a random walk with *iid* increments having a truncated Lévy PDF $p(x)$. For short time scales, $p(x)$ behaves like a stable law of index μ: the short term dynamics of the process is therefore correctly described by a Lévy flight which gives rise to the scale invariance with a self-similarity exponent $H = 1/\mu$ [1, 2, 11, 12]. However, since the variance of the distribution is finite, the Gaussian version of the Central Limit Theorem applies and for large time scales $T \gg T^* = \kappa\tau$, where κ is the kurtosis of the distribution $P_\tau(x)$. Thus at large time resolutions the distribution P_T will approach a Gaussian, which is consistent with the fact that the kurtosis tends to zero when $T \to \infty$ (Fig. 2). Between these two regimes -Lévy flight for short times and Gaussian diffusion for long times- is a crossover regime which is characterized by the appearance of fluctuations of the order $1/\alpha$.

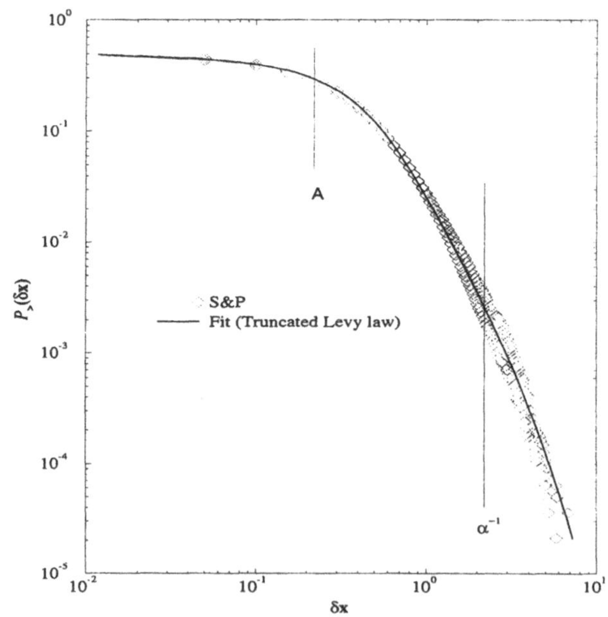

Fig. 3. — Rank ordering of 5-minute increments of S&P 500 index futures. The cumulative distribution function of price increments is well represented by an exponentially truncated Lévy distribution with finite variance.

5. Correlation and dependence

It is a well-known fact that price movements in liquid markets do not exhibit any significant autocorrelation: the autocorrelation function of the price changes

$$C(T) = \frac{\overline{\delta x_t \delta x_{t+T}} - \overline{\delta x_t}\ \overline{\delta x_{t+T}}}{var(\delta x_t)}$$

rapidly decays to zero in a few minutes: for $T \geq 15$ minutes it can be safely assumed to be zero for all practical purposes [20]. The absence of significant linear correlations in price increments and asset returns has been widely documented [22, 23] and often cited as support for the "Efficient Market Hypothesis" [24]. The absence of correlation is intuitively easy to understand: if price changes exhibit significant correlation, this correlation may be used to conceive a simple strategy with positive expected earnings; such strategies, termed *arbitrage*, will therefore tend to reduce correlations except for very short time scales, which represent the time the market takes to react to new information. This correlation time is typically several minutes for organized futures markets and even shorter for foreign exchange markets.

The fast decay of the correlation function implies the additivity of variances: for uncorrelated variables, the variance of the sum is the sum of the variances. The absence of linear correlation is thus consistent with the linear increase of the variance with respect to time scale.

However, the absence of serial correlation does not imply the independence of the increments [7]: for example the square or the absolute value of price changes exhibits slowly decaying serial correlations. Figure 4 displays the autocorrelation function $g(T)$ of the square of the increments, defined as:

$$g(T) = \frac{\overline{\delta x_t^2 \delta x_{t+T}^2} - \overline{\delta x_t^2}\ \overline{\delta x_{t+T}^2}}{var(\delta x^2)} = \frac{\overline{\delta x_t^2 \delta x_{t+T}^2} - \overline{\delta x_t^2}\ \overline{\delta x_{t+T}^2}}{\mu_4(\tau) - \sigma(\tau)^4}$$

for S&P 500 Futures.The slow decay of $g(T)$ (Figure 4) is well represented by a power law [7]:

$$g(k) \simeq \frac{g_0}{k^\alpha} \qquad \alpha = 0.37 \pm 0.037 \qquad g_0 = 0.08$$

This slow relaxation of the correlation function g indicates persistence in the *scale* of fluctuations. Other measures of the scale of fluctuations, such as the absolute value of the increments, exhibit the same type of persistence, a phenomenon which can be related to the "clustering of volatility", well known in the financial literature: a large price movement tends to be followed by another large price movement but not necessarily in the same direction.

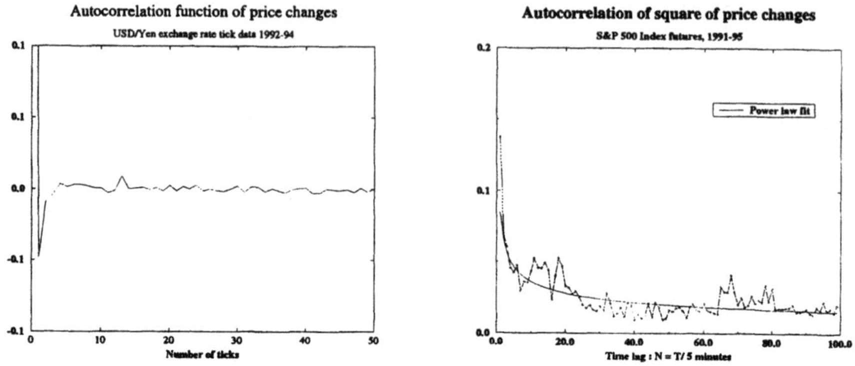

Fig. 4. — Left: Autocorrelation function of price increments for the US Dollar/ Yen exchange rate. Right: The autocorrelation function g of square of price increments of S&P futures exhibits a slow decay, well represented by a power law with exponent $\alpha \simeq 0.37$.

The persistent character of the scale of fluctuations may be quantitatively related to the anomalous scaling behavior i.e. slow decrease of the kurtosis as the time resolution increases. As a result, the convergence to the Gaussian is slowed down even more than in the case of a truncated Lévy flight with *iid* increments.

The presence of such subtle correlations points out the limits of a representation of market prices by a random walk model and the need to take into account finer effects due to nonlinear correlations and non-stationarity [25, 7]. These effects are especially important in options markets, where the behavior of volatility -the scale of price fluctuations- is crucial in determining the value of the option. It has been shown [25] that the correlations present in the volatility -as represented by the function g defined above- are accurately reflected in market prices for options.

6. Turbulence and finance

Recently some attempts have been made to draw analogies between scaling properties of foreign exchange rates and turbulent flows [26], based on some similarities between their probability density functions. Indeed, the statistical approach to the problem is the same, price increments playing the same role as velocity increments in the case of turbulence and a formal analogy may be very tempting. However, in the light of the above remarks, it is clear that such an analogy cannot be pushed too far: there is an essential difference between the two phenomena, namely, the presence of strong correlations between velocity increments, leading to Kolmogorov's famous $k^{-5/3}$ law [27], which are absent in the case of price increments [20, 28]. As remarked above, the autocorrelation function of price increments decreases rapidly to zero, resulting in a flat ("white-noise") spectrum (see figure below, taken from [20]). Translating into a power spectrum we obtain a dependence well approximated by ω^{-2}, which merely reflects the absence of correlation, as opposed to the $\omega^{-5/3}$ law for turbulent flows.

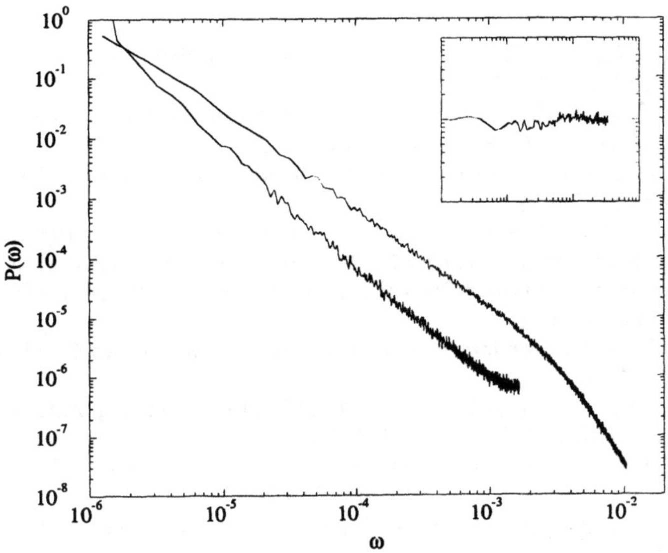

Fig. 5. — Fourier transform of the price autocorrelation function $\langle x(t)x(t+\tau)\rangle$ as a function of temporal frequency for the DEM-USD exchange rate (Oct 1991-Nov 1994) (*bottom curve*) as compared to the $k^{-5/3}$ power-law spectrum observed for the spatial velocity fluctuations in turbulent flows (the data were recorded by Y. Gagne and collaborators in a wind tunel experiment at $R_\lambda = 3050$) (top curve). Inset: Fourier transform of the price change autocorrelation function $\langle \Delta x(t)\Delta x(t+\tau)\rangle$, which is completely flat (and would behave as the $1/3$ $-th$ power of frequency in turbulent flows). Units are arbitrary.

7. Conclusion

The above example illustrates that although methods of statistical physics and in particular the concepts of scaling and scale invariance may be successfully applied in the study of market price fluctuations, sometimes revealing unsuspected properties of financial time series, analogies with problems in physics should be handled with care in order to avoid erroneous conclusions. Keeping this point in mind, the scientific study of financial markets proves to be a fascinating subject in itself, in particular for the physicist whose theoretical tools may prove to be useful in uncovering new properties and mechanisms in financial data.

• Acknowledgements:

We thank J.P. Aguilar for his helpful remarks.

References

[1] Mandelbrot, B. (1963) "The variation of certain speculative prices" *Journal of Business*, **XXXVI**, 392-417.
[2] Mandelbrot, B. & Taylor, H.M. (1967) "On the distribution of stock price differences " in *Operations Research*,**15**, 1057-1062.
[3] Pines, D. *et al.* (eds.) *The Economy as an evolving complex system* Santa Fe Institute, Addison-Wesley.
[4] Bachelier, L. (1900) "Théorie de la spéculation " *Annales Scientifiques de l'Ecole Normale Supérieure* , **III : 17**, 21-86. Bachelier's seminal work is also the first mathematical study of Brownian motion, preceding Einstein's famous work on the same subject.
[5] Cootner, P. (ed.) *The random character of stock market prices* Cambridge, MA: The MIT Press.
[6] Black, F. & Scholes, M. (1973) "The pricing of options and corporate liabilities" *Journal of Political Economy*, **81**, 635-654.
[7] Cont, R. (1997) "Scaling and correlation in financial time series" Science & Finance Working Paper 97-01 (cond-mat/9705075).
[8] Bouchaud, J.P. & Potters, M. (1997) *Théorie des risques financiers* Paris: Aléa Saclay (forthcoming).
[9] Gnedenko, B.V. & Kolmogorov, A.N. (1954) *Limit distributions for sums of independent random variables*, Addison-Wesley, Reading, MA.
[10] Bouchaud, J.P. & A. Georges (1990)*Physics Reports* **195**, 125.
[11] Mantegna, R.N. (1991) "Lévy walks and enhanced diffusion in the Milan stock exchange" *Physica A*, **179**, 232.
[12] Mantegna, R.N. & Stanley, H.E. (1995) "Scaling behavior of an economic index" *Nature*, **376**, 46-49.
[13] Zajdenweber, D. (1994) "Propriétés autosimilaires du CAC40" *Revue d'Economie Politique*, **104**, 408-434.
[14] Pictet O.V. *et al.* (1995) "Statistical study of foreign exchange rates, empirical evidence of a price change scaling law and intraday analysis" *Journal of Banking and Finance*, **14**, 1189-1208.
[15] Guillaume, D.M. *et al.* "From the bird's eye to the microscope: a survey of new stylized facts of the intra-day foreign exchange markets" Olsen & Associates Research Group working paper.
[16] Belkacem, L. (1996) *Processus stables et applications à la finance*, Thèse de Doctorat, Université Paris IX.
[17] Sornette, D., *et al.* (1996) "Rank-ordering statistics of extreme events : application to the distribution of large earthquakes", J.Geophys.Res. 101, 13883-13893.
[18] Cont, R. (1997) *Dynamical models of financial markets* (Doctoral thesis), Université de Paris XI.
[19] Mantegna, R.N. & Stanley, H.E. (1994) *Phys. Rev. Lett.*, **73**, 2946.
[20] Arnéodo,A. *et al.* (1996) "Comment on Turbulent Cascades in Foreign Exchange Markets" (preprint cond-mat/9607120) *Science & Finance* Working Paper 96-01.
[21] Koponen, I. (1995) *Phys. Rev. E* **52**, 1197.
[22] Fama, E.F. (1970) "Efficient capital markets: review of theory and empirical work" *Journal of Finance*,*25*, 383-417.

[23] Pagan, A. (1996) "The econometrics of financial markets" *Journal of Empirical Finance*,**3**, 15-102.

[24] Fama, E.F. (1965) "The behavior of stock market prices" *Journal of Business*,*38*, 34-105.

[25] Potters, M., Cont, R. & Bouchaud, J.P. (1996) "Financial markets as adaptive ecosystems" (preprint cond-mat/9609172) *Science & Finance* Working Paper 96-02.

[26] Ghashghaie, S. *et al.*(1996) "Tubulent cascades in foreign exchange markets" *Nature* **381** 767.

[27] Frisch, U. (1995) *Turbulence: The Legacy of A.N. Kolmogorov*, Cambridge University Press.

[28] Mantegna, R.N. & Stanley, H.E. (1996) *Nature*, **383**, 588-589.

[29] Feller, W. (1950) *Introduction to Probability theory and its applications*, **II**, 3rd ed., John Wiley & Sons, New York.

LECTURE 6

Hysteresis, Avalanches, and Barkhausen Noise

James P. Sethna, Olga Perkovic and Karin A. Dahmen*

Laboratory of Atomic and Solid State Physics, Cornell University, Ithaca, NY 14853-2501, U.S.A.
** Department of Physics, Harvard University, Cambridge, MA 02138, U.S.A.*

1. INTRODUCTION

We've been working on the crackling noise in hysteresis loops [1, 2, 3, 4, 5, 6]. Hysteresis occurs when you push and pull on a system with an external force, and the response lags behind the force. The hysteresis loop is the graph of force (say, an external magnetic field H) versus the response (say, the magnetization M of the material). In many materials, the hysteresis loop is not actually microscopically smooth: it is composed of small bursts, or avalanches (figure 1). In many first-order phase transitions, these bursts cause acoustic emission (crackling noise); in magnets, they are called Barkhausen noise.

Naturally, these pulses are associated with some kind of inhomogeneity or disorder in the material. Magnetic tapes are composed of small grains of iron oxide, and individual grains (when small enough) flip over as a unit, leading to a pulse in the magnetization. However, the pulses observed have a wide range of sizes: they can range over three to six decades in size in a typical experiment

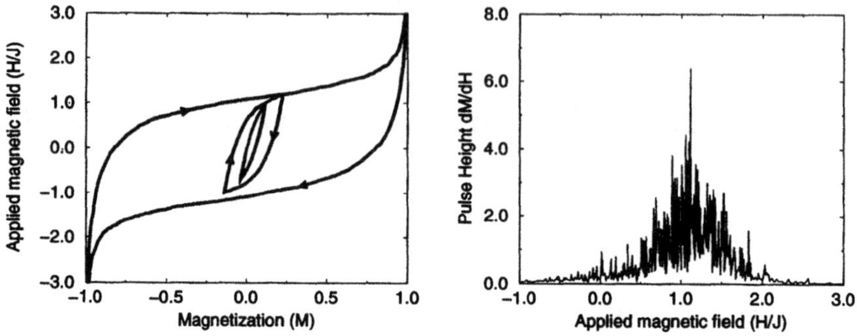

Fig. 1. — Left: A hysteresis loop in our model, showing the subloops. If you look carefully, you should be able to see small irregularities in the curve: these correspond to the avalanches causing the crackling noise. Right: The pulses in the upper branch of the outer hysteresis loop. Notice that the pulses become larger near $H = 1$ where the magnetization changes fastest with external field.

(figure 2). Since the grains in the material don't come in such a variety of sizes, one can conclude that many grains must be flipping at once, coupled together in a kind of avalanche.

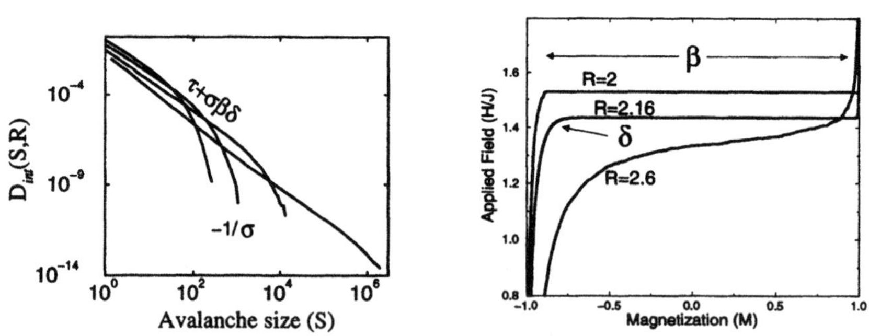

Fig. 2. — Left: The avalanche size distribution for our model in three dimensions, for disorders $R = 4$, 3.2, 2.6, and 2.25. The dashed line shows the expected behavior at the critical disorder $R_c \sim 2.16$ (a power law, $D_{int}(s, R_c) \propto s^{-(\tau+\sigma\beta\delta)}$). The curves are cut off after a few decades of scaling, at a size $S_{cutoff} \propto (R-R_c)^{-1/\sigma}$. Right: The jump in the avalanche size distribution ends at the critical disorder $R_c = 2.16$. The size of the jump scales as $\Delta M \propto (R-R_c)^{\beta}$. At R_c, the magnetization has a power-law form $H(M) \propto (M - M_c)^{\delta}$.

Having events of all sizes is not trivial! If the coupling between grains is weak compared to the disorder, the grains will tend to flip independently, leading to small avalanches. If the coupling is strong, a grain which flips will give a large kick to its neighbors, likely flipping several of them, who will flip several more — leading to one large avalanche. On the right in figure 2, we see that this is precisely what happens in our model. The large range of avalanches is associated with a critical value of the disorder R_c relative to the coupling: when the avalanches can't decide whether to be huge or small, they come in all sizes!

2. THE MODEL

To model these magnetic systems, we use a lattice of "spins" $S_i = \pm 1$, pointing up or down: each spin represents a domain or particle in the material. They are attached to their nearest neighbor (n.n.) spins by bonds of uniform strength J which we set for convenience to one; they are coupled to an external field $H(t)$ which we sweep from $-\infty$ to ∞. Finally, we model the inhomogeneities in the material with a random bias for up or down: at each site, we pick a random field h_i, distributed with a normal probability distribution $P(h) = \exp(-h^2/2R^2)/\sqrt{2\pi}R$. We call R the disorder: large R makes the distribution of h_i broad, makes the coupling between spins unimportant, and leads to smooth hysteresis loops and small avalanches. The energy of our system thus is

$$H = -\sum_{ij\ n.n.} JS_iS_j - \sum_i (H(t) + h_i)s_i. \tag{1}$$

Each spin flips as soon as its local external field $J\sum_{n.n.} s_j + H(t) + h_i$ changes sign: it then can kick over its neighbors if the resulting $2J$ change in their local fields is big enough. Thermal (and quantum) fluctuations aren't important for us, because the particles are large and (often by design) don't flip over spontaneously. Our model is called the random-field Ising model, and we simulate it at zero temperature [7, 8].

3. THE CRITICAL EXPONENTS

It's a remarkable fact that models like ours can accurately describe real systems near their transitions. The basic idea is a lot like hydrodynamics. All kinds of fluids look alike at long lengths and times, apart from their viscosity, density, and surface tension: despite rather different molecular structures and interactions, they all are described by the same equations for long distances and long times compared to the atomic scales. Similarly, a variety of hysteretic systems near the onset of a big jump (the "infinite avalanche") should be quantitatively describable by our simple model. This amazing property is called *universality*, and the family of models with common descriptions is called the *universality class*.

The most commonly measured universal quantities are the critical exponents. Many properties near the critical point have power-law scaling. This can be understood as a kind of self-universality: the system on one scale is quantitatively described by the same system at a different scale. Thus the properties of the system become scale invariant, and (in the usual way) develop power laws.

There are several critical exponents for our system that we measure and calculate. The most common is the power law giving the number of avalanches of a given size. This power law depends on whether you count all the avalanches in the hysteresis loop (the integrated avalanche size distribution plotted on the left in figure 2, whose power law is $\tau + \sigma\beta\delta$), or just measure them near the incipient jump in $M(H)$ (near which point the power law is τ). The exponents β, δ, and σ are also described in figure 2: they describe the shapes of the hysteresis loops and the cutoff in the avalanche size distribution as the disorder R is varied. The exponent ν describes the dependence of the size of the large avalanches on the distance from the critical point; the exponent z describes the lifetime of the large avalanches.

The first two columns of the left of figure 3 compare the two avalanche-size power-laws with experiments on a variety of materials [9]. While the scatter is large, the theory is well within the range of exponents measured. That doesn't mean we know the experiments are described by our theory: there might well be other universality classes with exponents not so far from ours... The other columns on the left of figure 3 represent combinations of exponents derived from different kinds of measurements: for example, the power spectrum of the noise from the hysteresis loops.

On the right-hand side of figure 3 we see the critical exponents in different spatial dimensions. Two dimensions might describe the behavior of magnetic tapes. Of course, dimensions greater than three have only theoretical interest! We'll see that we can learn something from high dimensions anyway...

4. THE EPSILON EXPANSION

What justifies us in thinking that our exponents and scaling is universal? How can we explain why we expect many systems to have exactly the same critical exponents (and scaling functions, and amplitude ratios, ...)? The theoretical justification of this was given by Leo Kadanoff, Ken Wilson, and Michael Fisher in the early 1970's, using what is called the *renormalization group* [11]. It's a bit technical, but theorists get unhappy unless they can point to a difficult calculation underlying their work.

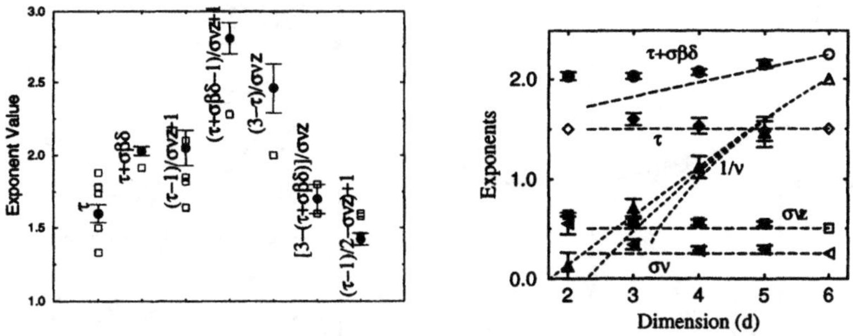

Fig. 3. — Left: Comparison of critical exponents with a variety of experiments [9], from [3]. Right: Exponents in various dimensions [3]. The dashed lines are from the ϵ-expansion [2,4,10].

The basic idea of the renormalization group is to think of the process of rescaling a system as a mapping from the space of all systems to itself! In statistical mechanics, one describes a system with a Hamiltonian: coarse-graining from one length scale to another maps Hamiltonian space into itself. In studying the period-doubling route for the onset of chaos, one describes the system with a mapping: rescaling the time by a factor of two is done by composing the mapping with itself. In our problem, we write a path integral for the probability of all histories for the system, and consider the effect of coarse-graining from one length-scale to another as a mapping taking one path-probability functional onto another [11].

This leap of abstraction is more useful than you might imagine. The subspace of all systems at their critical points must map onto itself. (If a system is on the verge of having an infinite avalanche, looking at it on a longer length scale won't change that.) Suppose one of the critical systems is a fixed-point under the mapping: suppose it has a "basin of attraction" of critical systems which flow towards it under coarse-graining. Then, on long length scales, all of these systems should look like the fixed point: the basin of attraction becomes the universality class.

Karin Dahmen [2,4] figured out how to implement this in a tangible calculation. In dimension d, each spin on a hypercubic lattice has $2d$ neighbors: in high dimensions, there are so many neighbors that one may just as well assume that every spin sees an average environment. This *mean-field theory* can be solved; it has a transition where the infinite avalanche first occurs, and that point leads to a fixed-point under the rescaling. In high dimensions (for us, dimensions $d > 6$) this fixed point is stable: all systems have the same critical properties as the mean-field fixed point. As the dimension decreases below $d = 6$, the mean-field fixed-point becomes unstable, but Karin did a pertur-

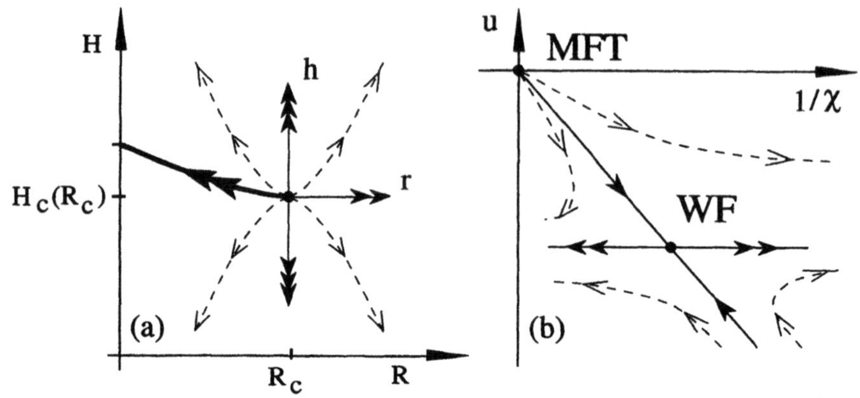

Fig. 4. — Left: Schematic phase diagram for our model, with arrows showing flows under coarse-graining. The dark line is $H_c(R)$, the external field at which the infinite avalanche occurs when the system is swept upwards from $H = -\infty$. Under coarse-graining, the effective external field $h = (H - H_c)/H$ grows fastest, and the effective disorder $r = (R - R_c)/R$ grows more slowly: all other directions are stable under coarse-graining. Right: Flow diagram near six dimensions, showing the mean-field fixed-point and the new, Wilson-Fisher fixed point. The mean-field fixed point is unstable below $d = 6$; at six dimensions, these two points merge.

bation theory in the dimension $\epsilon = 6 - d$ to find the new fixed-point to first order in ϵ (figure 4). The calculation is completely analogous to Ken Wilson's original calculation of the thermal, pure Ising model in $d = 4 - \epsilon$.

The results of Karin's calculation are shown in the right-hand part of figure 3. It's nice to see that the critical exponents approach their mean-field values as d approaches six, and that the ϵ-expansion captures the first corrections rather well. Indeed, the method works amazingly well, considering $\epsilon \geq 3$ in realistic problems! It so happens that the analogy to Ken Wilson's calculation is embarrassingly good: our calculation agrees with his to all orders in ϵ. Other people have calculated terms up to order ϵ^5 for the exponent ν in the pure, thermal Ising model [10]; we use their coefficients (in two higher dimensions) to predict $1/\nu$ in figure 3 right. There is one subtlety: the series for $\nu(\epsilon)$ doesn't converge without help (it's an asymptotic series), so we plot three different Borel resummations of the series for $1/\nu$ [10].

The fact that our series in $6 - \epsilon$ maps to Wilson's calculation in $4 - \epsilon$ is embarrassing for another reason. Our values for ν in $d = 3$ definitely do not agree with the pure, thermal Ising model in $d = 1$! This caused great anguish when it was first discovered in another context [10]. It's of course possible that we've erred in trying to perturb in a discrete variable like the dimension. The method seems to be working rather well, though, from figure 3. I personally believe it has something to do with the fact that the series doesn't converge for either problem: maybe we have one series trying to describe two different functions [3, 4, 12]?

5. WIDOM SCALING

It's important to stress that critical exponents are not the only predictions of the theory. I'd like to briefly discuss Ben Widom's discovery (later explained using the renormalization group): data for systems near criticality can be collapsed onto one another.

Consider again the avalanche size distribution (left figure 5, the same data as in left figure 2). Notice how the curves never quite lie along the dashed line, which I claimed was the power-law you would see if you were exactly at the critical point. Even when there are avalanches of size 10^6, the system still isn't exhibiting the critical exponents predicted! No wonder the experiments on the left of figure 3 don't agree with our theory: if you fit a power law to three decades of data, the exponent you get depends on $R - R_c$ and on which three decades you measure.

Does this mean our theory is useless? Not at all: our theory not only predicts the power laws, it also predicts the shape of the curves and the way they cut off. In particular, the avalanche size distribution is predicted to have the following form as $r = (R - R_c)/R \to 0$:

$$D_{int}(S, R) \sim S^{-(\tau+\sigma\beta\delta)} \mathcal{D}_{int}(s^{\sigma} r). \tag{2}$$

At the critical point $r = 0$: so long as $\mathcal{D}(0) \neq 0$, the distribution is a pure power law with power $\tau + \sigma\beta\delta$. Near the critical point, it starts to deviate from a pure power law when the argument of $\mathcal{D}$ becomes near one — that is, when the avalanche size $s \sim r^{-1/\sigma}$. Usually, the scaling function $\mathcal{D}$ is constant for small arguments, and dies away exponentially for large arguments. Again, we emphasize that the whole function $\mathcal{D}(x)$ is a universal property just as the critical exponents are.

We can make a plot of the scaling function, by taking our data and rescaling it:

$$\mathcal{D}_{int}(s^{\sigma}(R - R_c)/R) \sim S^{\tau+\sigma\beta\delta} D_{int}(S, R). \tag{3}$$

That is, if you plot $S^{\tau+\sigma\beta\delta} D_{int}(S, R)$ versus $s^{\sigma}(R - R_c)/R$, data taken at different R will all collapse onto the same curve. This is what Widom discovered in the early days of critical phenomena. The data collapse is shown in the inset to the left side of figure 5. The theory tells us further that any other system governed by the same universality class (one hopes someday a real experiment) will also rescale onto this same curve.

Now we can understand why our pure power-law is so elusive. The scaling curve almost vanishes at $S^{\sigma} r = 0$: it rises by about an order of magnitude

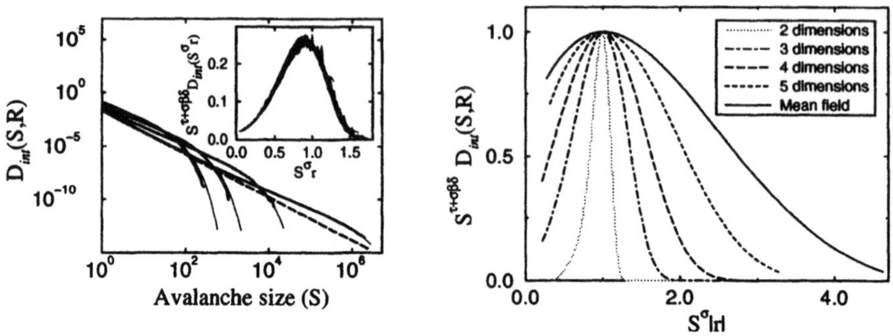

Fig. 5. — Left, main: The avalanche size distribution for our model in three dimensions, for disorders $R = 4$, 3.2, 2.6, and 2.25. The smooth curves going through the data are the scaling predictions of the theory. The dashed line shows the expected behavior at the critical disorder $R_c \sim 2.16$. Left, inset: The scaling collapses of these curves. The reason the slope of the avalanche size distribution converges so slowly is the large bump in this curve: it grows by an order of magnitude from its value at zero to the peak value. Right: The universal scaling curves in different dimensions. The explanation for the large bump in three dimensions is that the scaling curve $\mathcal{D}(0) = 0$ in two dimensions. We believe the exponent for the decay of the avalanche size distribution in two dimensions is shifted by one, because of this zero [5].

before dying exponentially. Indeed, each of the sets of data shows a bulge of about a factor of ten above the pure power law. (Why is the bulge so big? The right-hand side of figure 5 is our explanation: it's because we're so close to two dimensions, where $\mathcal{D}_{int}(0) = 0$.) We can work backward from the scaling curve and make predictions for the avalanche size distribution for each of the different disorders: the smooth curves in figure 5 are predictions of the scaling theory. The power law still isn't useful at $R = 2.25$, where $r = 0.04$, but the complete Widom scaling prediction is quite successful all the way out to $R = 4$, almost twice the critical value.

6. CONCLUSION

So, we have an understanding of why the noise pulses in magnets can span such a range of scales: they are near a critical point where the hysteresis loop develops a jump. We have a scaling description of the behavior near the critical point. We have a rough explanation of the experimental observations. We also think we understand why the measurements might fluctuate so far from our predictions: instead of varying a parameter and doing a scaling collapse, the experiments only measure an effective power-law. We recommend trying to get closer (or farther) from the critical point.

Why do the experiments see power laws? That is, why are the samples they use near the critical point? Unlike more traditional phase transitions, our model has a large *critical range*: 4% away from the critical point we have six decades of scaling, and a factor of two away we still have two decades. Perhaps there are mechanisms which tune the system precisely to the critical point, but it seems likely that the experimentalists could just be lucky and pick their sample inside this large range.

Finally, why is our critical range so large? This technically doesn't have a clean answer: the size of the critical range isn't a universal property! Universal questions aren't the only important ones, of course. I think there are three contributing factors. (1) The critical exponent $\nu = 1.42$ in our model, where in the three-dimensional Ising model it is 0.63. That means that getting twice as close to R_c makes the length spanned by an avalanche grow by a factor of 2.7, where getting twice as close to T_c for the Ising model makes the correlation length grow only by a factor 1.55. (2) The size S of an avalanche is more like a volume than a length. Six decades of scaling in S should be thought of as roughly two decades in length scale. (Actually, since the avalanches aren't space filling in three dimensions, the volume $S \sim \xi^{1/\sigma\nu} \sim \xi^{2.6}$, so six decades in size S gives 2.3 decades in the length scale ξ.) (3) Three dimensions is close to two dimensions. As you can see from the right side of figure five, the behavior in three dimensions is far removed from the mean-field behavior of the model in six and higher dimensions. The fluctuations are extremely important; in two dimensions, we believe that they almost completely dominate, perhaps even preventing an infinite avalanche from ever occurring! If $R_c = 0$ in two dimensions, then the critical regime must span to $\infty \times R_c$; no wonder in three dimensions that it spans to $2 \times R_c$.

There are still many important unsolved problems here, but it's clear that the traditional methods of critical phenomena — Widom scaling, the renormalization group, and the ϵ expansion — have been remarkably useful.

Acknowledgements

We acknowledge the support of DOE Grant #DE-FG02-88-ER45364 and NSF Grant #DMR-9118065. This work was partly conducted at the Cornell National Supercomputer Facility, funded in part by NSF, by NY State, and by IBM. Further pedagogical information (along with your own copy of the simulation, in Java) is available at
http://www.lassp.cornell.edu/sethna/hysteresis/hysteresis.html.

References

[1] J. P. Sethna, K. A. Dahmen, S. Kartha, J. A. Krumhansl, B. W. Roberts, and J. D. Shore, *Phys. Rev. Lett.*, **70**, 3347 (1993). J. P. Sethna, K. Dahmen, S. Kartha, J. A. Krumhansl, O. Perković, B. W. Roberts, and J. D. Shore (reply), *Phys. Rev. Lett.* **72**, 947 (1994).

[2] K. A. Dahmen and J. P. Sethna, *Phys. Rev. Lett.* **71**, 3222 (1993).

[3] O. Perković, K. A. Dahmen, and J. P. Sethna, *Phys. Rev. Lett.* **75**, 4528 (1995).

[4] K. A. Dahmen and J. P. Sethna, *Phys. Rev. B* **53**, 14872 (1996).

[5] O. Perković, K. A. Dahmen, and J. P. Sethna, Condensed-Matter Archive preprint #9609072.

[6] Olga Perković and James P. Sethna, *J. Appl. Phys.* **81**, 1590 (1997). Deepak Dhar, Prabodh Shukla, James P. Sethna, (submitted to *J. Phys. A*).

[7] Mark Robbins introduced this model, (allowing only connected spins to flip), in the study of the depinning transition (*e.g.* in fluid invasion): Hong Ji and Mark O. Robbins, *Phys. Rev. B* **46**, 14519 (1992) and references therein.

[8] Other models have been studied in a similar context: see C. M. Coram, A. E. Jacobs, N. Heinig, and K. B. Winterbon, *Phys. Rev. B* **40**, 6992 (1989); E. Vives and A. Planes, *Phys. Rev. B* **50**, 3839 (1994); E. Vives, J. Goicoechea, J. Ortín, and A. Planes, *Phys. Rev. E* **52**, R5 (1995).

[9] Barkhausen noise in different magnetic materials: Fe, alumel, metglass [P. J. Cote and L. V. Meisel, *Phys. Rev. Lett.* 67, 1334 (1991); L. V. Meisel and P. J. Cote, *Phys. Rev. B* 46, 10822 (1992)], NiS [K. Stierstadt and W. Boeckh, *Z. Physik* **186**, 154 (1965)], SiFe [G. Bertotti, G. Durin, and A. Magni *J. Appl. Phys.* **75**, 5490 (1994), H. Bittel *IEEE Trans. Magn.* **5**, 359 (1969), U. Lieneweg *IEEE Trans. Magn.* **10**, 118 (1974)], 81%NiFe [U. Lieneweg and W. Grosse-Nobis *Intern. J. Magnetism* **3**, 11 (1972)], AlSiFe [G. Bertotti, F. Fiorillo, and A. Montorsi *J. Appl. Phys.* **67**, 5574 (1990)], and FeNiCo [J.S. Urbach, R.C. Madison, and J.T. Markert, *Phys. Rev. Lett.* **75**, 276 (1995)]. The sample shapes were mostly wires. The quoted exponents were experimentally obtained from the pulse-area distribution in a small bin of the magnetic field H (exponent τ), the pulse-area distribution integrated over the entire hysteresis loop ($\tau + \sigma\beta\delta$), the distribution of pulse durations in a small bin of H ($(\tau - 1)/\sigma\nu z + 1$), the distribution of pulse durations integrated over the loop ($(\tau + \sigma\beta\delta - 1)/\sigma\nu z + 1$), the power spectrum of the pulses in a small bin of H ($(3 - \tau)/\sigma\nu z$), the power spectrum of the pulses

integrated over the hysteresis loop $((3-(\tau+\sigma\beta\delta))/\sigma\nu z)$, and the distribution of pulse energies in a small bin of H $((\tau-1)/(2-\sigma\nu z)+1)$. Notice that these experiments are mostly done in geometries which minimize the effects of demagnetization fields.

[10] H. Kleinert, J. Neu, V. Schulte-Frohlinde, K.G. Chetyrkin, and S. A. Larin (*Phys. Lett. B* **272**, 39 (1991) and erratum in *Phys. Lett. B* **319**, 545 (1993)); A. A. Vladimirov, D. I. Kazakov, and O. V. Tarasov, *Sov. Phys. JETP* **50** (3), 521 (1979) and references therein; J.C. LeGuillou and J. Zinn-Justin *Phys. Rev. B* **21**, 3976 (1980), *J. Physique Lett.* **46**, L137 (1985), *J. Physique* **48**, 19 (1987); J. Zinn-Justin "Quantum Field Theory and Critical Phenomena", 2nd edition, Clarendon Press, Oxford (1993)); G. Parisi, lectures given at the 1982 Les Houches summer school XXXIX "*Recent advances in field theory and statistical mechanics*" (North Holland), and references therein.

[11] The ϵ-expansion has a vast literature: recent works to study depinning transitions that inspired us include O. Narayan and D. S. Fisher, Phys. Rev. Lett. **68**, 3615 (1992), Phys. Rev. B **46**, 11520 (1992), and Phys. Rev. B **48**, 7030 (1993); T. Nattermann, S. Stepanow, L. H. Tang and H. Leschhorn, J. Phys. II France **2**, 1483 (1992)

[12] There is a calculation (M. Mézard and A. P. Young, *Europhysics Lett.* **18**, 653 (1992) which shows that the dimensional reduction result corresponds to a replica–symmetric solution in an expansion in an expansion in the inverse of the number of spin components M. They find in dimensions three through five that the replica–symmetry–breaking solution produces a change to order $1/M$. Thus the dimensional reduction corrections are not exponentially small as the number of spin components gets large. We have checked (Kamal Bhattacharya, unpublished) that their calculation near six dimensions is nonetheless compatible with exponentially small corrections in $\epsilon = 6 - d$, even at large M.

LECTURE 7

Burgers Turbulence and the Energy Landscape of Randomly Pinned Objects

J.-Ph. Bouchaud

Service de l'État Condensé, Centre d'Études de Saclay, Orme des Merisiers, 91191 Gif-sur-Yvette cedex, France

The physics of elastic objects pinned by random impurities is certainly one of the most topical current themes of statistical mechanics [1]. Two different general approaches have been proposed to describe the *statics* of these pinned manifolds, for which perturbation theory badly fails. The first one is the 'functional renormalisation group' (FRG) which aims at constructing the correlation function for the effective pinning potential acting on long wavelengths using renormalisation group (RG) ideas [3, 4]. The second is the variational replica method which combines a Gaussian trial Hamiltonian with 'replica symmetry breaking' (RSB) to obtain results in the low temperature, strongly pinned phase [5, 6, 7]. The aim of our work [8] was to understand the connections between these two approches. We show that both formalisms suggest the same physical picture: the effective, long wavelength pinning potential is a succession of parabolic wells of random depth, matching on singular points where the effective force (i.e. the derivative of the potential) is discontinuous. These discontinuities induce a singularity in the effective potential correlation function, and are encoded in the replica language by the RSB. The replica calculation furthermore provides an explicit construction of this effective (random) potential, and hence, in turn, information on the statistics of the depth of the potential minima. We hope that our construction can also shed light on the dynamics of such objects. For example, their relaxation can be analyzed in terms of hops between the different minima ('traps'), corresponding to metastable long wavelength configurations. The statistics of barrier heights control the trapping time distribution, and hence the low frequency response and its possible aging behaviour [9, 10].

The Hamiltonian we consider is the following:

$$\mathcal{H}(\{\vec{\phi}(\vec{x})\}) = \int d^D\vec{x} \left[\frac{c}{2} \left(\frac{\partial \vec{\phi}(\vec{x})}{\partial \vec{x}} \right)^2 + V_0(\vec{x}, \vec{\phi}(\vec{x})) \right], \tag{1}$$

where $\vec{x}$ is a D-dimensional vector labelling the internal coordinates of the object, and $\vec{\phi}(\vec{x})$ an N-dimensional vector giving the position in physical space of the point labelled $\vec{x}$. Various values of D and N actually correspond to interesting physical situations. For example, $D = 3$, $N = 2$ describes the elastic deformation of a vortex lattice, $D = 2$, $N = 1$ describes the problem of domain walls pinned by impurities in 3 dimensional space, while $D = 1$ corresponds to the so-called directed polymer (or single flux line) in a $N + 1$ dimensional space. The elastic modulus c measures the difficulty of distorting the structure, and $V_0(\vec{x}, \vec{\phi}(\vec{x}))$ is a random pinning potential, which we shall choose to be Gaussian with a short range connected correlation function:

$$\overline{V_0(\vec{x}, \vec{\phi}) V_0(\vec{x}', \vec{\phi}')}_c = N W \delta^D(\vec{x} - \vec{x}') R_0 \left(\frac{(\vec{\phi} - \vec{\phi}')^2}{N} \right), \tag{2}$$

where W measures the strength of the pinning potential. The scaling with N is chosen to ensure a correct $N \to \infty$ limit, in the sense that the pinning part of the free-energy is of order N, while each component of the pinning force $-\frac{\partial V_0}{\partial \vec{\phi}}$ remains of order 1, and R_0 is a certain short range function. One aim of the theory is to understand how the microscopic pinning potential will affect the elastic manifold on long length scales, relevant for macroscopic measurements.

The FRG method consists in writing down a recursion relation for the correlation function of the potential acting on 'slow' modes $\vec{\phi}_<$, after 'fast' modes $\vec{\phi}_>$ (corresponding to wavevectors in the high-momentum shell $[\Lambda/b, \Lambda]$) have been integrated out using perturbation theory. This procedure has been addressed in considerable detail in Ref. [4]. Assuming that the statistics of the effective potential remains Gaussian, the final result of the mode elimination plus rescaling of space and fields $\vec{x} \to b\vec{x}$ and $\phi \to b^\zeta \phi$ results in the following equation:

$$\partial_l R = (\epsilon - 4\zeta) R + \zeta \phi^i \partial_i R + \frac{1}{8\pi^2} \left[\frac{1}{2} \partial_i \partial_j R \partial_i \partial_j R - \partial_i \partial_j R \partial_i \partial_j R(0) \right]. \tag{3}$$

where $b = e^{dl}$, and i, j run from 1 to N. Iteration of this equation from the 'initial' condition $R(y) = R_0(y)$ converges towards the *fixed point* $R^*(y)$, describing the long wavelength properties, which has the singular small y expansion [4]

$$R^*(y) - R^*(0) = \epsilon y [a_1 - a_{3/2} \sqrt{y}] + \ldots, \tag{4}$$

where $\epsilon = 4 - D$. Another way of stating this result is in term of the effective *force* f acting on the manifold, defined as minus the derivative of the effective potential with respect to ϕ. The force correlation function then behaves as

$$\overline{[f^*(\phi) - f^*(\phi')]^2} = 12\epsilon a_{3/2}|\phi - \phi'|. \tag{5}$$

Together with the assumption of Gaussian statistics, this suggests that the effective force acting on the manifold behaves, for $N = 1$, as a *random walk* in ϕ space.

The replica approach is, in some sense, more ambitious, since it provides an explicit probabilistic construction of the effective disordered potential seen by the manifold. On the other hand, the method can only be controlled in the $N \to \infty$ limit, where a Gaussian variational Hamiltonian becomes exact. This method has been explained in full details on several occasions: one has to compute the nth moment of the partition function in the limit where $n \to 0$. The average of Z^n can be seen as the partition function of the following $n-$replica Hamiltonian:

$$\mathcal{H}_n = \frac{c}{2}\sum_{a=1}^{n} \int d^D\vec{x} \left(\frac{d\vec{\phi}^a}{d\vec{x}}\right)^2 - \frac{WN}{2}\sum_{a,b} \int d^D\vec{x} \exp\left[-\frac{(\vec{\phi}^a(\vec{x}) - \vec{\phi}^b(\vec{x}))^2}{2N\Delta^2}\right], \tag{6}$$

where an effective attraction between replicas has emerged from the disorder average. The idea is to treat this interaction using a Gaussian trial Hamiltonian:

$$\mathcal{H}_v \equiv \frac{1}{2}\sum_{a,b}\sum_{\vec{k}} \vec{\varphi}^a(-\vec{k})G_{ab}^{-1}(\vec{k})\vec{\varphi}^b(\vec{k}), \tag{7}$$

where $\varphi^a(\vec{k}) \equiv L^{-\frac{D}{2}} \int d^D\vec{x}\ \phi^a(\vec{x})e^{-i\vec{k}\cdot\vec{x}}$, and L is the 'linear' size of the manifold.

The trial free-energy obtained with $\mathcal{H}_v$ depends on G_{ab} and reads $\mathcal{F}_v[G] =< \mathcal{H}_n >_v - \frac{N}{2\beta} Tr \ln G$; the optimal matrix G is then determined by minimizing $\mathcal{F}_v[G]$, which leads to a set of self-consistent equations for G_{ab}. The structure of G_{ab} in replica space can be non trivial in the limit $n \to 0$, corresponding to 'replica symmetry breaking'. The physical meaning of this procedure has already been described in detail in [5, 11, 2], and we shall come back to it later. From the knowledge of G_{ab}, one can reconstruct the statistics of the large scale pinning potential.

The solution of the self-consistent equations for $G_{ab}(\vec{k})$ was discussed in [5, 6, 8]. Let us specialize to the case $N = \infty$, and introduce two important physical quantities, namely:

• The Larkin-Ovchinnikov length ξ_{LO} separating a 'weakly distorted' regime for $|\vec{x}| < \xi_{LO}$, where all the displacements induced by the random potential are small compared to the correlation length of the potential Δ, from a strongly distorted regime. Simple dimensional arguments lead to [12, 6]

$$\xi_{LO} \equiv \left(\frac{c^2\Delta^4}{\hat{W}}\right)^{\frac{1}{4-D}}, \tag{8}$$

where $\hat{W}$ is a rescaled potential strength, defined as $\hat{W} = (2\pi)^2 W/(4-D)$. It is easily seen from the study of the linearised, random force problem, that a 'weak disorder' regime with non trivial wandering exponent only exists when $\hat{W}$ is small enough.

• A 'Reynolds' number Re (this terminology comes from the analogy with Burgers' equation [2]), defined as the ratio of the elastic energy stored in a volume ξ_{LO}^D to the temperature $\frac{1}{\beta}$. We shall define Re as

$$Re \equiv \beta \mathcal{C}(D) \hat{W}^{\frac{2-D}{4-D}} (c\Delta^2)^{\frac{D}{4-D}}, \tag{9}$$

where $\mathcal{C}(D)$ is a dimension dependent number. Note that for $D = 4 - \epsilon$ with ϵ small, $\mathcal{C} = \frac{\epsilon^2}{2}$.

We shall only consider the case of low temperature and weak disorder, so that $Re \gg 1$ and $\xi_{LO} \gg a$, where $a = \frac{2\pi}{\Lambda}$ is the small scale lattice constant which regularizes the integrals over $\vec{k}$. Under these conditions, the replica symmetry has to be 'fully' broken [5, 8]. The final result for the correlation function of the pinning potential acting on a small Fourier mode reads:

$$R_{RSB}(y) - R_{RSB}(0) = -\epsilon \hat{W} \frac{y}{2\Delta^2} \left(1 - \frac{\sqrt{\pi}}{2} \frac{\sqrt{y}}{\Delta \xi_{LO}^{D/2}} \right), \tag{10}$$

with $y = (\vec{\varphi}_0 - \vec{\varphi}_0')^2$. Quite remarkably, Eq. (10) has the same form as the FRG result, Eq. (4), provided $\hat{W}$ is chosen in such a way that ξ_{LO} remains fixed as $D \to 4$. This $y^{\frac{3}{2}}$ behaviour was first obtained within a replica theory in [2] in the case $D = 1$ (corresponding to Burgers' turbulence), and, as discussed now, correpsonds to the presence of 'shocks'.

Let us first discuss the replica construction of the effective potential in the simpler case $D = 1$ where a 'one step' solution holds [5, 2]. In this case, one has:

$$V_\Omega^*(\varphi) = -\frac{1}{\beta} \ln \left[\sum_\alpha e^{-\beta F_\alpha - \frac{(\varphi - \varphi_\alpha)^2}{u_c \Delta^2}} \right], \tag{11}$$

where α label the 'states', centered around φ_α and of free-energy F_α, both depending on the 'sample' Ω. The major prediction of the replica theory is that the F_α are exponentially distributed for 'deep' states [13] , i.e:

$$\rho(F_\alpha) \propto_{F_\alpha \to -\infty} \exp(-\beta u_c |F_\alpha|). \tag{12}$$

where u_c parametrizes the shape of the matrix G_{ab}. The full distribution of the effective force $\frac{\partial V_\Omega^*}{\partial \vec{\varphi}_0}$ (corresponding to the velocity in the Burgers problem) was analyzed in detail in [2]. Using the turbulence language, it was found that the velocity field organizes in a 'froth-like' structure of $N-1$ dimensional shocks of vanishing width in the limit $Re \to \infty$. Correspondingly, the potential is made of parabolas matching at angular points – the shocks. The singular behaviour of the force-force correlation function, Eq. (5), is due to the fact that with a probability proportional to the 'distance' $|\vec{\varphi}_0 - \vec{\varphi}_0'|$, there is a shock which gives a *finite* contribution to $\vec{f}(\vec{\varphi}_0) - \vec{f}(\vec{\varphi}_0')$. This means in particular that all the moments $\overline{|\vec{f}(\vec{\varphi}_0) - \vec{f}(\vec{\varphi}_0')|^p}$ grow as $|\vec{\varphi}_0 - \vec{\varphi}_0'|$ for $p \geq 1$, instead of $|\vec{\varphi}_0 - \vec{\varphi}_0'|^{\frac{p}{2}}$ as for Gaussian statistics. It is not clear how this strong departure from Gaussian statistics can be incorporated in an FRG treatment. The relation with Burgers' equation is not coincidental and actually quite interesting. Keeping $N = 1$ for simplicity, consider a toy model for the FRG mode elimination in which the renormalized effective potential is defined as

$$\beta V_R(\varphi_<) = -\ln\left[\int d\varphi_> e^{-\beta[\frac{c\Lambda^2}{2}\varphi_>^2 + V_0(\varphi_< + \varphi_>)]}\right]. \tag{13}$$

This means that $V_R(\varphi_<)$ is precisely the Cole-Hopf solution of the Burgers equation [15]:

$$\frac{\partial V(\varphi,t)}{\partial t} = \frac{1}{2\beta c\Lambda^2}\frac{\partial^2 V(\varphi,t)}{\partial\varphi^2} - \frac{c\Lambda^2}{2}\left(\frac{\partial V(\varphi,t)}{\partial\varphi}\right)^2 \tag{14}$$

with

$$V(\varphi, t=0) = V_0(\varphi) \qquad V_R(\varphi) = V(\varphi, t=1). \tag{15}$$

As is well known [15, 16], a random set of initial conditions (here the bare pinning potential acting on φ) develops shocks which separates as time grows, between which the 'potential' $V(\varphi)$ has a parabolic shape. Elimination of fast modes in a disordered system thus naturally generates a 'scalloped' potential, with singular points (which are smoothed out at finite temperature or finite *Re*) separating potential wells – the famous 'states' appearing in the replica theory.

In the case of continuous RSB, the effective potential is recursively constructed via a set of 'Matrioshka doll' Gaussians. For each length scale ℓ, one can define a characteristic value of the parameter $u(\ell)$ which plays the role of u_c in Eq. (12) and sets the scale of the energy fluctuations. The large scale structure of the effective potential is thus a succession of parabolas of depth $\propto \ell^\theta$, where θ is the so-called energy exponent, but this envelope structure is decorated by hierarchically imbedded parabolas corresponding to all the smaller length scales, between ℓ and ξ_{LO}, beyond which the shocks disappear, since one enters into the effectively replica symmetric random force regime.

We have thus shown that both the FRG and RSB suggest quite an appealing physical picture: the phase-space of the system is, on large length scales, divided into 'cells' corresponding to favourable configurations where the potential is locally parabolic, and whose depth is exponentially distributed. These cells are themselves subdivided into smaller cells, corresponding to larger length scales, etc.. This hierarchical construction is similar to the one usually advocated for the phase space of spin-glasses [17], based on Parisi's RSB solution of the Sherrington-Kirkpatrick model [5, 10]. The enormous advantage of random manifolds is that this construction can be directly performed in physical space.

An important consequence of this construction is that it allows us to discuss the dynamical properties for finite N [18]. In the case of a one-step RSB, one can directly calculate from Eq. (11), the distribution of the height of the *barriers* ΔE between two neighbouring wells, and finds that it decays exponentially as $\exp(-\beta u_c \Delta E)$. It is interesting to notice that the barriers thus behave in the same way as energy depths [20], a point recently studied in detail for randomly pinned lines in [21]. A natural picture for the dynamics is thus to imagine that the manifolds jumps from well to well, each of which representing a long-lived conformation of the manifold. Such a picture is corroborated by recent numerical simulations in $D = 1$, $N = 1$ [22]. The lifetime of each 'trap' is activated $\tau \simeq \tau_0 \exp(\beta \Delta E)$, and is thus distributed as a power-law $\tau^{-1-u(k)}$ for large τ, where the exponent $u(k) \propto k^\theta$ depends on the 'size' of the jump (i.e. the mode involved in the change of conformation), small $u(k)$ corresponding to large wavelengths. Then, as emphasized in [9] where precisely the same 'trap' picture was advocated for spin-glasses, the dynamics becomes non stationary and aging effects appear at low temperatures and/or long-wavelengths such that $u(k) < 1$. For example, the response of the manifold to a spatially modulated external field is expected to behave, for $t \ll t_w$, as $(\frac{t}{t_w})^{1-u(k)}$, where t_w is the time elapsed since the quench from high temperature. Correspondingly, the a.c. response should behave, for $\omega t_w \gg 1$, as $(\omega t_w)^{u(k)-1}$, again much in the same way as observed in spin-glasses [9]. For finite N however, one may expect that the exponential distribution of deep states ceases to be valid outside the scaling region, i.e. for $\Delta E >>> \frac{1}{\beta u(k)}$ [9, 14]. This will lead to 'interrupted aging' for modes such that $\ln t_w >>> u(k)^{-1}$. These equilibrated modes thereafter only contribute to the stationnary part of the response (or correlation).

It is thus rather satisfactory that the 'traps' appear naturally in the context of pinned manifolds through the replica description, and that this picture actually complement the 'droplet' construction. It would of course be gratifying to understand precisely how these ideas could be extended to finite dimensional spin-glasses.

Acknowledgments I want like to thank Leon Balents and Marc Mézard, with whom the content of this paper was elaborated.

References

[1] D. S. Fisher, M.P.A. Fisher, D. A. Huse, Phys Rev B 43, 130 (1991); G. Blatter, M. V. Feigel'man, V. B. Geshkenbin, A. I. Larkin and V.M. Vinokur, Rev. Mod. Phys. **66** (1994) 4; M. Kardar, D. Ertas, in 'Scale

Invariance, Interfaces and Non-equilibrium dynamics', NATO-ASI; T. Halpin-Healey and Y.C. Zhang; Phys. Rep. **254** (1995) 217
[2] J.P. Bouchaud, M. Mézard, G. Parisi, Phys. Rev. **E 52** (1995) 3656
[3] D.S. Fisher, Phys. Rev. Lett. **56** (1986) 1964
[4] L. Balents, D. S. Fisher, Phys. Rev. **B 48** (1993) 5949
[5] M. Mézard, G. Parisi, J. Physique I **1** 809 (1991); J.Phys. **A23** L1229 (1990)
[6] J.P. Bouchaud, M. Mézard, J. Yedidia, Phys. Rev B **46** 14 686 (1992),
[7] T. Giamarchi, P. Le Doussal, Phys. Rev. **B 52** 1242 (1995).
[8] L. Balents, J.P. Bouchaud, M. Mézard, J. Physique I 6 (1996) 1007.
[9] J.P. Bouchaud, *J. Physique I (Paris)* **2** (1992) 1705; J.P. Bouchaud, D.S. Dean, *J. Physique I (Paris)* **5** (1995) 265. See also: C. Monthus, J.P. Bouchaud, preprint cond-mat 9601012, to appear in J. Phys. A.
[10] V.S. Dotsenko, M. V. Feigel'man, L.B. Ioffe, Spin-Glasses and related problems, Soviet Scientific Reviews, vol. 15 (Harwood, 1990)
[11] M. Mézard, G. Parisi, M.A. Virasoro, "Spin Glass Theory and Beyond", (World Scientific, Singapore 1987)
[12] A. Larkin, Y. Ovchinnikov, J. Low. Temp. Phys. **34** (1979) 409
[13] This can actually be understood within the general context of extreme event statistics [14].
[14] J.P. Bouchaud, M. Mézard, in preparation.
[15] J. M. Burgers, 'The Non-Linear Diffusion Equation', D. Reidel Pub. Co. (1974)
[16] S. Kida, J. Fluid. Mech. **93** (1979) 337
[17] For experimental arguments in favour of a hierarchical organisation of phase-space, see: E. Vincent, J. Hammann, M. Ocio, p. 207 in "Recent Progress in Random Magnets", D.H. Ryan Editor, (World Scientific Pub. Co. Pte. Ltd, Singapore 1992).
[18] Infinite N properties have been thoroughly investigated in [19] but might belong to a different dynamical class since activated effects disappear, at least for short range correlations.
[19] L. Cugliandolo, J. Kurchan, Phys. Rev. Lett. **71** (1993) 173; L. Cugliandolo, P. Le Doussal, *Phys. Rev.* **E 53** (1996) 1525. S. Franz, M. Mézard, Europhys. Lett. **26** (1994) 209, Physica **A209** (1994) 1. For a recent review, see J.P. Bouchaud, L. Cugliandolo, J. Kurchan, M. Mézard, *Out of Equilibrium dynamics in spin-glasses and other glassy systems*, to appear in "Spin-glasses and Random Fields", P. Young Editor.
[20] L. B. Ioffe, V.M. Vinokur, J. Phys. C **20**, 6149 (1987).
[21] B. Drossel, M. Kardar, Phys. Rev. **E52** 4841 (1995).
[22] H. Yoshino, *Off-Equilibrium Dynamics of 1+1 dimensional Directed Polymer in Random Media*, preprint cond-mat 9510024.

LECTURE 8

Power Laws and Scale Invariance in Physical Metallurgy

Y. Bréchet and M. Lebyodkin

L.T.P.C.M., I.N.P. Grenoble, B.P. 75,
38402 Saint-Martin-d'Hères, France

1. INTRODUCTION

Physical metallurgy, and more generally materials science has a somewhat ambiguous status with respect to the topic of this school. On one hand materials scientists are definitely interested in scale invariance and numerous power laws exist in physical metallurgy, either to describe microstructures, or to model the relations between microstructures and properties. By nature, the activity of a materials scientist is to build bridges between different scales, and to explain macroscopic behaviour using microscopic ingredients. On the other hand, universality is not precisely the aim of materials scientists, and they are more interested in knowing the macroscopic consequences of a microscopic phenomenon, or conversely to derive from macroscopic observations information about the micromechanisms. A characteristic and almost caricatural example of this difference in attitude between the two communities - physics and materials science- occured concerning the structure of fracture surfaces: when it became clear that the self affine structure was the same irrespectively of the fracture mechanisms, physicist were overexcited by the universality of the result, whereas materials scientist were very disappointed! However, beside the different status in the two fields of the "generic vs specific" distinction, materials science provides numerous examples of scaling or power law behaviours, for which materials scientists have derived explanations relying on detailed mechanisms, but which appear to be applicable in a much wider range of situation that the ones implied by the hypothesis used in their derivation. The purpose of this lecture is to provide an introduction to the field of physical metallurgy (§2), to illustrate the concepts of power laws and scale invariance by examples taken from the field (§3), and to outline open questions in physical metallurgy for which tools developed in statistical physics might be useful (§4).

2. THE ACTORS AND THE LENGTH SCALES IN PHYSICAL METALLURGY

2.1. The actors of physical metallurgy

Metals and alloys are usually in a crystalline form, but their crystalline structure has some defects [1-3] which are responsible for most of the mechanical properties which make these materials attractive. These defects can be of structural nature, or of chemical nature. The structural defects are deviation from the crystalline regularity. The simplest structural defects are point defects. Those point defects have a thermodynamic equilibrium concentration and are interacting either through their elastic field, or because of the perturbation introduced in local electronic densities. The second class of structural defects are dislocation lines. Their role is crucial in plasticity, they are always out of thermodynamic equilibrium and are interacting elastically with a force inversely proportional to their separation distance. The third class of structural defects are grain boundaries which separate regions of the material with different cristalline orientations. These planar defects are also out of equilibrium thermodynamically, and present basically no long range stress field. Point defects are moving according to a random walk responsible for diffusion of species inside the crystal in the solid state. Dislocation lines couple to an applied stress (Peach Kolher force) and can either glide on the plane corresponding to their line direction and to the Burgers vectors, or climb out of this plane if they interact with point defects. Grain boundaries will move under capillary forces associated to their curvature. Dislocation lines, as well as grain boundaries can be prevented to move or slowed down in their migration due to "chemical defects". These obstacles to defect migration are of crucial importance to applications because they are the basic mechanisms responsible for structural hardening and grain size control, which give for instance, their attractive properties to aeronautic alloys. Metals are very seldom used as pure elements. They are alloyed and the interaction between the elements in an alloy lead to regions in the (Temperature T, composition c) plane where the thermodynamically stable situation is to be two phased. The transition between the one phase situation and the two phase regime may take place by different mechanisms labelled diffusive (i.e. involving diffusion) or displacive (with no long range atomic transport). The possibility of the transformations is indicated by the phase diagram whereas the kinetics of these transformations is described by the so called (TTT) curve (time, temperature, transformation). For practical consequences, the volume fraction transformed is not the only relevant parameter: the scale fo the obtained structure is also of central importance. When a transformation takes place, part of the available free energy is consumed creating interface between the two phases: the more energy available, the more interface can be created, and the finer will be the structure. As a "rule of thumb", the deeper below the stability limit of the phase diagram the alloy is treated, the finer will be the obtained microstructure: the typical length scale being $\Delta G/\gamma$ where ΔG is the available free energy and γ the interface energy.

2.2. A microstructure viewpoint on mechanical properties

At temperature below $T_m/3$, when submitted to a stress, a metal deforms [1-3]. First it deforms in a linear reversible manner. This elastic regime reflects the reaction of the perfect harmonic crystal to the applied stress, and the *strain* which is this response occurs immediately. When the applied stress exceeds a limiting value called the yield stress, dislocations start to unpin from obstacles and a plastic *strain rate* results: while deformation proceeds, dislocation multiply, impinge each other and the deformation will stop at a given strain due to this "self pinning" of dislocations. An increase in stress is needed to deform the metal further: this phenomenon is called "work hardening". When temperature is higher than $T_m/3$, recovery processes take place and the result of an applied stress is a constant strain rate (creep test). When the stress is too small to unpin dislocations, creep can occur directly by diffusion of point defects. The various possible mechanisms are summarised in the so called "Ashby map" [1]. Beside the monotonic loading mode, materials can be tested in fatigue [4]. In this test, a stress amplitude, or a strain amplitude is applied alternatively in tension and compression. Even if the range of loads is much lower than the load for fracture in a monotomic test, failure will finally occur by "fatigue" of the material. Fracture can also be studied on notched specimens: if an acute crack of length a is introduced in a material loaded with a stress σ, it can be shown that it will break when the applied stress σ is such as $\sigma(\pi a)^{1/2}$ reaches a critical value K_c called "toughness". Fracture has recieved a lot of interest in the physics community in recent years [7]. It might be worth stating a few basic ideas about the mechanisms of fracture [3]. Fracture can occur either in a brittle manner, or in a ductile manner. In the brittle case, the only energy needed to propagate a crack is the creation of new surface, and the crack remains acute. In the ductile case, plastic dissipation takes place and the work dissipated that way is 3 orders of magnitude larger than the work involved in creating new surfaces, moreover the propagating crack is no longer acute but is "blunted" by plasticity. These difference results in values of toughness which can be 50 times larger than in the brittle case. The fracture surfaces are also drastically different: brittle fracture exhibits "cleavage rivers" whereas ductile fracture show "dimple patterns". Seing these stricking morphological differences, and the related consequences on the macroscopic properties such as K_c, it is indeed very puzzling to observe that the self affine characteristics are equivalent, and this "universality" certainly is still waiting for an explanation [8].

2.3. Changing scales

The different "actors", either structural defects of chemical defects, span the whole spectrum of length scales form the atomic level (10^{-10} m) to the structural component level (10 m) [6]. One needs to superimpose on this picture three other length scales controlling either the transformations, or the mechanical properties: the diffusion length scale $L_d = D/V$ (D being the diffusion coefficient and V the interface velocity in a tranformation, the thermodynamic length scale $L_t = \gamma/\Delta G$ (already mentioned), and the micromechanical length scale $L_m = \mu b/\sigma$ (μ being the shear modulus, b the atomic spacing and σ the applied stress). L_d controls the extension of the diffusion field ahead of the moving interface, L_t is the natural length scale which appears at the beginning of a phase transformation and L_d is the length scale below which dislocation dislocation interactions is dominant compared to their interaction with the applied stress field. The field of physical metallurgy goes from 10^{-10} m to 10^{-2} m, above which the relay is taken by continuum mechanics. The ideal situation would be to derive from process parameters (solidification, rolling, heat treatments) the characteristics of the microstructures, then deriving from this microstructure local yield or fracture criteria, then translate these criteria at the macroscopic level (sample behaviour) into a constitutive equation (relation between σ, ε and ε) which can finally be included in a finite element code to calculate a structure.

3. SCALE INVARIANCE AND POWER LAWS IN PHYSICAL METALLURGY: EXAMPLES

There are no example in materials science, to our knowledge, where time or length scales are absent. Scale invariance in metallurgy means invariance of the distribution function of a quantity, normalised to its average value. The really interesting question is now the departure from this scaling behaviour under certain circomstances. Similarly, power laws occur frequently, but they can usually be explained by very simple arguments in which scale invariance plays no role. However, these power laws are more "robust" than their derivation, and therefore their significance may be deeper than the classical arguments suggest. The examples that follow aim at giving a representative but not a complete picture of the field.

3.1. Examples taken from phase transformations

Diffusive phase transformations in alloys conserve the order parameter which is the composition. For an homogeneous bulk precipitation where transport takes place by diffusion, the predicted asymptotic $t^{1/3}$ scaling law for the average size of particles R after a time t spent in the coarsening regime (i.e. when the driving force is only the reduction in surface energy) is well verified experimentally. The distribution in size aroud this average value is also invariant [4]. However, the predicted LSW distribution is far too sharp compared to the experimental one (Fig. 1). LSW results was proven for a dilute precipitation of spherical particles. Finite dilution effects have been studied extensively, they change the prefactor of the power law but not the exponent [9]. It is worth noticing that dendrite coarsening, which has the same solute conservation requirement, but a different morphology (both positive and negative curvatures are present) follow the same scaling law: the $t^{1/3}$ behaviour is in fact very "generic". However, when elastic interaction between precipitates are present (due to coherency stresses) structures are stabilised at a given size, coarsening stops, and the size distribution sharpens: scale invariance is no longer valid. So the state of the art is the following: asymptotic coarsening kinetics are well understood in the absence of elastic interactions. In spite of numerous studies on the behaviour of a pair of interacting precipitates [10], the collective effect of elastic interactions on coarsening kinetics and on particle distributions is still to be understood.

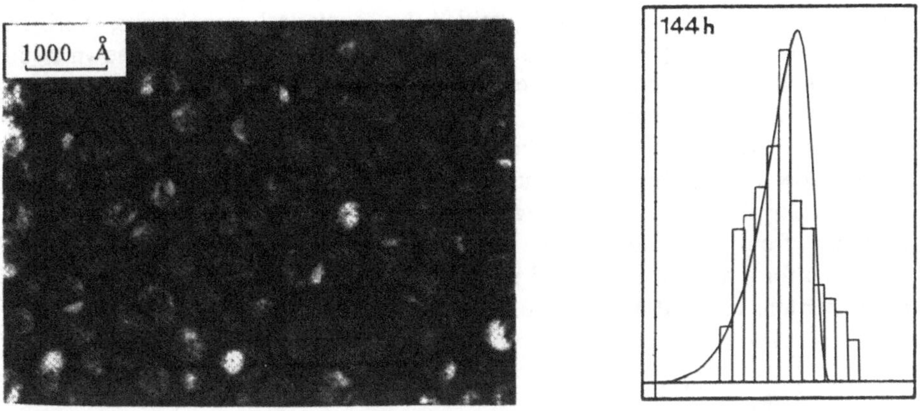

Figure 1: a) continuous precipitation in Al-Li alloys b) experimental size distribution and LSW theory

Another class of precipitation of interest with respect to scaling, in the one which takes place by interface diffusion at a moving grain boundary [11]. The structure is lamellar (just like a eutectic, but in the solid phase). The system undercooled at a given temperature selects both a length scale and a velocity. When both the parent and the daughter phase have the same structure (discontinuous precipitation) this selection mechanism is now well understood due to the different kinetics of lamellar and interlamellar growth [12] which are bound to grow cooperatively (which selects the velocity). The spacing between lamellae however shows a well define scaling behaviour [13] still not understood (Fig. 2). When the two phases have different structures (eutectoid reaction) the selection rule is still controversial.

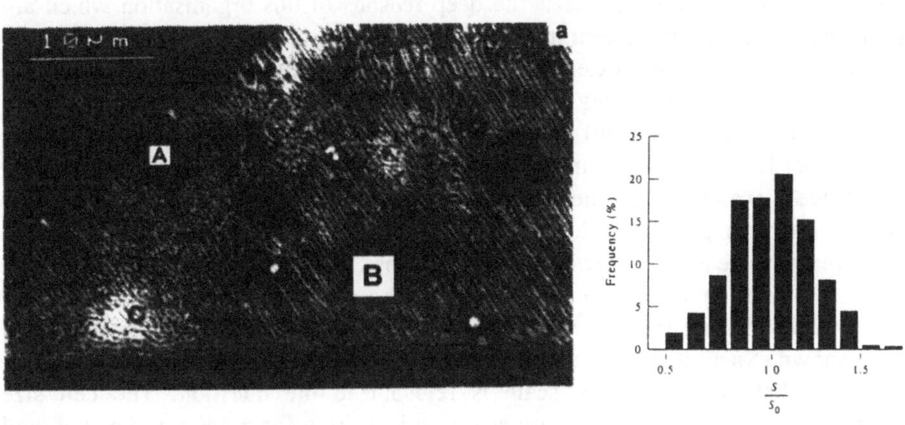

Figure 2: a) discontinuous precipitation in Mg-Al alloys b) spacing distribution

3.2. Organisation of structural defects

Grain growth, a phenomenon completely driven by capillarity, has also been extensively studied over the years. A $t^{1/2}$ behaviour for the average size has been predicted on simple arguments, verified by numerical simulations, verified experimentally on very pure metals. The size distribution of grains is shown to be Log Normal [14]. An interesting feature which has recieved little attention in the statistical physics community is precisely the deviation from this "standard behaviour". Grain boundaries can be pinned by a second phase (Zener pinning) or can be slowed down by the viscous drag associated with clouds of solute atoms (Cahn-Lucke drag). The result is a stagnation of the average grain size, and a departure from the Log normal distribution. The interesting question to be addressed for a metallurgist would be "what does this departure from the standard behaviour tells us about the nature of the pinning force or the drag force".

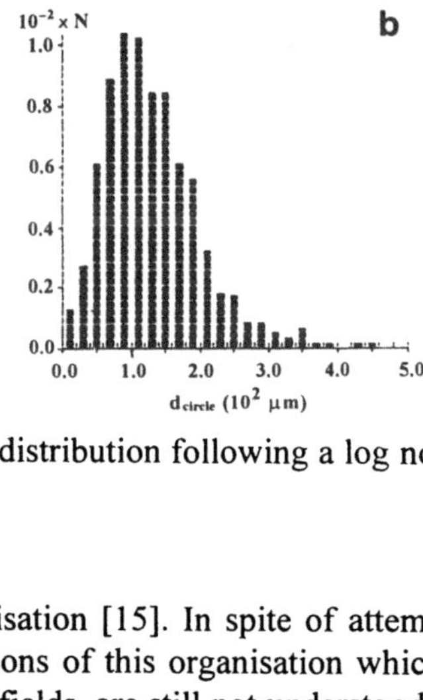

Figure 3: a) Grain structure in Mg b) grain size distribution following a log normal law

Dislocation have a stricking tendancy to organisation [15]. In spite of attemps to adapt "reaction diffusion models" the deep reasons of this organisation which are probably related to self screening of dislocation fields, are still not understood. The actual trend is to perform extensive 3D computer simulations to treat the collective effects. Very accurate scaling laws are observed: the flow stress scales as $\rho^{1/2}$ (ρ being the dislocation density) no matter how organised is the dislocation structure. This power law is simple to understand: the stress to escape obstacles separated by a distance L scales as 1/L, and the distance between dislocations acting as pinning centers for other pinning dislocations is on average $\rho^{-1/2}$. The average cell size is observed to scale as the inverse of the flow stress: this again can be understood by dimensional arguments assuming that $\rho^{-1/2}$ is the only length scale of the system. The problem is that i) these reasoning tell nothing about the prefactor, ii) it has been shown experimentally that the screening radius does not scale with $\rho^{-1/2}$ which indicates that another length scale is relevant to the question. The cell size distribution, and its possible invariance, is still waiting for a patient, devoted, and somewhat masochistic researcher to be determined experimentally.

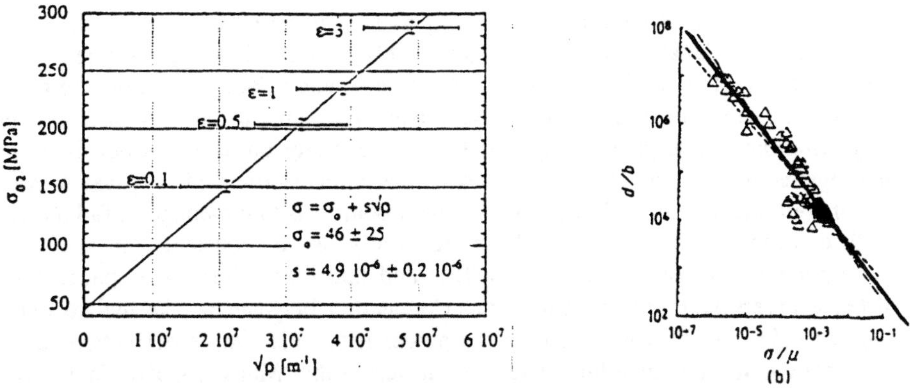

Figure 4: a) flow stress vs dislocation density b) cell size vs flow stress

3.3. Scaling laws for mechanical behaviour

Mechanical behaviour of alloys also provide numerous example of power laws. For instance the yield stress of a polycristal scales with $d^{-1/2}$ where d is the grain size. This law, known as Hall-Petch law, is one of the best established one in mechanical metallurgy. The classical derivation [16] proposed in textbooks involves pile ups of dislocations, formally equivalent to a crack, and concentrating the applied stress by a factor proportional to $d^{1/2}$ thus leading immediately to the power law. However, it is difficult to be convinced by such an explanation since Hall Petch law is valid even in pure Aluminium in which pile ups have never been observed, is valid not only for the yield stress but also for the flow stress. Moreover, the hardening by subgrains (whose only difference with grains is to be at least partially transparent to dislocations) leads to a scaling in 1/d. Clearly Hall Petch law is much more robust than its current derivation.

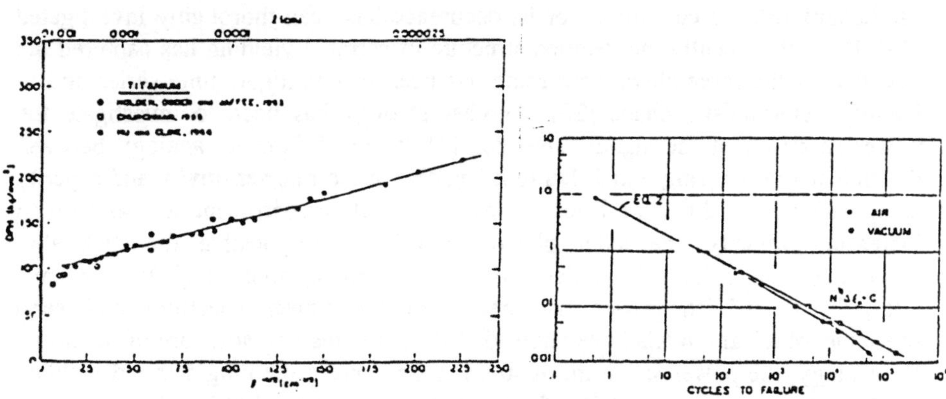

Figure 5: a) Hall Petch relation in Titanium b) Manson Coffin law in fatigue

A similar and somewhat frustrating experience occured to the author while providing an explanation for the very general Manson Coffin law in fatigue [17]. It is observed that the number of cycles to fracture N_f under fatigue loading at a given plastic strain amplitude $\Delta\varepsilon$ scales as $\Delta\varepsilon^{-2}$. This law seems to be very universal on polycristalline single phase alloys, and more generally on alloys which have no heterogeneous sites for crack nucleation. The argument proposed [17] was making use of well know facts in fatigue but the main point was that damage in fatigue is a surface event, that cracks nucleate randomly at a constant rate at sites corresponding to strain localisation (PSB) and that these sites were spaced by a distance inversely proportionnal to $\Delta\varepsilon$. Connection between cracks had to occur geometrically and thus was derived the power law. This simple derivation seems quite convincing for metallic alloys... but it occurs that Manson Coffin law is still valid for rubber for which none of the previous hypothesis is fulfilled. Then one is left with a doubt, either that Manson Coffin law is much deeper than the advanced reason which tried to incorporate all the known facts on fatigue, or that it is valid also in elastomers for completely different reasons.

An interesting example of scaling is observed in plastic instabilities. In normal situations, an increase in strain or in strain rate hardens the material. However a phenomenon called dynamical strain ageing (DSA) can lead to a negative strain rate sensitivity (SRS). This means that the higher the strain rate, the lower the stress necessary to produce it. One can easily be convinced that such a situation is not stable, and indeed the stress-strain curve exhibits some serrated yielding and plasticity is strongly heterogeneous, appearing under the form of propagative bands [18]. This phenomenon, known as Portevin Le Chatelier effect, is due to dynamic interaction between dislocations and diffusing solute atoms. In the intermediate regime where both the diffusion time for solute atoms and the waiting time for dislocations are comparable, which corresponds to a range of temperature and of strain rate, dynamical strain ageing takes place: the lower is the strain rate, the longer is the residence time of dislocations, the larger will be the solute content and therefore the stress to depin the dislocations. The serrated yielding corresponding to this unstable regime has been extensively studied as an example of relaxation oscillations [5], the conditions for its occurence has been thoroughly investigated [19]. But only recently the detailed structure of serrated yielding has captured the attention of the researchers. One route has been to investigate time series and to identify deterministic chaos [20]. Another strategy has been to investigate the power spectrum of the signal. The idea [5] stemmed from an analogy between dynamical strain ageing which led to a negative strain rate sensitivity and asperity creep which in solid friction leads to a dynamic friction coefficient decreasing with increasing velocity, the latter phenomenon being responsible for stick slip. Numerous studies have been devoted to stick slip phenomena [21]: they have allowed to establish power law nature of the power spectrum and semi phenomenological models have allowed to describe this fine structure of the noise. By analogy, the power spectrum of stress strain curve presenting serrated yielding has been studied experimentally (for a detailed account see [22]) and a power law behaviour over two decades of stress jumps amplitude was found only in a limited range of strain rate and temperature. The reason for this "non universality" can be easily understood: the system can be schematised as a collection of slices, each of them having a constitutive equation with a negative SRS leading to relaxation oscillations. A simple computer model [23] was built on this idea. As it is, the system has a natural time scale which is related to the periodicity of the relaxation oscillations. If the slices are strongly coupled, the size distribution of jumps will be bell shaped, whereas when they are weakly coupled a power law behaviour is observed. Seeking a power law is interesting, imposing a universal power law is missing a deep feature of the phenomenon. This simple example also shows that a whole range of complex behaviour concerning the spatial nature of plastic waves can be obtained with no further ingredient than the ones listed above. Last but not least, in order to obtain the observed exponent, when a power law is observed, the coupling between layers has been shown to be of the order of an elastic modulus. The information contained in the fine structure of the noise allows us to identify the mechanism by which spatial coupling between band enters: the elastic incompatibilities stemming from differently deforming slices. This example can be taken as a paradigm of what a materials scientist can expect from this type of studies: less a quest for universality than a new insight into the specificity of his system.

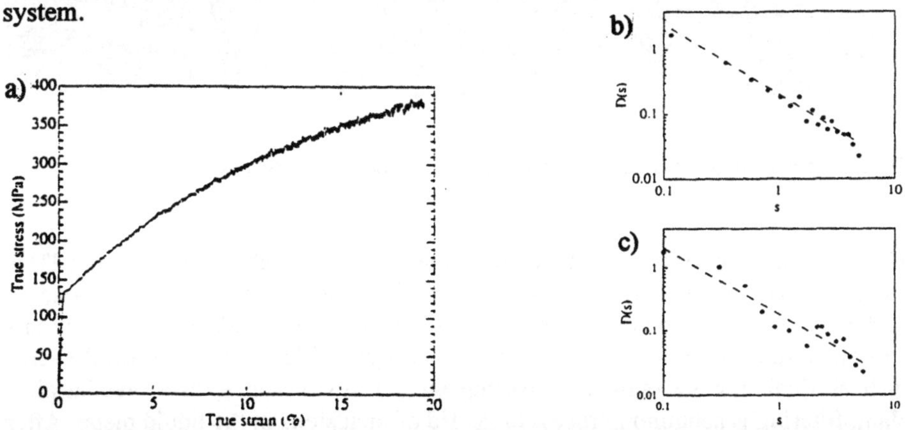

Figure 6: Portevin Le Chatelier effect. a) Serrated yielding b) scaling behaviour for the size distribution of stress drops c) scaling behaviour obtained from a "spring block" computer model

4. SUGGESTIONS FOR FUTHER WORK AND NEED OF NEW APPROACHES

Some open questions in materials science which would benefit greatly from the input of methods developed in the framework of scale invariance studies. One could say in a somewhat provocative manner that the materials scientist will be more interested in scaling breakdown, revealing a typical scale of the system, rather than in scaling invariance itself. The tools developed by mathematicians and physicists to describe scale invariance could be of great interest for situations, either in experimental observation, or in computer simulations, where structures are emerging from an apparently chaotic background. For instance, each time a dislocation arrives at a surface, it creates a step of a unit Burgers vector. The development of Atomic Force Microscopy allows one to measure the step height distribution which is a direct measure of avalanche sizes: certainly a careful analysis of these slip lines statistics wouls help undestanding fundamental aspects of plasticity. In recent years, intensive computer simulations [6] have been developed to deal with various problems in physical metallurgy. In particular the problem of dislocation self organisation has led to the development of a 3D dislocation population simulation. The strains attaignable are still rather limited, but the dislocation structures developed start to shown an incipient organisation, which is also visible on the stress field mapping. Using wavelet analysis on this complex problem could be a very interesting tool.

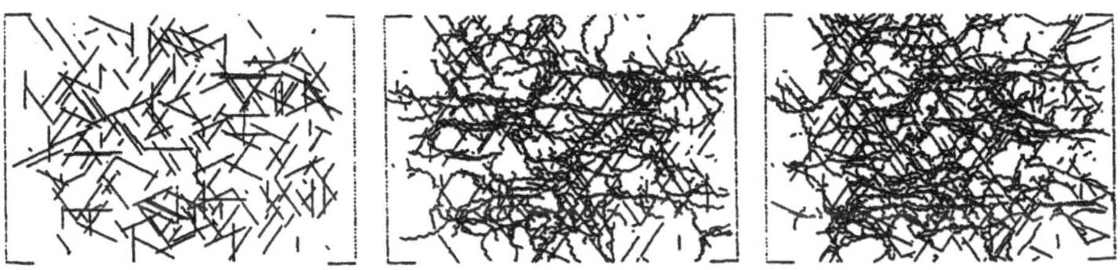

Figure 7: Simulated dislocation structures for increasing strains [24]

Materials science provides numerous situations in which a "noisy signal" is observed. The whole field of non destructive control (acoustic emission, or Barkhausen noise) would certainly benefit from the methods presented in this school. A special interest should be put on analysis able to predict critical events such as final fracture (similar to earthquakes...). For instance, in a metallurgical plant, filtering is acommon process to get rid of inclusions in the liquid metal. After a while, the filter will get stuck with particles and the necessary metallostatic pressure therefore the height of liquid metal above the filter increasing, this situation will lead to a catastrophic spilling of liquid metal. It is well known from the technical people in foundry that this catastrophy is "announced" by weird fluctuations in the liquid metal level. Another example of "noisy signal" recently studied in detail from the viewpoint of deterministic chaos concerns the electrochemical noise [25] which is a signature of corrosion of a metal. On the economical side, materials science can provide to physicists interested in economy relatively simple situations to model. It is well known that Aluminium consumption and production is somewhat erratic and this behaviour is likely to be related to the existence of this metal under three forms: primary production, material under use, and recycled materials.

5. CONCLUSION

Power laws and invariant distributions of rescaled variables occur frequently in materials science. These power laws can often be deduced from simple arguments, but are also often more general than these derivations would suggest. Situations were information is hidden in a "noisy signal" are also frequent and the application of the tools developed to characterize scale invariance seem very promising.

References

General bibliography on materials science : introductory references

[1] M. Ashby and D. Jones, Engineering materials Vols. 1. and 2 (Pergamon press, 1986).
[2] P. Haasen, Physical metallurgy (Cambridge University Press, 1992).
[3] "Materials Science and Technology", R. Cahn, P. Haasen, E. Kramers Eds., Vol. 4, "Plasticity and fracture", H. Mughrabi Ed., VCH Pub. (1993).
[4] "Materials Science and Technology", R. Cahn, P. Haasen, E. Kramers Eds., Vol. 5, "Phase transformations", P. Haasen Ed., VCH Pub. (1993).
[5] "Non Linear phenomena in materials science", L. Kubin, G. Martin Eds. (Transtech), Vol. 1 (1987), Vol. 2 (1992), Vol. 3 (1994).
[6] "Computer simulations in materials science", H. Kirchner, L. Kubin, V. Pontikis Eds. (Kluwer Academic Publisher, 1995).

Specialised references

[7] H. Hermann and S. Roux, "Statistical models for fracture " (North Holland, 1990).
[8] E. Bouchaud, in "Dislocations", *Transtech* (1993) 353-367.
[9] P. Voorhees, *J. Stat. Phys.* **38** (1985) 231-252.
[10] W. Johnson, *Met. Trans.* (1983) 2219-2227.
[11] G. Purdy, "Interfacial reactions ", in reference [4].
[12] L. Klinger, Y. Bréchet and G. Purdy, *Acta Materialia* (1997) in press.
[13] D. Duly, Y. Bréchet and M.C. Cheynet, *Acta Materialia* **42** (1994) 3843-3855.
[14] F. Humphreys, "Recrystallisation" (Elsevier, 1995).
[15] L. Kubin, "Dislocation patterning" in reference [3].
[16] J. Friedel, "Dislocations" (Pergamon, 1966).
[17] D. Sornette, Y. Bréchet and T. Magnin, *Europhys. Lett.* **20** (1992) 433-438.
[18] Y. Estrin and L. Kubin, "Continuum Models fo materials with microstructure", H. Mulhaus Ed. (Wiley and Sons, 1995) 395-435.
[19] Y. Estrin and L. Kubin, *Acta. Mat.* **38** (1990) 697-709.
[20] G. Ananthakrishna, C. Freyssengeas, L. Kubin and J. Vergnols, *Scr. Mat.* (1996).

[21] T. Baumberger, *Solid State Comm.* (1997) under press.
[22] M. Lebyodkin,Y. Bréchet,Y. Estrin and L. Kubin, *Acta. Mat.* **44** (1996) 4531-4541.
[23] M. Lebyodkin,Y. Bréchet,Y. Estrin and L. Kubin, *Phys. Rev. Lett.* **74** (1995) 4758-62.
[24] M. Verdier, PhD, Institut National Polytechnique de Grenoble (1997).
[25] H. Mayet, PhD, Institut National Polytechnique de Grenoble (1997).

LECTURE 9

Scale Invariance in Fluids with Anticorrelated Entropy-Specific Volume Fluctuations

H.E. Stanley, S.T. Harrington, P.H. Poole* and F. Sciortino**

Center for Polymer Studies and Physics Department,
Boston University, Boston, MA 02215, U.S.A.
** Department of Applied Mathematics, University of Western Ontario,*
London, Ontario N6A 5B7, Canada
*** Dipartimento di Fisica, Università di Roma*
La Sapienza, 00185 Rome, Italy

1. INTRODUCTION

Although water has been the topic of considerable research over the past 100 years, few would dare to claim that its peculiar properties are completely understood [1, 2]. Standard liquid theories fail to explain its dynamical and thermodynamic properties, which differ from those of most other liquids [3, 4, 5, 6]. Many workers believe that a central role is played by the water molecule's ability to form hydrogen bonds and to create a tetrahedrally coordinated open network in both the solid and the liquid phases [7, 8, 9, 10, 11, 12, 13].

In experiments, it is generally observed that the peculiar properties of water become more pronounced when the liquid is supercooled or stretched [5, 6]. In particular, thermodynamic response functions such as the specific heat, isothermal compressibility and thermal expansivity, as well as various relaxation times, seem to exhibit a power law type singularity at a certain temperature $T_A \approx 228$K, similar to the behavior commonly encountered when approaching a second order phase transition [6, 14, 15, 16].

2. THE HYPOTHESIS

Science can progress by making hypotheses, and then testing them. Accordingly, in this talk we will examine the hypothesis that the unusual properties of liquid water are related to the existence of a "second" critical point. First we present the hypothesis, then we explain why this hypothesis is worthy of critical examination, and finally we briefly review some of the facts that are consistent with the hypothesis.

Water has a well-studied critical point C with coordinates (T_C, P_C, ρ_C), where (to one significant figure) $T_C \approx 600$K is the critical temperature, $P_C \approx 20$MPa is the critical pressure, and $\rho_C \approx 0.3$ is the critical density characterizing C. We hypothesize that there may exist a second critical point C' with coordinates $(T_{C'}, P_{C'}, \rho_{C'})$, where (also to one significant figure) $T_{C'} \approx 200$K, $P_{C'} \approx 100$MPa, and $\rho_{C'} \approx 1$.

Just as C separates a 1-fluid region (a single phase) above T_C from a 2-fluid region (two phases) below, so also C' separates a 1-liquid region above $T_{C'}$ from a 2-liquid region below $T_{C'}$. The two fluid phases that can co-exist below C—called liquid and gas—differ in volume per molecule $\bar{V}$. The specific volume difference between liquid and gas phases is a suitable "order parameter," which continuously approaches zero as one approaches C displaying a power law singularity characterized by the critical exponent β. Similarly, the two liquid phases that can co-exist below C' differ in specific volume; we shall call these two liquid phases high-density liquid (HDL) and low-density liquid (LDL). Again, this density difference, because it approaches zero as one approaches C', is an order parameter characterizing the liquid-liquid critical point.

The phases differ not only in their volume per molecule $\bar{V}$; they also differ in the entropy per molecule $\bar{S}$. Below T_C, the gas has the larger entropy, while below $T_{C'}$ the HDL phase has the larger entropy. Hence, using the Clausius-Clapeyron relation, the coexistence line in the PT phase diagram has positive slope for C but negative slope for C'.

3. WHY CARE?

One might think it senseless to invest much energy studying the critical point, as it is only a single point in an entire phase diagram. However we know that this "one little point" exerts an immense influence on properties at points in the phase diagram that are far away. For example, there are significant deviations in response functions from their corresponding values for a non-interacting fluid at temperatures more than twice the value of the critical temperature T_C.

For water, response functions are also anomalous at temperatures roughly twice the temperature $T_{C'}$. For example, the isothermal compressibility K_T is anomalously *large* at temperatures as high as 400 K (100°C); K_T actually passes through a minimum at 319K (46°C), below which it increases by almost a factor of two as temperature is further decreased. Similarly, the coefficient of thermal expansion α_P is anomalously *small* for temperatures as high as 400 K, and actually decreases to zero at 277 K (4°C), below which it continues decreases at a continually faster rate. Finally, the constant-pressure specific heat C_P is anomalously large for temperatures as high as 400 K, a fact of tremendous practical importance for using water as a coolant in, e.g., a car engine. The dynamic properties of water are also anomalous, with pronounced non-Arrhenius (often power law) behavior typically occurring at low temperature. Thus if there were a critical point at, say, 200K, then the influence of this critical point could be of relevance to the anomalies observed in water at much higher temperatures.

4. WHY BELIEVE?

In the remainder of this talk, we shall discuss some tentative results that are not inconsistent with the critical point hypothesis. First we present an intuitive plausibility argument, then we discuss evidence from simulations and experiments.

4.1. Plausibility Arguments.

A non-interacting gas has no critical point, but a gas with arbitrarily weak attractive interactions does since at sufficiently small temperature, the ratio of the interaction to kT will become sufficiently significant to condense the liquid out of the gas. That all interacting gases display a critical point below which a distinct liquid phase appears was not always appreciated. Indeed, in the early years of this century one spoke of "permanent gases"—to describe gases that had never been liquefied. Helium is an example of what was once thought to be a permanent gas [17].

Nowadays, we understand that permanent gases cannot exist since all molecules exert some attractive interaction, and at sufficiently low temperature this attractive interaction will make a significant contribution. To make the argument more concrete, one can picture droplets of lower specific volume $\bar{V}$ forming in a single-component fluid. Once the interaction between molecules is fixed (and P is fixed at some value above P_C), then the only remaining control parameter is T; as T decreases the high-density droplets increase in number and size and eventually below T_C they coalesce as a distinct liquid phase.

Water differs from most liquids due to the presence of a line of maximum density (TMD line) in the PT phase diagram. This TMD is physically very significant, as it divides the entire PT phase diagram into two regions with remarkably different properties: the coefficient of thermal expansion—which is proportional to the thermal average ("correlation function") $\langle\delta\bar{V}\delta\bar{S}\rangle$—is *negative* on the low-temperature side of the TMD line, while it is *positive* on the high-temperature side. Here $\bar{V}$ is the volume per molecule, $\bar{S}$ the entropy per molecule, and the δX notation indicates the departure of a quantity X from its mean value.

That $\langle\delta\bar{V}\delta\bar{S}\rangle$ is negative is a thermodynamic necessity given the presence of a TMD line. What microscopic phenomenon causes it? One not implausible explanation [7] is related to the presence of local regions of the hydrogen bond network that are characterized by four "good" hydrogen bonds—and these local regions can be considered as droplets just as the high-density droplets in a gas above C. Stated more formally: the sensitivity of hydrogen bonds to the orientation of the molecules forming it encourages local regions to form that are partially ordered in the sense that if there is a region of the water network where each molecule has four "good" (strong) hydrogen bonds, then the local entropy is lower (so $\delta\bar{S} < 0$) and the local specific volume is larger (so $\delta\bar{V} > 0$), so the contribution to $\langle\delta\bar{V}\delta\bar{S}\rangle$ is negative for such regions.

As the temperature is lowered, there is no *a priori* reason why the "droplets" characterized by negative values of $\delta\bar{V}\delta\bar{S}$ should not increase in number and size, just as the droplets associated with a normal phase transition increase in number, since all water molecules exert mutual interactions on one another, and these interactions—because of their sensitivity to orientation and well as distance—favor the open clusters characterized by $\delta\bar{S}\delta\bar{V} < 0$. It is thus plausible that at sufficiently low temperature these orientation-sensitive interactions will make a larger and larger contribution, and at sufficiently low temperature (and for sufficiently low pressure), a new phase—having roughly the density of the fully hydrogen bonded network—will "condense" out of the one-fluid region.

This intuitive picture has received striking support from a recent generalization of the van der Waals theory. Specifically, Poole et al. [18] allow each water molecule to be in many bonding states, only one of which corresponds to a "good" quality hydrogen bond (with a larger number of states corresponding to "poor" quality bonds). To build in this feature, Poole et al. adopt the approach of Sastry and co-workers [19, 20] and assume that there are $\Omega \gg 1$ configurations of a weak bond, all having $\epsilon = 0$, and only *a single configuration* in which the HB is strong with $\epsilon = \epsilon_{\rm HB}$. Thus the thermal behavior of the HBs is represented by independent $(\Omega + 1)$–state systems, each described by a partition function $Z = \Omega + \exp(-\epsilon_{\rm HB}/kT)$. Poole et al. find that for small values of the parameter $\epsilon_{\rm HB}$, there is no critical point (but rather a re-entrant spinodal of the form first conjectured by Speedy [21]). However for $\epsilon_{\rm HB}$ above a threshold (about 16 kJ/mol), a critical point appears.

The possibility of a second critical point has received recent support by phenomenological analysis of Ponyatovskii and colleagues[22] and by lattice gas models[20, 23]. Also, Roberts and co-workers [24] have shown that simulation results for a microscopic "water-like" Hamiltonian confirms the presence of a second phase transition, previously deduced from approximate calculations[23].

4.2. Evidence from simulations.

Simulation studies of liquid water have a rich history and have contributed greatly to our understanding of the subject. In fact, this year is the 25th anniversary of the introduction of the ST2 ("Stillinger-2") potential, in which water is represented by a central point from which emanate 4 arms—two carrying positive charges to represent the two protons associated with each water molecule, and two carrying negative charges to represent the two lone electron pairs [25]. The central points interact via a Lennard-Jones potential, while the point charges and the arms interact via a Coulomb potential. Thus every pair of waterlike particles has $4^2 + 1 = 17$ interaction terms. Corresponding to the rather "cumbersome" nature of such a potential is the fact that most studies are limited to extremely small systems—a typical number being $N = 6^3 = 216$ waterlike particles. Recently some studies have considered larger systems, but the typical size rarely exceeds $N = 12^3 = 1728$. It is hoped that by using fast multipole methods one can begin to simulate much larger systems[26].

One way to obtain less cumbersome simulations is to simplify the intermolecular potential. To this end, the simpler TIP4P potential [27] and the much simpler SPC/E potential [28] have enjoyed considerable popularity. However the opposite direction is also under active investigation: simulating more realistic potentials, such as polarizable potentials[29]. The researcher is left with the perplexing problem of which model potential to adopt!

With these caveats, let us very briefly summarize some recent work that might be interpreted as being consistent with (or at least not contradicting) the hypothesis that a HDL-LDL critical point C' exists. We emphasize that most of this work has not reached the stage that it can be interpreted as "evidence" favoring the hypothesis, so we also outline appropriate avenues where future work may strengthen the argumentation.

4.3. Does $1/K_T^{max}$ extrapolate to zero at $(T_{C'}, P_{C'})$?

The compressibility K_T diverges at a second order critical point. Thus, we expect $1/K_T^{\max}$ to extrapolate to zero at the "new" HDL-LDL critical point C', exactly as it does for the "old" liquid-gas critical point C. Recent ST2 calculations [30] are consistent with a plausible extrapolation to a single point in the phase diagram at which $K_T^{\max} = \infty$. The caveat is that one can never know that a given quantity is approaching infinity—it could as well just be approaching a very large number. Indeed, the possibility has been raised, and seriously discussed, that there is no genuine singularity [31]; this possibility will be discussed briefly at the end of this lecture.

4.4. Is there a "kink" in the P_ρ isotherms for sufficiently low temperature?

If there is a critical point, then we expect to find a kink in the $P\rho$ isotherms when T is below $T_{C'}$. Indeed, such a kink appears to exist for the ST2 potential, at a temperature of 235K but not at a temperature of 280 K, consistent with $T_{C'}$ somewhere between 235 K and 280 K. This finding, originally made for simulations of 216 ST2 particles [32, 33], has very recently been strikingly confirmed for a system 8 times larger [34]. An analogous kink has not been found for the TIP4P potential, but a prominent inflection occurs at the lowest temperature studied—suggesting that such a kink may be developing. Work is underway testing for inflections and possible kinks for other water potentials in three, and also in two, dimensions [?, 35].

4.5. Is there a unique structure of the liquid near the kink point?

If there exists a critical point C', then we would expect a two–phase coexistence region below C'. To investigate the possible structural difference between these two phases, Sciortino et al [30] have studied the structure of the liquid at a temperature just below the estimated value of $T_{C'}$ at two values of ρ on the two sides of $\rho_{C'}$. They find that the structure of the liquid state of ST2 at $\rho = 1.05$ g/cm^3 is similar to the experimental data on high-density amorphous (HDA) solid water, while the structure of the structure at $\rho = 0.92$ g/cm^3 resembles the data on low-density amorphous (LDA) solid water. The correspondence between the HDA ice phase and ST2 water just above $\rho_{C'}$, and between the LDA phase and ST2 water just below $\rho_{C'}$ suggests that the two phases that become critical at C' in ST2 water are related to the known HDA and LDA phases of amorphous ice[36].

4.6. Does the coordination number approach four C' is approached?

Sciortino et al [30] have studied the coordination number $N_{\rm nn}$ of the ST2 liquid as a function of T and V, where $N_{\rm nn}$ is the average number of nearest–neighbors found in the first coordination shell of an O atom. For the high–T isotherms, their results show that a 4–coordinated "LDL"-like configuration is approached at negative P, in agreement with previous simulations of Geiger and co–workers [37]. For $T \leq 273$ K, N_{nn} also approaches 4 at positive P. That is, if T is low enough, it appears that a 4-coordinated network can form in liquid water even for $P > 0$. This result is consistent with an experimental study of the evolution of the structure function $S(Q)$ as water is supercooled at atmospheric pressure, in which it was found that the structure tends toward that of the LDA ice [36].

4.7. Is it possible that two apparent "phases" may coexist below C'?

Convincing evidence for a HDL-LDL critical point C' would be the presence of two coexisting phases below C'. This search is the focus of ongoing work. One can, e.g., partition the water molecules into two groups ("red" and "blue" molecules), those with fewer than the average number of nearest neighbors and those with more than the average and find that the red molecules and the blue molecules segregate to opposite sites of the 18 A box in which they are residing. These preliminary investigations at a temperature somewhat below $T_{C'}$ do not prove phase coexistence [34, 38], but work is underway to establish this possibility. In particular, one must first rule out the likelihood that the two "phases" are merely large fluctuations due to a large correlation length (because near a critical point there should be fluctuations of all sizes and shapes, while the sample separating into two distinct regions is rather different). Also, one must seek to find the phase separation occurring in much larger systems. To be conclusive, firstly one must demonstrate that phase separation occurs in a much larger system, and secondly one must study systematically the time dependence of $S(Q)$ as one quenches into the two-phase region from a large value of temperature.

Separate calculations of the weighted correlation function $h(r)$ for the two tentatively identified HDL and LDL phases suggest similarities with experimental results on the two amorphous solid phases HDA and LDA[38]. Additional work remains to be done to establish this point.

4.8. Do fluctuations appear on all time scales?

For the ST2 potential, a histogram of hydrogen bond lifetimes reveals power law behavior over as much as two decades, with the region of "scale free behavior" extending over a larger time domain as T is decreased [39]. For the TIP4P potential, no calculations have yet been carried out, but for the SPC/E potential, non-Arrhenius behavior has also been found at high temperatures [40]. At low temperatures, it is possible that power law behavior is found [26, 34]. An important caveat in interpreting these results is that this scale free behavior is exactly what one would expect if the hydrogen bonded network were regarded as possessing defects (corresponding to molecules with fewer than four good bonds), and these defects were allowed to diffuse randomly[41, 42]. Possibly some of these ambiguities will be resolved by applying to this problem Sasai's "local structure index" that permits one to study in some detail the local dynamics[43].

4.9. Is there "critical slowing down" of a characteristic time scale?

For the ST2 potential, the characteristic value of hydrogen bond lifetime, defined as the value of time at which the power law distribution of bond lifetimes is cut off by an exponential, depends sensitively on temperature and in fact is consistent with a power law divergence as T approaches $T_{C'}$ [39]. The temperature dependence of the cutoff has not been studied for other potentials.

Appearing to diverge at roughly the same temperature is a less ambiguous measure of the characteristic time—the inverse of the self-diffusion coefficient D [44]. This slowing down of the dynamics is consistent with what one expects near a critical point. Specifically, $1/D$ strongly increases as $N_{nn} \to 4$. Consistent with this picture, it was found [45, 46] that additional nearest neighbors beyond 4 have a "catalytic" effect on the mobility of the central molecule, in that they lower the local energy barrier of the molecular exchanges that are the microscopic basis of diffusion, demonstrating the importance for molecular mobility of molecular environments having more than four nearest neighbors.

Because of the relation between $1/D$ and $(N_{nn} - 4)$, the manner in which $N_{nn} \to 4$ is also significant. At high T the decrease of N_{nn} with P is relatively uniform. However, as T decreases, N_{nn} is observed to vary more and more abruptly from a high-coordinated structure ($N_{nn} > 6$) to $N_{nn} \simeq 4$. It should be possible to collapse this family of curves onto a single "scaling function" if the two axes are divided by appropriate powers of $T - T_{C'}$; these tests are underway.

4.10. Is the characteristic dynamics of each "phase" different?

We can identify molecules as "red"/"blue" if they are in a region of locally high/low density for a specified amount of time (say 100 ps). Looking at the mean square displacement of the red and blue "phases," we see that the red molecules (corresponding to high densities) move much further than blue molecules (corresponding to low densities)[34, 38]. The nature of transport in each phase is under active investigation, particularly in light of recent proposals for the nature of the anomalous dynamics taking place in low-temperature water [47].

4.11. Is there evidence for a HDL-LDL critical point from independent simulations?

Recently, Tanaka independently found supporting evidence of a critical point by simulations for the TIP4P potential [48]. Tanaka's value of the critical temperature $T_{C'}$ agrees with the earlier estimates, but his critical pressure $P_{C'}$ occurs at roughly atmospheric pressure, or perhaps at negative pressures [48]. The resolution between the two different values of $P_{C'}$ is an open question that will hopefully be resolved shortly.

4.12. Evidence from experiments.

Before discussing experiments, two very important statements must be made. The first is that the numerical values of thermodynamic variables arising from computer simulations are always rather different from the true values. It is customary to correct for this fact by comparing the values of T and P at the temperature of maximum density to the experimental values. In the case of ST2, e.g., the TMD $\rho = 1.00$ g cm^{-3} appears at T = (277+35)K and P = 82 MPa, so ST2 at these conditions overestimates the temperature by 35 K and the pressure by 82 MPa. If a correction of a similar magnitude were to hold in the supercooled region, then we would anticipate that the experimental coordinates of the HDL-LDL critical point would be approximately $T_{C'} \approx 200$ K and $P_{C'} \approx 100$ MPa. The magnitude of the correction depends upon the potential used, so simulations using the TIP4P and SPC/E potentials can be compared with ST2 or with experiment only if a correction appropriate to that potential is first made.

The second statement concerns the presence of an impenetrable "Berlin wall": the line $T_H(P)$ of homogeneous nucleation temperatures [1, 4]. By careful analysis of experimental data above $T_H(P)$, Speedy and Angell pioneered the view that some sort of singular behavior is occurring in water at a temperature $T_s(P)$ some 5-10 degrees below $T_H(P)$. Regardless of the pressure, they found that the identical value of $T_s(P)$ could be used to fit the behavior of quite distinct equilibrium and dynamic quantities.

With these two caveats, one can begin to discuss relevant experimental work. Three general types of experiments have been performed, the first corresponding to a single isobar at 0.1 MPa, the second to a family of isobars up to 600 MPa, and the third to a family of isotherms below $T_{C'}$.

4.13. Density fluctuations along the P = 0.1 MPa isobar.

The correlation length ξ for density fluctuations should increase close to a critical point; this quantity has recently been measured along a P=0.1 MPa isobar [49], down to quite low temperatures (239K). A gentle increase in correlation length was found, but no indication of a divergence, consistent with the possibility that the HDL-LDL critical point, if it exists, lies at a much higher pressure.

4.14. Structure along isobars up to P = 600 MPa.

Bellissent-Funel and Bosio have recently undertaken a detailed structural study of D_2O using neutron scattering to study the effect of decreasing the temperature on the correlation function[36]. As paths in the PT phase diagram, they have chosen a family of isobars ranging in pressure up to 600 MPa (well above the HDL-LDL critical point of about 100 MPa). They plot the temperature dependence of the first peak position Q_o of the structure factor for each isobar. They find that for the 0.1 MPa isobar, Q_o approaches 1.7 A^{-1}—the value for LDA, low-density amorphous ice. In contrast, for the 465 and 600 MPa isobars, Q_o approaches a 30% larger value, 2.2 A^{-1}—the value for HDA, high-density amorphous ice. For the 260 MPa isobar, $Q_o \to 2.0A^{-1}$, as if the sample were a two-phase mixture of HDA and LDA.

4.15. Reversible conversion of LDA to HDA with pressure.

Since the HDL-LDL critical point occurs below $T_H(P)$, it is not possible to probe the two phases experimentally. However two analogous solid amorphous phases of H_2O have been studied extensively by Mishima and co-workers[50, 51, 52]. In particular, Mishima has recently succeeded in converting the LDA phase to the HDA phase on increasing the pressure, and then reversing this conversion by lowering the pressure. The jump in density was measured for a range of temperatures from 77K to 140 K, and the density jump (when HDA is compressed to LDA) was found to occur at roughly 200 MPa. Moreover, the magnitude of the density jump decreases as the temperature is raised, just as would occur if instead of making measurements on the HDA and LDA amorphous solid phases, one were instead considering the HDL and LDL liquid phases. These results are corroborated by independently performed computer simulations using both the ST2 and TIP4P intermolecular potentials[53].

If we assume that HDA and LDA ice are the glasses formed from the two liquid phases discussed above, then the HDA–LDA transition can be interpreted in terms of an abrupt change from one microstate in the phase space of the high–density liquid, to a microstate in the phase space of the low–density liquid. The experimentally detected HDA–LDA transition line would then be the extension into the glassy regime of the line of first–order phase transitions separating the HDL and LDL phases.

5. DISCUSSION

Thus far in this talk, we have presented some tentative evidence that is consistent with the hypothesis that there exists a second HDL-LDL critical point C', in addition to the well-known liquid-gas critical point C. There are also some facts, which we next present, that do not favor the existence of C'.

First, the vast experimental work of Angell and co-workers [4, 6] is equally consistent with a line of spinodal singularities or with a HLD-LDL critical point, and the generalized van der Waals theory is equally consistent with both possibilities [18].

Moreover, the observed polymorphism of amorphous *solid* water which was invoked to support the possibility of a line of liquid–liquid phase transitions terminating in a critical point supports, but certainly does not prove, the existence of a phase transition in the behavior of the supercooled *liquid*. Merely a large but finite value of K_T might be sufficient to account for abrupt density changes in the amorphous ices in the glass regime [54, 55].

Very recently, the dynamics of SPC/E water have been found to be well described by mode coupling theory, with no need to invoke the presence of a critical point[47]. For ST2 water, on the other hand, the situation may be somewhat different[34, 38, 56].

Many open questions remain, and many experimental results are of potential relevance to the task of answering these questions. For example, the collective excitations that occur at low frequency and large wave vector [57] have been the object of much recent work [58, 59]. Also, the full implications of a re-entrant spinodal possibility [60, 61, 62, 63] may not have been fully incorporated into our understanding of water. The minimum exhibited by the *adiabatic* compressibility in atmospheric pressure experiments [64] is of potential relevance since a maximum in any quantity that would be singular at a spinodal may be relevant to placing the critical point at a larger pressure than atmospheric.

Acknowledgments

We thank C. A. Angell, S. S. Borick, S. V. Buldyrev, S.-H. Chen, L. C. Cruz, P. G. Debenedetti, P. Gallo, T. Grande, P. F. McMillan, P. Ray, C. J. Roberts, R. Sadr-Lahijany, S. Sastry, R. J. Speedy, F. W. Starr, F. H. Stillinger, H. Tanaka, P. Tartaglia, B. Widom and R. Zhang for enlightening discussions and seminal help in various aspects of this work, and NSF, BP, GNSM, and NSERC (Canada) for financial support.

References

[1] *Water: A Comprehensive Treatise*, Vol. 1–7, F. Franks, ed. (Plenum Press, NY, 1972);
Water Science Reviews, Vol. 1–4, F. Franks, ed. (Cambridge University Press, Cambridge, 1985).

[2] P. G. Debenedetti, *Metastable Liquids* (Princeton Univ Press, Princeton, 1997).

[3] G. S. Kell, in *Water: A Comprehensive Treatise*, Vol. 1, F. Franks, ed. (Plenum Press, NY, 1972), pp. 363–412.

[4] C. A. Angell, in *Water: A Comprehensive Treatise*, Vol. 7, F. Franks, ed. (Plenum Press, NY, 1980), pp. 1–81.

[5] E. W. Lang and H.-D. Lüdemann, *Angew. Chem. Int. Ed. Engl.* **21**, 315–329 (1982).

[6] C. A. Angell, *Ann. Rev. Phys. Chem.* **34**, 593–630 (1983).

[7] H. E. Stanley, J. Phys. A **12**, L329-L337 (1979);
H. E. Stanley and J. Teixeira, *J. Chem. Phys.* **73**, 3404-3422 (1980);
H. E. Stanley, J. Teixeira, A. Geiger, and R. L. Blumberg, *Physica A* **106**, 260–277 (1981).

[8] A. Geiger, F. H. Stillinger and A. Rahman, *J. Chem. Phys.* **70**, 4185–4193 (1979).

[9] F. H. Stillinger, *Science* **209**, 451–457 (1980);
F. H. Stillinger and T. A. Weber, *Phys. Rev. A* **25** 978–989 (1982);

F. H. Stillinger and T. A. Weber, *Phys. Rev. A* **28** 2408–2416 (1983); F. H. Stillinger and T. A. Weber, *J. Phys. Chem.* **87**, 2833–2840 (1983); F. H. Stillinger and T. A. Weber, *J. Chem. Phys.* **80**, 4434–4437 (1984).

[10] A. Geiger and H. E. Stanley, *Phys. Rev. Lett.* **49** 1749–1752 (1982); H. E. Stanley, R. L. Blumberg, and A. Geiger, *Phys. Rev. B* **28** 1626–1629 (1983); H. E. Stanley, R. L. Blumberg, A. Geiger, P. Mausbach and J. Teixeira, J. de Physique **45**, C7[3]–C7[12] (1984).

[11] A. Geiger and H. E. Stanley, *Phys. Rev. Lett.* **49**, 1895-1898 (1982); R. L. Blumberg, H. E. Stanley, A. Geiger and P. Mausbach, *J. Chem. Phys.* **80**, 5230–5241 (1984).

[12] A. Geiger, P. Mausbach, J. Schnitker, R. L. Blumberg, and H. E. Stanley, *J. Physique* **45**, C7[13]–C7[30] (1984).

[13] C. A. Angell and V. Rodgers, *J. Chem. Phys.* **80**, 6245 (1984).

[14] C. A. Angell, J. Shuppert, and J. C. Tucker, *J. Phys. Chem.* **77**, 3092–3099 (1973); H. Kanno and C. A. Angell, J. Chem. Phys. **70**, 4008 (1979); H. Kanno and C. A. Angell, J. Chem. Phys. **73**, 1940 (1980).

[15] R. J. Speedy and C. A. Angell, *J. Chem. Phys.* **65**, 851–858 (1976).

[16] C. A. Angell, M. Oguni, and W. J. Sichina, *J. Phys. Chem.* **86**, 998–1002 (1982).

[17] K. Mendelssohn, *The Quest for Absolute Zero: The Meaning of Low-Temperature Physics* (McGraw, New York, 1966), p. 42ff.

[18] P. H. Poole, F. Sciortino, T. Grande, H. E. Stanley and C. A. Angell, *Phys. Rev. Lett.* **73**, 1632–1635 (1994); P. H. Poole, F. Sciortino, T. Grande, H. E. Stanley and C. A. Angell, "Mean-Field Approach to Water Structure," *Phys. Rev. E* (submitted).

[19] S. Sastry, F. Sciortino and H.E. Stanley, *J. Chem. Phys.* **98**, 9863–9872 (1993).

[20] S. S. Borick, P. G. Debenedetti and S. Sastry, *J. Phys. Chem.* **99**, 3781 (1995); S. S. Borick and P. G. Debenedetti, *J. Phys. Chem.* **97**, 6292–6303 (1993).

[21] R. J. Speedy, *J. Phys. Chem.* **86**, 3002–3005 (1982).

[22] E. G. Ponyatovskii, V. V. Sinitsyn and T. A. Pozdnyakova, *JETP Lett.* **60** 360–364 (1994).

[23] C. J. Roberts and P. G. Debenedetti, J. Chem. Phys. **105**, 658–672 (1996).

[24] C. J. Roberts, A. Z. Panagiotopoulos and P. G. Debenedetti, Phys. Rev. Lett. **77**, 4386–4389 (1996).

[25] F. H. Stillinger and A. Rahman, *J. Chem. Phys.* **57**, 1281–1292 (1972).

[26] F. W. Starr et al (preprint)

[27] W. L. Jorgensen, J. Chandrasekhar, J. Madura, R. W. Impey and M. Klein, *J. Chem. Phys.* **79**, 926–935 (1983).

[28] H. J. C. Berendsen, J. R. Grigera and T. P. Straatsma, *J. Phys. Chem.* **91**, 6269–6271 (1987).

[29] U. Niesar, G. Corongiu, E. Clementi, G.R. Kneller, and D. Bhattacharya, *J. Phys. Chem.* **94**, 7949 (1990).
[30] F. Sciortino, P. H. Poole, U. Essmann and H. E. Stanley, *Phys. Rev. E* **55**, 727 (1997).
[31] S. Sastry, P. G. Debenedetti, F. Sciortino, and H. E. Stanley, *Phys. Rev. E* **53**, 6144–6154 (1996).
[32] P. H. Poole, F. Sciortino, U. Essmann, H. E. Stanley, *Nature* **360**, 324–328 (1992); C. Sorenson. *Nature* **360**, 303-304 (1992).
[33] P. H. Poole, F. Sciortino, U. Essmann and H. E. Stanley, *Phys. Rev. E* **48**, 3799–3817 (1993)
[34] S. T. Harrington, R. Zhang, P. H. Poole, F. Sciortino and H. E. Stanley, Phys. Rev. Lett. **78**, 2409–2412 (1997).
[35] L. Cruz et al. (preprint).
[36] M. C. Bellissent-Funel and L. Bosio, *J. Chem. Phys.* **102**, 3727–3735 (1995).
[37] A. Geiger, P. Mausbach and J. Schnitker, in *Water and Aqueous Solutions*, edited by G. W. Neilson and J. E. Enderby (Adam Hilger, Bristol, 1986) p. 15.
[38] R. Zhang et al (preprint)
[39] F. Sciortino, P. H. Poole, H. E. Stanley and S. Havlin, *Phys. Rev. Lett.* **64**, 1686–1689 (1990).
[40] A. Luzar and D. Chandler, *Phys. Rev. Lett.* **76**, 928-931 (1996); A. Luzar and D. Chandler, *Nature* **379**, 55–57 (1996).
[41] H. Larralde, F. Sciortino and H. E. Stanley, "Restructuring the Hydrogen Bond Network of Water" (preprint); H. Larralde, Ph. D. Thesis (Boston University, 1993)
[42] F. Sciortino, A. Geiger and H. E. Stanley, *Phys. Rev. Lett.* **65**, 3452–3455 (1990).
[43] E. Shiratani and M. Sasai, *J. Chem. Phys.* **104**, 7671-7680 (1996).
[44] P. H. Poole, F. Sciortino, U. Essmann, M. Hemmati, H. E. Stanley and C. A. Angell, in: *Hydrogen Bond Networks* Eds M.-C. Bellissent-Funel and J. C. Dore. [Proc. 1993 Cargèse Conf.] (Kluwer Academic, Dordrecht, 1994), pp. 53–60.
[45] F. Sciortino, A. Geiger and H. E. Stanley, *Nature* **354**, 218–221 (1991).
[46] F. Sciortino, A. Geiger and H. E. Stanley, *J. Chem. Phys.* **96**, 3857–3865 (1992).
[47] P. Gallo, F. Sciortino, P. Tartaglia, and S.-H. Chen, Phys. Rev. Lett. **76**, 2730–2733 (1996); F. Sciortino, P. Gallo, P. Tartaglia, and S.-H. Chen, *Phys. Rev. E* **54**, 6331–6346 (1996).
[48] H. Tanaka, *Nature* **380**, 328–331 (1996); H. Tanaka, *J. Chem. Phys.* **105**, 5099–5111 (1996); R. J. Speedy, *Nature* **380**, 289–290 (1996).
[49] Y. Xie, K. F. Ludwig, Jr., G. Morales, D. E. Hare and C. M. Sorensen, Phys. Rev.Lett. **71**, 2050–2053 (1993).

[50] O. Mishima, *J. Chem. Phys.* **100**, 5910-5912 (1994).
[51] O. Mishima, K. Takemura and K. Aoki, *Science* **254**, 406-408 (1991).
[52] O. Mishima and H. E. Stanley, "Metastable melting lines of ice phases," (abstract);
O. Mishima, *Nature* **384**, 546 (1996);
O. Mishima, L. D. Calvert and E. Whalley, *Nature* **310**, 393–395 (1984);
O. Mishima, L. D. Calvert and E. Whalley, *Nature* **314**, 76–78 (1985).
[53] P. H. Poole, U. Essmann, F. Sciortino and H. E. Stanley, *Phys. Rev. E* **48**, 4605–4610 (1993),
[54] P. H. Poole, T. Grande, F. Sciortino, H. E. Stanley and C. A. Angell, *J. Comp. Mat. Sci.* **4**, 373–382 (1995).
[55] F. Sciortino, U. Essmann, H. E. Stanley, M. Hemmati, J. Shao, G. H. Wolf, and C. A. Angell, *Phys. Rev. E* **52**, 6484–6491 (1995).
[56] D. Paschek and A. Geiger, "Simulation study on the nature of the diffusive motion in supercooled water," preprint.
[57] J. Teixeira, M.-C. Bellissent-Funel, S.-H. Chen and B. Dorner, *Phys. Rev. Lett.* **54**, 2681–2683 (1985).
[58] F. Sciortino and S. Sastry, *J. Chem. Phys.* **100**, 3881–3893 (1994).
[59] S. Sastry, F. Sciortino and H. E. Stanley, *J. Chem. Phys.* **95**, 7775–7776 (1991).
[60] P. G. Debenedetti and M. C. D'Antonio, *J. Chem. Phys.* **84**, 3339–3345 (1986).
[61] P. G. Debenedetti and M. C. D'Antonio, *J. Chem. Phys.* **85**, 4005–4010 (1986).
[62] M. C. D'Antonio and P. G. Debenedetti, *J. Chem. Phys.* **86**, 2229–2235 (1987).
[63] P. G. Debenedetti and M. C. D'Antonio, *AIChE J.* **34**, 447–455 (1988).
[64] E. Trinh and R. E. Apfel, J. Chem. Phys. **69**, 4245–4251 (1978);
E. Trinh and R. E. Apfel, *J. Chem. Phys.* **72**, 6731–6735 (1980).

LECTURE 10

Scale Invariance(s) in the Cosmic Matter Distribution ?

M. Lachièze-Rey

C.E. Saclay, 91191 Gif-sur-Yvette, France

1. Introduction

The conclusion talk of this conference by Procaccia pointed out the absence of meaning of such expressions as *scale invariant systems* or *scale invariant processes* in physics: only scale invariant *quantities* (mathematical functions) or *equations* can be defined with any sense. It is thus irrelevant to invoke a possible scale invariance of the cosmic matter distribution. But a scale invariance has been claimed to appear in some quantities used for its description, or in the equations governing its formation or evolution.

Structure formation is thought to be the result of a very long term process having extended over the whole cosmic period, the *gravitational instability*. Although additional phenomena may also have played a role, especially in the initial and final phases, this process is mainly driven by gravitational interactions, which have no preferred scale. For this reason it appears reasonable to expect some scale invariance in the matter distribution.

On the one hand, signs of scale invariance have been claimed to be observed in the (discontinuous) distribution of objects, mainly galaxies and galaxy clusters, from the study of correlation functions or related statistical indicators. I report them in section 2, and the possible links with dynamics in section 3. On the other hand, scale invariance or self-similarity has been claimed to characterize the (continuous) matter distribution *inside* astrophysical objects like galaxies or galaxy clusters: the shape of their density profiles follows approximatively a *power law*. This is reported in section 4.

2. The large scale matter distribution

Although we see only the *visible* component of the matter, under the form of stars, gas, dust,..., many evidences indicate the presence of large quantities of dark invisible matter. From the scale free character of gravity, we expect a possible scale invariance in the total (dark + luminous) mass distribution. But this latter differs from that of luminous matter although signs of scale invariance have been searched, and reported, in the distribution of visible matter. The populations of galaxies, or others luminous object, do not reflect that of total matter. Moreover, different classes of objects are differently clustered, even galaxies of different luminosities. The ratio between the local abundance of a given type of objects - e.g. galaxies of a given luminosity in a given spectral domain - , and the (unknown) abundance of total mass at the same point is called the *bias*. For a given category of objects, the bias depends on many factors (type of objects, position, scale, ...), that we cannot observe, and that no model is able to predict.

2.1. Typical scales

Gravitational interactions have no preferred scale. Cosmology has however some typical spatial scales, which vary with time.

- The *curvature radius of space* increases proportionally to the scale factor $R(t)$ of the cosmic expansion. A constant scale $c/\sqrt{\Lambda}$ is associated with the cosmological constant Λ (c is the velocity of light). But none of these scales is thought to have played a role in the cosmic matter distribution.
- The cosmic evolution is characterized, at each moment t_i of the cosmic history, by the horizon length $L_H(t_i) \approx c\, t_i$, which measures the size of a causally connected region of the universe. A process occurring at an instant t_i may thus imprint the scale $L_H(t_i)$ onto the resulting systems.
- We will see, in addition, that the process of structure formation through gravitational instability allows, in the cosmic context, to define a characteristic scale $L_{NL}(t_i)$ at any moment t_i (NL holds for non linear, see below).
- Finally, non gravitational processes, which have played a role in structure formation, have also imprinted typical scales. Damping of the fluctuations in early phases may have a non–gravitational origin. Cooling and dissipation have occurred in the final phases of structure formation.

On the other hand, the cosmic matter appears distributed in a hierarchy of astronomical units with typical scales : *galaxies*, have typical dimensions of a few tens of kpc, although the scale generally associated to them is their averaged separation, $\approx$ 1 Mpc, also the scale of the region which was once occupied by the matter presently condensed in the galaxy. Their masses $\approx 10^{11} - 10^{12} M_\odot$ correspond to a density contrast $\delta \equiv \frac{\delta\rho}{<\rho>} \approx 10^5$. Galaxies are often associated in *pairs*, *groups* (up to 100 members), and *clusters*. Clusters of galaxies, gathering up to 1000 members (the *richness* of a cluster is a measure of the number of galaxies that it contains), have masses $10^{14} - 10^{15} M_\odot$, and scales around 5 Mpc. These well defined systems, with a density contrast $\delta \approx 100$, offer a special interest because they seem to be the largest collapsed systems in the universe. *Superclusters* reach scales of tens of Mpc, and masses in the range $10^{15} - 10^{16} M_\odot$, with density contrast around a few units. We do not know about the existence of larger structures beyond.

The characteristic dimensions of this hierarchy (which continues towards the small scales) result from the interplay of of gravitational instability in various regimes (thus involving the various scales defined above, also possibly with a memory from initial conditions) and other interactions like dissipative radiation during the collapse of structures. We do not know precisely its origin. According to various claims, the existence of this discrete hierarchy could give room to some scale invariance.

2.2. Correlation functions

A considerable attention has been paid to the spatial distribution of galaxies, considered as the building blocks of the Universe. For this task, a great importance has been given to correlation functions, in relation to the common thought that the distribution of galaxies (or similar objects) is a random realization of an underlying statistical process, characterized by a well defined probability distribution function. In this sense, the measured correlation functions would represent the autocorrelation functions of the underlying matter field. Recent results suggest however that this may not be the case.

2.2.1. Galaxy – galaxy correlation functions

The 2-point correlation function $\xi(r)$ was originally defined (see, e.g., Peebles, 1990) through

$$1 + \xi(r) = \frac{1/V \int_V n(\mathbf{x})\, n(\mathbf{x}+\mathbf{r})\, d^3x}{\left(1/V \int_V n(\mathbf{x})\, d^3x\right)^2}, \tag{1}$$

where $n(\mathbf{x})$ is a local density of galaxies. It has been first estimated from the 2-dimensional galaxy distribution, via a deprojection procedure. More recently it has been measured directly from 3-d distributions, with the drawback that this concerns redshift space rather than physical space.

The classical result is that, in first approximation, $\xi_{gg}(r)$ is well fitted by a power law $(r/C_{gg})^{-\gamma_{gg}}$, with $\gamma_{gg} \approx 1.8$ and $C_{gg} \approx 5$ Mpc. For historical reasons in astrophysics, C_{gg} is called a *correlation length*. This formulation may suggest erroneously that C_{gg} is a true physical length, although here the correlation function is precisely scale-free (not an exponential, for instance). This conventional view, and the interpretation of C_{gg} as a correlation length, have been challenged in the last years, as we detail below. But the presence of this power law, considered to be valid in the range $0.1\ h^{-1}$ Mpc to $10\ h^{-1}$ Mpc, has been claimed to be a first sign of scale invariance.

2.2.2. The cluster – cluster correlation function

The cluster – cluster correlation function ξ_{cc} shows a similar power law behavior, with $\xi_{cc}(r) \approx 20\ \xi_{gg}(r)$. More generally, the distributions of galaxy clusters with different richnesses, and of superclusters, are all compatible with power law correlations with the same exponent, and different amplitudes, expressed by $C_{gg} \approx$ 5 Mpc, $C_{cc} \approx 25$ Mpc (which in fact includes different categories according to the richness), and $C_{ss} \approx 60$ Mpc. There is no agreement concerning the significance of these relations. Various authors have attempted to express them by an unique function thanks to some rescaling.

Szalay & Schramm (1985) proposed that

$$\xi_i(r) = C^{te}\ (r/R_i)^{-\gamma}, \tag{2}$$

valid for a sample i of cosmic objects of size R_i. This same and unique formula describes well galaxies and clusters of different richnesses, but not superclusters. With this interpretation, galaxies and clusters would be considered to share the same rescaled correlation function, the difference being due to the different sizes of the catalogs in which they are measured.

On the other hand, Bahcall & Burgett (1986) proposed the different formula

$$\xi_i(r) = C^{te}\ (r\ < n_i >^{1/3})^{-\gamma}, \tag{3}$$

where the scaling goes with the mean interparticle distance $n_i^{-1/3}$. This relation holds for clusters of different richnesses and superclusters, with $C/ < n >^{-1/3} \approx$ 0.56, but not for galaxies, for which $C_g/ < n_g >^{-1/3}$ is 3 times higher. This idea was suggested by the following formula concerning the correlation of the peaks of a random gaussian distribution with correlation function ξ, and dispersion σ, the peaks being defined above a given threshold ν:

$$\xi_\nu(r, S) \approx exp[\nu^2\ \xi(r, S)/\sigma^2] - 1, \tag{4}$$

for $\nu >> 1$, where S is the order of the mean separation between the peaks. According to this analogy, galaxy clusters would be simply considered as statistical peaks in the galaxy distribution. This conception disagrees however with other observational facts: clusters are not zones where there is a random excess of galaxies, but are real, gravitationnally linked, objects, with precise physical characteristics and histories.

More recently, Montuori et al. (1997) have argued that the use of ξ is not justified, and that the correlation lengths so defined have no physical meaning and only reflect the dimensions of the catalogs. According to this approach, which is also that of Pietronero (see below), galaxies and galaxy clusters share the same statistical properties.

2.2.3. Higher order correlations

Three-point correlation functions, more difficult to measure, appear very intense. They are written as $\xi_{123} = \xi_{12} + \xi_{13} + \xi_{23} + \zeta_{123}$, which defines the reduced part ζ. A non zero value of ζ proves that the distribution is non Gaussian. It is interesting to mention the Kirkwood formula which applies to the distribution of the peaks of a Gaussian field:

$$\zeta = Q\ [\xi_{12}\,\xi_{23} + \xi_{12}\,\xi_{13} + \xi_{23}\,\xi_{31} + \xi_{12}\,\xi_{23}\,\xi_{31}].$$

The observed results for galaxies are however compatible with :

$$\zeta = Q\ [\xi_{12}\,\xi_{23} + \xi_{12}\,\xi_{13} + \xi_{23}\,\xi_{31}],$$

with $Q = 0.7 - 1$. Similar results have been measured for the distributions of galaxy clusters or quasars, but with $Q = 0.2 - 0.6$.

One defines similarly the four-point correlation functions, with a reduced part η. Observation suggest that η can be written as

$$\eta = R_a\ [\xi_{12}\,\xi_{23}\,\xi_{34} + ...] + R_b\ [\xi_{12}\,\xi_{13}\,\xi_{14} + ...].$$

These results lead to the empirical suggestion for the nth order correlation function:

$$\xi^{(n)}_{reduced} = [\xi^{(2)}]^{(n-1)},$$

originally proposed by Sharp (1981).

It is a general tendency that the (reduced) correlations become stronger when the order increases. Thus, low order correlations contain a very limited information concerning the matter distribution. The difficulty to measure these higher order correlations has motivated the introduction of various statistical indicators to characterize the galaxy distribution, like nearest neighbor distance distributions, percolation thresholds, topological genus, Void probability function, Counting probabilities, fractal dimensions, wavelet transform coefficients, ... Void and Count probabilities have exhibited a special importance concerning a possible scale invariance.

2.3. Void and Count probability functions

The count probability $P_N(V)$ is defined as the probability that a volume V of given shape (i.e., spherical), randomly placed in the sample, contains N galaxies. For $N = 0$, $P_0(V)$ is the Void probability function (VPF) and plays a special role. In fact $P_0(n;V)$ depends also on the average density n of the sample: at given V, the denser the sample, the smaller P_0.

For a random Poisson distribution, calculations lead to $P_0(n;V) = exp(-nV)$. This suggests to define

$$\chi(n;V) = -logP_0(n;V)/nV,$$

with $\chi = 1$ for a random Poisson distribution. Interestingly, $P_0(n;V)$ plays the role of a generating function for the P_N's through

$$P_N(n,V) = \frac{(-n)^N}{N!}\frac{d^N}{dn^N}P_0(n;V). \tag{5}$$

This implies that all statistical information is contained in $P_0(n;V)$, although this is difficult to use in practice (this requires different realizations of the same statistical process with different densities ; this is possible in some simulations; also partially, and artificially by undersampling, the real sample). Also, P_0 is linked to the correlation functions of all orders by

$$\chi = 1 + \sum_i \frac{(-n)^i}{i!}\int\int \ldots dV_i \ldots \xi^{(i)}(x_1,\ldots,x_i,\ldots). \tag{6}$$

These formulae have suggested that P_0 could verify some scale invariance, or rather scaling relation, in the sense that $\chi(n,V)$ (which is a priori a function of the 2 variables n and V) writes as a function of the "scaling variable" $q(n,V) = nV < \xi(V) >$ only, where $< \xi(V) >$ is the correlation function averaged in the volume V:

$$\chi = \chi[q(n,V)]. \tag{7}$$

This scaling relation has been checked for the galaxy distribution (Maurogordato & Lachièze-Rey, 1987, and references therein) and also for the cluster distribution (Cappi et al. 1991) : when n changes, χ remains a function of q only. Quite surprisingly, it holds also between galaxies and clusters: both distribution are described by the same function $\chi(q)$. No dynamical models gives a reason why it should be so ! Similar relations exist with the counting probabilities.

2.4. Difficulties with correlation functions

The power law form of the 2-point correlation function, the scaling relation for the VPF's, the rescaling between galaxies clusters and superclusters, have been occasionally interpreted as signs of scale invariance. However, they all rely on the use of correlation functions, which presents serious difficulties. On the first hand, edge effects in the catalogs are very severe, and strongly limit the validity of the analyses (see, e.g., McCauley 1997).

More seriously, it has been recognized for a long time (see, e.g., Maurogordato & Lachièze-Rey, 1991) that the evaluation of ξ, for a given population, depends on the size of the sampled volume. This strongly questions the relevance of its use. This is due in part to the difficulty for choosing a value n of the averaged density, necessary to estimate ξ in a galaxy catalog (see Maurogordato & Lachièze-Rey 1991). The use of an *intrinsic* density, defined as the ratio $n = N/V$ of the number of galaxies in the sample divided by its volume, implies a zero value for the integral of ξ in the catalog. It results that different values of $\xi(r)$ are measured for *the same* galaxy distribution, in catalogs of different sizes, in contrary to what is expected to characterize clustering properties.

The alternative use of an *extrinsic* density leads to important difficulties also. This latter is defined as an average value $n =< n >$, estimated independently of the catalog, and in particular of its size. Firstly, how to define such a value ? The lacunarity, the filling factors, and thus the averaged density of the galaxy distribution vary with the size of the sample, so that we have no access to a fair value for n. Even if it were the case, such a normalization would not be convenient, as a pedagogic example can show: consider a distribution with fluctuations at scale 50 Mpc only, with no power at lower scales (5 Mpc in particular). From this distribution, extract 2 catalogs of size $\approx$ 50 Mpc : C^+ in a dense fluctuation, and C^- in an underdense one. Measurements of ξ(5 Mpc) give a positive value in C^+, and a negative one in C^-, in contradiction with the fact that there is absolutely no structure at 5 Mpc. Clearly, the correlation function fails to express the correct result!

All estimations of ξ are thus biased. This is at the origin of the controversy between Pietronero (1997) and Davis (1997), and of the attempts to describe the galaxy distribution as a fractal. How relevant can be a correlation length which increases with the size of the catalog ? How could a correlation length $\approx$ 5 Mpc be reconciled with the large voids, or the correlations between galaxy clusters ? ... It appears more and more difficult to give any relevance to this quantity. This is advocated by Pietronero (1997, and references therein), which promotes the use of $\Gamma(r) =< n > (\xi(r) + 1)$ as a better measure of low order clustering. Results expressed with $\Gamma(r)$ appear much more stable, in the sense that they do not depend on the position and size of the sample, as does ξ. It is well fitted by an unique power law, with slope $\gamma \approx -1$, over the range 1 to 150 h^{-1} Mpc. This has be interpreted a a sign of fractal behavior and/or of scale invariance.

In any case, all results resting on a value of the mean density, or of the correlation function should be taken with the greatest care. There is no agreement about the slope of a possible power law, 1.7 or 1, nor about the value, and even the existence, of a correlation length. Interestingly, McCauley (1997) made the remark that almost any function looks like a power law when plotted over a limited range, and may thus be erroneously interpreted as scale invariant.

2.4.1. Fractal or multifractal dimensions ?

There have been many claims that the galaxy distribution could present a fractal character. This originated from the remark that the measured galaxy density varies with the scale, and from the power law behavior of ξ. A fractal *correlation dimension* was claimed : $d = 3 - \gamma \approx 1.2$ for the conventional view, or $d \approx 2$ for Pietronero (1997). It is well known, however, that any distribution could appear similar to a fractal, when examined over a limited range. It seems that the relevant interrogation is not if the galaxy distribution can or not be described by a fractal distribution, but if the answer to this question presents any interest: the numerous claims about fractality have presently not suggested any interesting prediction or explanation.

Davis (1997) interprets all difficulties with ξ as effects of bad sampling or luminosity segregation. Even if this explanation were sufficient, this would precisely sign the failure of the approach with ξ. Conversely, Pietronero (1997) claims that the galaxy distribution is well described by a fractal with dimension $D = 3 - \gamma \approx 2$ over a very wide range.

The power law of the correlation function implies that, *on average*, the density around a galaxy decreases proportionally to r^{3-D}. Independently of any interpretation, the number of galaxies $n_i(r^i)$ at a distance r^i from a galaxy i is a function of r^i. For a fractal distribution, $n_i(r^i)$ would be a power law, with the same exponent for all galaxies. McCauley (1997) clearly showed that this is not the case, and thus that the galaxy distribution is not fractal.

This leaves the possibility that it is multifractal. Various authors have measured the slopes of the functions $n_i(r^i)$, occasionally claiming a multifractal character. Results are contradictory, measurements procedures appear often incorrect, undersampling effects neglected, and claims of multifractallity not founded (see, e.g., Dubrulle & Lachièze-Rey, 1994, or McCauley, 1997 for critical discussions). Interestingly, McCauley (1997) remarked that there is no direct link between the fact that all local distributions $n_i(r^i)$ are power laws (with different exponents) and that the averaged function i.e., the correlation function is a power law. This was also analysed by Dubrulle & Lachièze-Rey (1994) with respect to the presence of prefactors varying from one galaxy to the other.

To conclude, the reported signs of scale invariance appear due to the breaking of the validity of the conventional approach, rather than reflecting a real scale invariant process, which in any case has not been found. Similar critics holds for the use of correlation functions, and for a possible fractal or multifractal character. Efforts will continue to define and measure indicators able to express a possible global scale invariance. An other interesting track is to examine the local properties of fluctuations at different scales, in particular to check if they can be themselves self-similar.

3. Scale invariance and dynamics

3.1. The power spectrum of fluctuations

Before turning to this other aspect of scale invariance, it may be worth recalling an other way in which global scale invariance in the matter distribution has been invoked, i.e., through the *power spectrum of fluctuations*. From a density field $\rho(\mathbf{x})$, we define the density fluctuation $\delta\rho(\mathbf{x}) = \rho(\mathbf{x}) - < \rho >$, and the density contrast $\delta(\mathbf{x}) = \delta\rho(\mathbf{x}) / < \rho >$.

The field $\delta(\mathbf{x})$ is usually decomposed in (spatial) Fourier components $\delta_{\mathbf{k}}$, the vector $\mathbf{k}$ characterizing the spatial scale $L = 2\pi/k$. The modes $\delta_{\mathbf{k}}$ are complex quantities :

$$\delta_{\mathbf{k}} = \frac{1}{V} \int_V d^3x \ \rho(\mathbf{x}) \ \exp(i\mathbf{k}.\mathbf{x})$$

(one finds sometimes other normalizations, with 2π factors at different places).

The cosmic isotropy implies that the average modes depend only on the module $k = \mid \mathbf{k} \mid$. The average level of fluctuations, at scale L, is characterized by the Power fluctuation spectrum

$$P(k) = <\mid \delta_{\mathbf{k}} \mid^2> = \int_{|\mathbf{k}|=k} d^3k \ \delta_k^2, \qquad (8)$$

or, more correctly,

$$P(k) \ \delta_D(\mathbf{k} - \mathbf{k}') = < \delta_{\mathbf{k}} \delta_{\mathbf{k}'} >,$$

the average being over all modes with module $k = 2\pi/L$. In the case of Gaussian statistics, the power spectrum contains all information about fluctuations.

A *power law spectrum* is often considered, defined as $P(k) = A \ k^n$. Convergence requirements forbid the spectrum to be a constant power law at all scales: $n < -3$ at large scale would imply an inhomogeneous Universe; and $n > -3$ at small scales would imply so intense fluctuations that the universe should be filled with black holes. The initial value of n is a very important parameter for the models of structure formation. A very interesting possibility (among others) is that of a *scale invariant* spectrum , also called "Harrison-Zeldovich" spectrum (Harrison 1970), corresponding to $n = 1$. It has the remarkable property that the fluctuations of the gravitatinal potential keep the same level at all scales. It has been claimed to be the "most natural", or to result from some "models" of inflation. It is marginally compatible with the COBE results.

3.2. Gravitational instability : standard views

The standard view for structure formation in the Universe is that of gravitational instability, characterized by the following successive stages:

- Creation of initial fluctuations early in the Universe, characterized by their statistics (mainly their initial power spectrum), their adiabatic or isothermal character, and the nature of the matter involved (baryonic, *cold dark matter*, *hot dark matter*, other ..., a mixture).
- A relativistic linear evolution, prior to recombination, from which the spectrum and characteristics of the cosmic microwave background fluctuations results.
- A linear post-recombination amplification of the fluctuations, leading to the apparition of very large structures.
- The non linear gravitational evolution, during which matter accretes onto the first formed structures.
- The formation, breaking and merging of richer structure elements.

- The isolation, from the Hubble flow, of structures at all scales. This is believed to lead to the formation of *halos*, i.e. those almost spherical mass distributions of matter which are believed to surround galaxies, and which are observed, at larger scales, in clusters.
- The dissipative formation of disks, of stars, leading to the apparition of galaxies.

To follow globally the evolution, one can examine the deformation of the power spectrum. Following conventional views, the statistics is initially almost Gaussian, so that the spectrum contains almost all statistical information. But this becomes less and less true. Beside this global evolution, it is also necessary to follow the evolution more locally, under the form of the formation of pre-halos or halos. During *phase mixing*, the distribution function (positions and velocities) becomes smooth. During *violent relaxation*, the distribution function of particles evolves in the time dependent (self consistent) gravitational potential. It may erase quickly (in a few *dynamical times*) the memory of initial conditions. A secular evolution, named *gravothermal catastrophe* follows.

Unfortunately, we do not know with generality to which distributions these processes lead, although the observed profiles of globular clusters, elliptical galaxies, galaxy clusters probably result from them. And we do not know how to make the link between the two approaches, global and local. The global approach extends the linear and quasi-linear formalism through a *perturbative approach*, using the power spectrum. The second one is completely local, with probably a loss of the memory of initial conditions. In particular, we do not know how to make the link between the initial spectrum of fluctuations and the halo profiles (for instance).

3.3. A time dependent scale for gravitational clustering

Although gravitational interactions offer no preferred scale, the process of gravitational instability in the cosmic context allows us to define a time dependent characteristic scale, which helps us to understand the matter distribution. To each cosmological period defined by its redshift z, it is possible to associate the non linear scale $L_{NL}(z) = 1/k_{NL}(z)$. It is defined as the scale for which the linear rms density fluctuation $\delta(L_{NL})$ just becomes non linear, and thus begins to collapse with an increased efficiency. This is defined by requiring that $\delta(L_{NL})$ is greater than a predefined threshold, δ_{NL}, resulting from calculations of collapse in a spherical idealized situation.

Then $L_{NL}(z)$ provides a characteristic size for structure formation at each z, that of those objects which form at redshift z. For hierarchical models $L_{NL}(z)$ decreases with z (increases with time) : smaller halos formed earlier !

There is a (recent) agreement that the pattern of halo formation does not depend very much on the details of the cosmological context, and on the peculiar model. However the timings, and thus the function $L_{NL}(z)$ depend on the cosmology and of the model for structure formation: for instance, since structure formation is less efficient in a $\Omega < 1$ universe (i.e., in which the averaged density is lower than the critical value for recollapse), it has to proceed earlier in order to match the present distribution.

3.4. A global scale invariance ?

Most models assume that the initial spectrum of fluctuations was scale invariant. Since the linear evolution preserves the initial conditions (with a time dependent rescaling), this original scale invariance could still appear in the spectrum of the fluctuations of the cosmic microwave background (compatible with the COBE results), or in the very large scales power spectrum of the matter distribution, not presently known.

The possible apparition of scale invariance during the non linear development of fluctuations is a very difficult subject. A huge number of numerical simulations of gravitational collapse have been made, with various codes, resolutions, cosmological models, initial conditions, authors, ... It is however very difficult to extract from them well established results, which can be confirmed on a physical basis. Scaling invariance of the VPF's (or count probabilities) appears in some numerical simulations, at least in a certain scale range, and at some stage of the evolution. For instance, the simulated matter distribution from *cold dark matter* models (which, by the way, are invalidated for other reasons) presents a scale invariant character, as defined previously from non linear indicators. However, no specific model, even purely descriptive, accounts for all the statistical properties of the resulting distribution, and the possible origin of such a scale invariance is not understood.

To deal analytically with scale invariance, Peacock & Dodds (1996, and references therein) suggested empirical scaling formulae for the growth of correlation functions. First, they calculate what would be the evolution of fluctuations in the linear approximation, and denote the result with L-index. Then they make the empirical assumption that the real (non linear) power spectrum $P_{NL}(k)$ can be obtained from this approximation: its value at some scale k_{NL} depends on the value of the *linear* power spectrum $P_L(k)$ at a different scale k_L, related to k_{NL} through

$$k_L = [1 + P_{NL}(k_{NL})]^{-1/3} k_{NL}. \tag{9}$$

The non linear spectrum at scale k_{NL} is assumed to be

$$P_{NL}(k_{NL}) = f_{NL}[P_L(k_L)], \tag{10}$$

where the non linear function f_{NL} is obtained by a fit on simulation results. This purely empirical relation, with no demonstration or dynamical foundation, appears approximately verified in some simulations.

3.5. The hierarchical assumption

It was suggested, initially by Peebles (1980, Hierarchical assumption), that the n-point correlation function can be written as a sum of products of $(n-1)$

2-points correlation functions (with a summation over the possible branching trees). This can be written in a symbolic way as

$$\xi_n = Q_n [\xi_2]^{n-1}. \tag{11}$$

If ξ_2 is a power law with exponent γ, this hierarchical assumption implies the scaling relation

$$\xi_n(S\mathbf{r}_1, S\mathbf{r}_2, \ldots, S\mathbf{r}_n) = S^{\gamma(n-1)}\ \xi_n(\mathbf{r}_1, \mathbf{r}_2, \ldots, \mathbf{r}_n), \tag{12}$$

where S denotes a scale transformation. This latter scaling law may be taken as a weaker assumption leading to similar predictions. Such relations apparently hold in galaxy catalogs, for the low order correlation functions. However they are impossible to check for higher orders. It was shown by White (1979) that these relations imply the scale invariance for the VPF in the sense defined above, also present in galaxy catalogs.

Hierarchical assumption or scaling hypothesis are thus plausible. However, estimations of the relevant coefficients, by different authors and different methods, disagree. Given also the difficulties with correlation functions mentioned above, scale invariance cannot be considered to be clearly established. Such criteria may be so weak that any distribution would seem to be scale invariant according to them. In any case, in the absence of any explanation or justification, a possible scale invariance, the hierarchical assumption or the scaling hypothesis remain at a purely phenomenological level.

4. Scale invariance in local collapse ?

Locally, the massive matter is distributed in halos, around spiral or elliptical galaxies, and galaxy clusters. Their observed profiles are generally approximately fitted by power laws; moreover, some regularities in the halo parameters appear inside a given population, and even between populations. This has motivated analytical and numerical work to search for the origin of a possible self–similarity or scale invariance during the formation of a halo, or in the shape of its density profile (in which case, it should result from the non linear dynamics, since initial conditions are not preserved in the non linear collapse).

Different numerical works have analysed the gravitational collapse (Euler – Poisson or Vlasov – Poisson system) of isolated perturbations, most often with spherical symmetry. In some cases, a scale invariant regime, with a power law density profile, is established after orbit crossing (after linear and quasi linear development). Most often, some kind of scale invariance may be recognized in the resulting mass distribution. In general, it is not indicated why scale invariance does or does not occur.

Bertschinger (1985, see also Fillmore & Goldreich, 1984 and Hoffman, 1988) proposed for instance a self similar infall model. He proved that secondary infall of matter, onto already formed galaxies, lead to a self-similar evolution, corresponding to precise halo profiles with $\rho \propto r^{-2} - r^{-3}$, apparently resulting from phase mixing. Starting from a mass fluctuation with constant density, he calculated the external density profile of collapsing matter. In this very simple and peculiar case, he found that the evolution (collisional or not) follows a scaling solution: expressed with correctly scaled variables, it obeys an universal law, whose validity extends with time further and further from the central condensation. Henriksen & Widrow (1996) showed however that these solutions are unstable, the instability leading to a true violent relaxation. Their simulations also showed a self similar phase occurring during the collapse, with characteristics depending on the initial profile. However, since self similarity (a scaling symmetry) is *imposed* to the evolution, it is not easy to estimate how general are these conclusions.

More recently, Navarro, Frenk and White (1997) showed that the dark matter halo profiles, resulting from numerical simulations, have the same universal shape

$$\frac{\rho}{\rho_{crit}} = \frac{\delta_c}{\frac{r}{r_s}(1+\frac{r}{r_s})^2}.$$

For a given halo, this formula applies over at least two decades. The two parameters – the characteristic density δ_c and a characteristic scale r_s, appear strongly correlated, so that the massive halos form a one parameter family, which can be chosen as the halo mass M, proportional to $\delta_c r_s^3$. Low-mass halos are denser and have greater concentration c. Having collapsed earlier (see above), they are less evolved: the characteristic density of a halo appears to be proportional to the average density of the Universe at the time where it was assembled.

These results, roughly confirmed by Huss, Jain & Steinmetz (1997), are claimed to be independent of the values of the cosmological parameters, and of the shape of the power spectrum. They would imply that there is only one type of (dynamical) halo, depending on one unique parameter, for instance the total mass. This mass $M_*(z)$ only depends, in first approximation, on the redshift of formation of the halo and thus reflects its age.

It is not clear presently if this family of profiles will correctly fit the real halos. Rotation curves of elliptical galaxies rather suggest an r^{-2} profile at large distances. Various authors have proposed a very wide variety of profiles for halos of different kinds (galaxy or cluster size).

5. Conclusions

Concerning the large scale distribution of matter, it is clear that the good statistical tool has not yet been found. The classical correlation functions, and derived indicators, clearly appear unable to express its characteristics, in particular a possible scale invariance, or a fractal character. The new indicator Γ introduces by Pietronero appears more convincing but the controversy is presently not resolved. It is not clear how relevant are the signs of scale invariance found in the galaxy distribution or in the simulations. In any case, they lack any physical interpretation and remain at the phenomenological level. The proposed phenomenological descriptions, like for instance the hierarchical hypothesis, have not allowed us to to identify a possible mechanism which could lead to scale invariance.

More locally, it seems now well established that some formation (or infall) models of the halos lead to a self similar behavior, or to a scale invariant distribution. Questions remain about their generic character and their stability. Also, the possible underlying mechanisms are not understood. The recent results, suggesting that massive halos form a one parameter family, depending only on the age of formation, are very encouraging. They remain to be confirmed by dynamical arguments, and confronted to the real Universe.

References

[1] Bahcall N. A. & Burgett W. S. 1986, Ap.J. 300, L35
[2] Bertschinger E., 1985, Ap.J.S. 58, 39
[3] Cappi A., Maurogordato S. & Lachièze-Rey M. 1991, A& A 243, 28
[4] Davis M. 1997, preprint astro-ph 9610149
[5] Dubrulle B. & Lachièze-Rey M. 1994, A&A 289,667
[6] Fillmore J. A. & Goldreich P., 1984, Ap.J. 281,1
[7] Harrison E. R. 1970, Phys. Rev. D1, 2726
[8] Henriksen R. N. & Widrow L. M. 1996, preprint astro-ph/9612023
[9] Hoffman Y., 1988, Ap.J., 328,489
[10] Huss A., Jain B. & Steinmetz M., 1997, preprint astro-ph/9703014
[11] Maurogordato S. & Lachièze-Rey M. 1987, Ap.J. 320,13
[12] Maurogordato S. & Lachièze-Rey M. 1991, Ap.J. 369,30
[13] McCauley J.L. 1997, astro-ph 9703046
[14] Montuori M., Sylos-Labini F. & Amici A. 1997, preprint astro-ph/9705123
[15] Navarro J. F., Frenk C. S. & White S. D. M. 1997, preprint astro-ph/9611107
[16] Peacock J. A. & Dodds S. J. 1996, MNRAS 280, L19
[17] Peebles P. J. E. 1980, *The large scale structure of the universe*, Princeton University Press
[18] Pietronero L., Montuori M. & Sylos Romani F. 1997, astro-ph 9611197
[19] Szalay A. S. & Schramm D. N. 1985, Nature 314,718
[20] Sharp N. A. 1981, MNRAS 195, 857
[21] White S. D. M., 1979, MNRAS 186,145

LECTURE 11

Scale Invariance of the Avalanche Effect in Phase Transition Modulated Mantle Mixing

S.L. Butler and W.R. Peltier

Department of Physics, University of Toronto, Toronto, Ontario M5S 1A7, Canada

1. INTRODUCTION

As is well known, a wide variety of forced dissipative systems achieve a statistical equilibrium through restorative events which exhibit scale invariance, evidenced by power law distributions of both amplitude and duration with frequency. Geophysical examples include the distribution of earthquake seismic moments in a given region, the well known Gutenberg-Richter Law (see for example Main [1], Knopoff [2]), as well as cut-offs of river meanders (Stolum [3]). In the discussion to follow, we report a further geophysical example of this generic phenomenon, one which is of planetary scale: this concerns the distribution of mass flux events that develop across a pressure induced phase transition in a model of thermal convection in the Earth's mantle.

It is now rather universally accepted that the driving mechanism of surface plate tectonics is thermal convection in this region of the Earth. Sources of heat which drive this circulation include internal heating due to radioactive decay of of K, U, and Th, basal heating due to the flux of heat from the planet's molten iron outer core, and "effective heating" associated with the secular cooling of the Earth as a whole. Though the mantle is made of solid rock, creep processes within and between crystal grains allow the polycrystalline material to flow like a liquid on sufficiently long time scales. An outstanding question, however, concerns the issue as to whether the mantle convects throughout its entire depth as a single radial cell or whether the upper and lower mantle convect separately. Considerable theoretical effort has been invested in recent years to investigate a variety of physical effects that could control the extent to which the circulation is layered.

As a result of increasing pressure with depth in the mantle, a phase transition occurs at roughly 670 km depth in which the dominant mineral at that depth, spinel, transforms into a mixture of perovskite and magnesiowűstite. This phase transition is endothermic and it defines a horizon which separates the lower mantle from the transition zone and upper mantle. (The transition zone in the mantle comprises the region between 400 km and 670 km depth.) In our model (Solheim and Peltier [4],[5]) and in other similar models, e.g. Machetel and Weber [6], Peltier and Solheim [7], Tackley et al. [8], Honda et al. [9], Solheim and Peltier [10],[11], Tackley et al. [12], Peltier [13], this phase transition has been shown to be capable of acting as a barrier to radial mass transport, inducing a partially layered style of convection, if the overall convective vigor is sufficiently great. In a given simulation, the circulation has been shown to be almost perfectly layered with only weak mass flux events occurring for many hundreds of millions of years of model time. These periods have been shown to be separated by a small number of very large mass flux events in which a great deal of mixing occurs between the lower mantle and upper mantle and transition zone. For present purposes all mass flux events crossing the phase transition will be referred to as "avalanches" and it is the scale invariance of these avalanche events that we will demonstrate in what follows.

2. MANTLE CONVECTION WITH PHASE TRANSITIONS

The physical mechanism through which the endothermic phase transition at 670 km depth inhibits radial flow is associated with the buoyancy force resulting from deflection of the phase boundary due to local fluctuations in temperature at the position of the phase transition. For example, when cold material sinks from the surface to the depth of the phase transition, it is associated with a local decrease in temperature. Since the phase transition is endothermic, it has a negative Clapeyron slope and cooling requires that the phase boundary sink to a greater depth where the pressure is increased if the transition is to remain in thermodynamic equilibrium. Since the deeper phase is more dense than the shallower phase, the downward deflection of the phase boundary will result in an upward buoyancy force on the sinking material, tending to retard its motion. Latent heat will also be absorbed by material descending through the phase transition, leading to a further cooling of the phase transition region and a further deflection of the phase boundary. Of course, latent heat induced cooling will also tend to accelerate material downward, due to the material's own thermal expansivity. However, detailed analysis has shown this effect to be secondary to effects due to phase boundary deflection if the convection is sufficiently vigorous (Solheim and Peltier [9]).

3. A THEORETICAL MODEL

The influence of equilibrium phase transitions on mantle mixing may be assessed by solving the equations of fluid dynamics in the infinite Prandtl number limit, using appropriate finite difference approximations to the model equations. We will discuss results obtained in axi-symmetric spherical geometry and subject to the anelastic approximation which will allow us to include the effects of a number of depth dependent properties. The influence of both the 670 km phase transition and the exothermic phase transition at 400 km depth are included. The phase transition from Olivine to Spinel which occurs at 400 km depth is included in order to maintain fidelity to the real Earth. Its effect is opposite to that of the 670 km phase transition in that a phase boundary deflection tends to increase the intensity of the flow, but its net effect on the convective circulation have been shown to be far less significant (e.g. Solheim and Peltier [10]).

Explicitly, the governing equations consist of those for mass conservation (continuity),

$$\nabla \cdot (\rho_r \mathbf{u}) = 0, \tag{1}$$

momentum conservation,

$$0 = -\rho g \hat{\mathbf{r}} - \nabla p + \alpha_0 T_c \{\nabla \times (\eta \nabla \times \mathbf{u}) + 4/3 \nabla(\eta \nabla \cdot \mathbf{u})\}, \tag{2}$$

energy conservation,

$$DT/Dt - (\tau_0/\alpha_0 T_c c_p) D/Dt(l_1\Gamma_1 + l_2\Gamma_2) = (\kappa/Ra)[\nabla^2 T + (1/k)(\partial k/\partial r)(\partial T/\partial r)]$$
$$+ \mu/(c_p Ra) + \tau_0 \Phi/(\rho_r c_p) - \tau u_r (T + T_{top}) \tag{3}$$

and an equation of state

$$\rho = \rho_r\{1 - \alpha_0 T_c \alpha (T - T_r) + 1/K_T(p - p_r)\} + \Delta_1(\Gamma_1 - \Gamma_{r_1}) + \Delta_2(\Gamma_2 - \Gamma_{r_2}). \tag{4}$$

In the above system, subscript r denotes a depth dependent quantity whereas r itself represents the radial coordinate, $\mathbf{u}$ the velocity field, and p, ρ, and T the pressure, density, and temperature fields respectively. α, c_p, κ, k, K_T denote the thermal expansivity, heat capacity, thermal diffusivity, thermal conductivity, and the isothermal compressibility, while g represents the acceleration due to gravity. Subscript 0 is employed to denote a constant reference value used for nondimensionalization purposes while T_c and T_{top} respectively represent the temperature difference across the mantle and the temperature at the Earth's surface. Φ represents the heat generated by viscous dissipation. The nondimensional parameters in the above system of equations include the Rayleigh number,

$$Ra = \alpha_0 T_c g_0 d^3 \rho_0 / \kappa_0 \eta_0, \tag{5}$$

the dissipation numbers

$$\tau_0 = g_0 \alpha_0 d / c_p; \quad \tau_r = \tau_0 g \alpha / c_p, \tag{6}$$

and the nondimensional constant internal heating rate per unit mass

$$\mu = \rho_0 \chi d^2 / \kappa_0 T_c. \tag{7}$$

In the definitions of these parameters, d represents the thickness of the mantle, whereas χ is the dimensional constant heating rate per unit mass used to represent heating due to radioactive decay processes.

Thermal convection is driven by buoyancy forces which result from lateral variations in density which, in our model, result from changes in temperature, pressure, and phase as displayed in equation 4. These buoyancy forces cause discrepancies between gravitational and pressure gradient forces which are balanced by viscous forces (Equation 2). One will note that in the infinite Prandtl number limit, the inertial term is dropped from the momentum equation due to the extremely viscous nature of the fluid. Heat generated due to viscous dissipation feeds back on the temperature field (Equation 3) as do effects due to latent heat associated with the phase transitions, heat advection and conduction, internal heating and temperature changes associated with adiabatic expansion and compression. One will also note that the anelastic approximation results in neglect of the $\partial\rho/\partial t$ term in the mass conservation equation. This has the effect of removing acoustic waves which are many orders of magnitude faster than the motions of mantle material in which we are interested.

Effects due to the phase transitions are included through the phase density functions

$$\Gamma_i = 1/2\{1 + tanh[(r_{pi} - r)h_i/d]\} \tag{8}$$

in which subscript 1 denotes the phase transition at 400 km depth while subscript 2 denotes the 670 km phase transition. h_i is a width parameter representing the thickness of the divariant phase loop of the phase transition. In equation 3, the l_i represent latent heat per unit mass while in equation 4 Δ_i represents the difference in density between the upper and lower phases. r_{pi} represents the position of the phase boundary and is a function of the polar coordinate and of time. Its value must be determined in each angular grid column and for each time step. This may be accomplished by assuming a constant Clapeyron slope. If the transition is in thermodynamic equilibrium then the phase boundary must occur at a depth where the pressure is given by

$$p = p_{0i} + \beta_i T. \tag{9}$$

In the above p_{0i} is the zero temperature pressure of the coexistence curve for phase transition i. The Clapeyron slopes, β_i, for the 400 and 670 km phase transitions are taken to be 3 MPa/K and -2.8MPa/K respectively (e.g. Chopelas et al. [14]). The latent heats are related to the Clapeyron slopes through the Clausius-Clapeyron relation

$$\beta_i = \rho^2 l_i / [\Delta_i (T + T_{top})]. \tag{10}$$

To solve equations 1-4 the momentum equation is first transformed into stream-function-vorticity form. It is also useful to ensure that only a single time derivative appears in equation 3 and this may be achieved by introducing a transformed temperature, A, such that:

$$A = T - \tau_0/(\alpha_0 T_c c_p)(l_1\Gamma_1 + l_2\Gamma_2) \tag{11}$$

Equations 9 then reduce to the coupled set of nonlinear algebraic equations

$$\beta_1\{A + T_{top} + \tau_0/(2\alpha_0 T_c c_p)[l_1 + l_2 + l_2 tanh(r_{p2} - r_{p1})h_2]\} + p_0 - p_r = 0 \tag{12}$$

$$\beta_2\{A + T_{top} + \tau_0/(2\alpha_0 T_c c_p)[l_1 + l_2 + l_1 tanh((r_{p1} - r_{p2})h_1)\} + p_0 - p_r = 0 \tag{13}$$

These equations must be solved for the r_{pi} at each time step and at each spatial position. The boundary conditions applied for the purpose of model integration are that the tangential stress and radial velocity vanish at the surface and at the core mantle boundary (CMB). The CMB and surface are also taken to be isothermal. For a more detailed description of the model, the interested reader is directed to Solheim and Peltier [10].

4. RESULTS

In figure (1) we display a grey scale contour plot of the temperature field in the model mantle; a large (45^0 from right) and smaller (95^0 from right) avalanche can be seen. The positions of the 670 and 400 km phase transitions are indicated and one can clearly see that some down-going plumes are blocked by the 670 km phase transition.

Figure (2) shows a typical time series for the mass flux crossing the 670 km horizon (dotted line), averaged over the polar coordinate, as well as the similarly averaged mass flux crossing the 400 km horizon (solid line) for a constant viscosity model with $Ra = 10^7$ and $\mu = 10$. These parameters result in a core mantle boundary temperature which is approximately correct for the modern earth but deliver a surface heat flow which is somewhat too large. These conditions may, however, be representative of the early earth. It will be noted that the mass flux is highly time dependent and that there are prolonged periods during which the mass flux crossing the 400 km horizon is considerably greater than that across the deeper 670 km horizon, indicating a layering of the circulation. This layering is interrupted by large mass flux events, the last

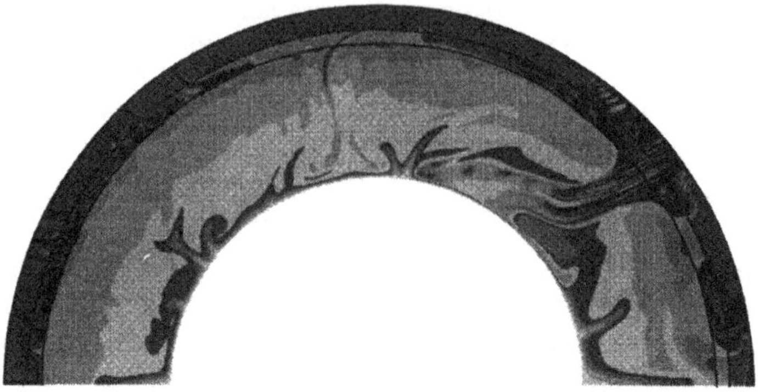

Figure 1: A grey scale contour plot of a 'snap-shot' of the temperature field in the mantle.

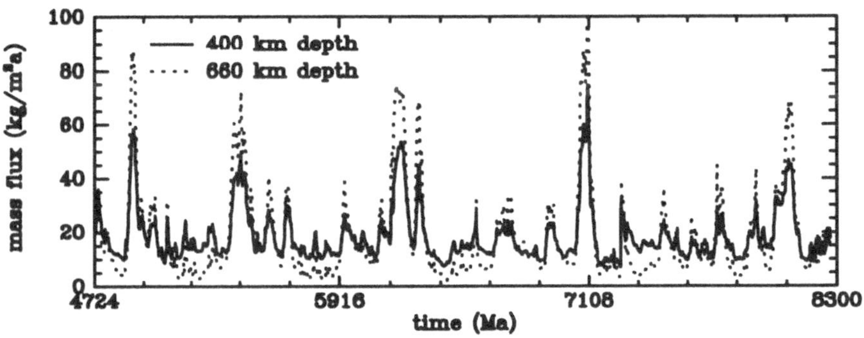

Figure 2: Time series of the mass flux averaged over the polar angle.

of which (8076 Ma) is shown in the temperature field in figure (1). In figure (3) we display the mass flux across the 670 km horizon averaged over both position and time as a function of time for a time span covering 3.3 billion years of model time. The average is clearly converging to a value around 18 kg/(m^2 year), indicating that the system has reached a statistically stationary state. Figure (4) shows the mass flux across the 670 km horizon as a function of angular position for a few "instants" of time, including a very large avalanche event. Avalanches of many different sizes can, however, be seen. It will also be noted that associated with the largest downgoing avalanche are two large upward mass flux events. This clearly illustrates that at least for the largest events, mass is locally conserved whereas it was previously believed that diffuse upwellings balanced the mass flux associated with mantle avalanches. Since there are large upward mass flux events associated with the large downwellings,

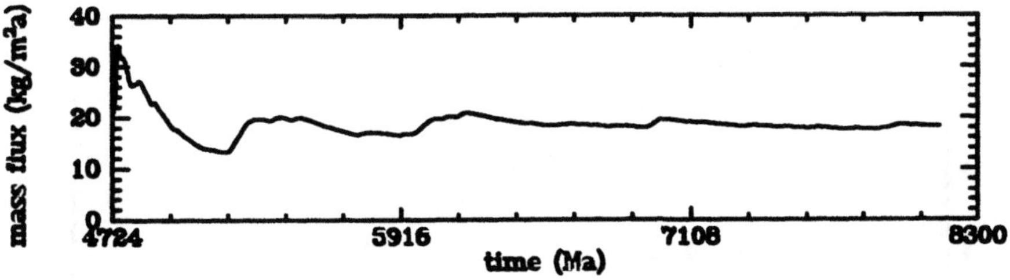

Figure 3: The mass flux crossing the 670 km horizon averaged over both space and time.

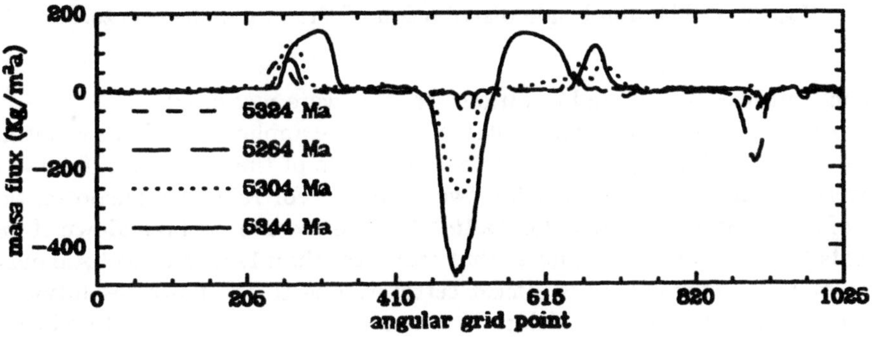

Figure 4: The mass flux as a function of polar angle, crossing the 670 km horizon for four different instances in time.

one might consider the term avalanche to be a misnomer; we will continue to use it in what follows, however.

In order to assess the possible scale invariance of these avalanche events, it was necessary to count them and to measure their magnitudes. This was accomplished for a time interval that consisted of the last 940 Ma of the time series shown in figure 1. In order to count the avalanches, at each time step of the model integration, zero crossings of mass flux were detected. The peak of the mass flux between each of these zero crossings was then found and was compared to the zero crossings of the previous time step in order to determine whether the event was the same as one that existed previously. If so, the mass flux at that time step was added to the previously accumulated total for the avalanche in order to derive a total integrated mass for a given avalanche. If the peak mass flux did not lie between zero crossings of mass flux from the previous time step, the event was considered to constitute a new avalanche. An avalanche was considered to have ended if the associated mass flux dropped below a certain threshold. In this way over 15 000 avalanches were counted and

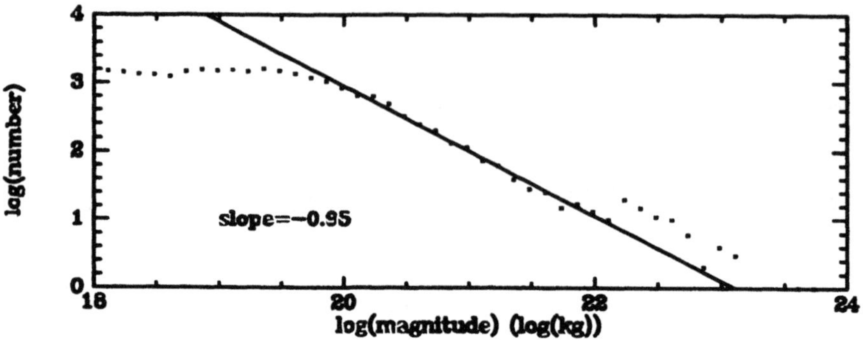

Figure 5: The amplitude distribution of the avalanche events.

their magnitudes (in kg) and durations (in Ma) were measured.

Figure (5) presents the results obtained from the application of this counting procedure. It will be observed that the amplitude of the avalanches follows a power law relationship over roughly two decades. For reference purposes, the mass of the mantle in our model is 4×10^{24} kg. We observe the breakdown of the power law behaviour for avalanches that carry less than 1/10 000 the total mass of the mantle. This cut off is almost certainly due to the finite resolution of the finite difference grid on which the avalanche process is resolved. In a future numerical experiment it will be interesting to further increase the spatial resolution in order to determine whether or not scaling will then persist to smaller amplitude events. Such an experiment would also provide a useful check on the error in our calculated scaling exponent. The power law relation actually seems to underpredict the largest avalanches, making this cut-off incompatible with the finite size effects that one might expect. This anomaly is most likely due to insufficient counting time. The time scale separating the largest avalanche events is roughly 500 Ma, a time scale which could be related to the characteristic timescale of the supercontinent cycle (Peltier et al. [15]). The entire counting time in this experiment was only 940 Ma and we may have, as a result, over counted the largest avalanches. The other possibility is that there is a different physical mechanism that is responsible for the largest avalanches; a phenomenon similar to characteristic earthquakes, if in fact these exist (See for example Knopoff [2], Schwartz and Coppersmith [17]). Further evidence for characteristic avalanches comes from the value of the scaling exponent. Since it has magnitude less than one, the distribution is not normalizeable at infinity and as such the largest avalanches must be characteristic avalanches with scale invariance existing only for the smaller events (Sornette [16]). Resolution of the issue as to the existence of "characteristic avalanches" will require a longer counting time both to determine whether more large events occur than the power law would predict and to reduce the error on the observed scaling exponent. Also interesting is the issue as to which avalanches carry the largest

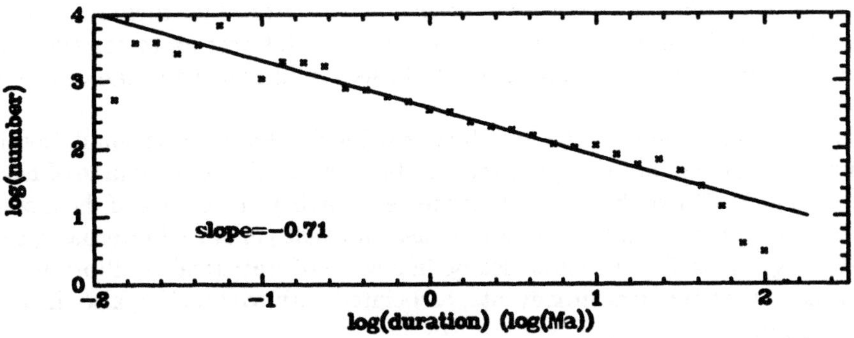

Figure 6: The duration distribution of the avalanche events.

amount of mass. If in fact the power law exponent has magnitude less than 1, then the total mass flux across the 670 km horizon is dominated by the largest avalanches.

In figure (6) we note that the duration of the avalanches also follows a power law decay with slope approximately -0.71. The power law is once again stable for roughly two decades. In this case, the long duration cut-off probably results from the finite counting time of the experiment, with the longest avalanches lasting a significant fraction of the total counting time. The scatter at short durations is probably due to the finite time step that was taken during the model run, with the shortest lived avalanches being not much longer than a single time step.

5. DISCUSSION AND CONCLUSIONS

The results presented in figures (5) and (6) provide strong evidence that the avalanche effect in our model of thermal convection in the presence of an endothermic phase transition exhibits a scale invariance over a wide range of magnitude and duration. That it does so, without requiring tuning of the control parameters, argues that the system displays self-organized criticality. (e.g. Bak et al. [18])

Besides increasing the counting time in order to determine whether 'characteristic avalanches' exist, it would also be interesting to perform a similar counting experiment on a system with a different set of control parameters. Will the slopes of the power laws be found to be invariant? This would indicate an underlying symmetry in the models. If so, what could be learned from these symmetries concerning thermal convection in the presence of an endothermic phase transitions? Furthermore, our spherical axi-symmetric model represents the phase transitions by solving explicitly for phase boundary position as a function of polar angle and time; other models represent the phase transitions by employing an effective thermal expansivity which includes a jump at the position of the phase transition. It would be interesting to see whether a model constructed on this approximate basis would reveal the same scaling behaviour.

It is also interesting that the heat flux crossing 670 km for this simulation is dominated by advection, implying that avalanches are the main source of heat flux crossing the 670 km horizon. Once the heat leaving the surface of the model is equal to the sum of that entering the base and that generated internally, the heat flowing across 670 km must also be in a state of statistical equilibrium. It is very likely that the heat flux events, associated with avalanches, also display scale invariance.

Finally, it is useful to reflect on the nature of the critical state in this system. In the model, hot material rises as thermal plumes from below the endothermic phase boundary whereas cold material sinks as thermal plumes from above. Both flows are strongly inhibited by the phase boundary and a thermal boundary layer is built up across this position (Solheim and Peltier [10], Peltier [13]). It has been shown that the largest avalanches can be understood as developing out of linear instability of this thermal boundary layer and that the onset of these avalanches depends on a relatively small number of parameters that characterize the mean circulation (Butler and Peltier [19]). In particular, the length scale governing the size of the largest avalanches is the depth of the entire mantle. It may be that the largest avalanches arise from a mechanism that is distinct from that which causes the smaller events. If true, this would argue for the existence of 'characteristic avalanches'. The other, more likely, possibility is that avalanches of all sizes arise from the same instability mechanism but that smaller events depend on local conditions not captured by mean flow characteristics. The length scale governing the size of these smaller avalanches must be controlled by local conditions in their vicinity.

References

[1] Main I., *Reviews of Geophys.* **34** (1996) 433-462.
[2] Knopoff L., this volume, Lecture 12.
[3] Stolum H., *Science* **271** (1996) 1710-1713.
[4] Solheim L.P., and Peltier W.R., *Geophys. Astrophys. Fluid Dyn.* **53** (1990) 205-255.
[5] Solheim L.P., and Peltier W.R., *Can. J. Earth Sci.* **30(5)** (1993) 881-892.
[6] Machetel P. and Weber P., *Nature* **350** (1991) 55-57.
[7] Peltier W.R. and Solheim L.P., *Geophys.. Res. Lett.* **19** (1991) 321-324.
[8] Tackley P.J., Stevenson D.J., Glatzmaier G.A., and Schubert G., *Nature* **361** (1993) 699-704.
[9] Honda S., Yuen D.A., Balachandar S., and Reuteler D., *Science* **259** (1993) 1308-1311.

[10] Solheim L.P. and Peltier W.R., *J. Geophys. Res.* **99** (1994) 6997-7018.
[11] Solheim L.P. and Peltier W.R., *J. Geophys. Res.* **99** (1994) 15877-15901.
[12] Tackley P.J., Stevenson D.J., Glatzmaier G.A., and Schubert G., *J. Geophys. Res.* **99** (1994) 15877-15901.
[13] Peltier W.R., *Philos. Trans. Roy. Soc. London Ser. A* **354** (1996) 1425-1447.
[14] Chopelas A., Boehler R. and Ko T., *Phys. Chem. Miner.* **21** (1994) 351-359.
[15] Peltier W.R., Butler S.L. and Solheim L.P., The Earth's Deep Interior (Gordon and Breach, Newark N.J., 1996) pp. 405-430.
[16] Sornette D. private communication.
[17] Schwartz D.P. and Coppersmith K.J., *J. Geophys. Res.* **89** (1984) 5681-5698.
[18] Bak P., Tang C. and Wiesenfeld K., *Phys. Rev. A* **38** (1988) 364-373.
[19] Butler S. and Peltier W.R., *J. Geophys. Res.* **102** (1997) 2731-2749.

LECTURE 12

Scale Invariance of Earthquakes

L. Knopoff

Department of Physics and Astronomy and Institute of Geophysics and Planetary Physics, University of California, Los Angeles, U.S.A.

1. EARTHQUAKE ENERGY DISTRIBUTIONS

If the model of self-organized criticality (SOC) yields power law distributions [1-5], does it follow that the well-known power-law distribution of earthquake energies [6-8] imply that the earthquake process is SOC? I show below that the earth is not SOC and that there are alternatives to understand the power law.

The distributions of the energies are derived from analysis of earthquake catalogs, which are usually listings of five parameters, namely the hypocenters – which are the locations in three dimensions of the points of initiation of earthquake fractures–, the times of their initiation, and their magnitudes. The magnitude M of an earthquake is proportional to the base ten logarithm of the energy radiated to distant sites. There are two hints here that dynamical processes are important in the physics of the self-organization: 1) earthquakes are fractures – and fractures cannot develop instantly –, and 2) earthquakes radiate vibratory wave motions of stress and displacement to distant points. We cannot ignore radiation since it is the way we have of knowing that most earthquakes have actually taken place. Radiation during the brief seconds of the earthquake is a potent influence on stress redistribution.

On a standard picture, the nature of fracture requires that we start with an elastic medium that has been prestressed up to some critical fracture threshold; at this time fracture initiates locally and the zone of fracture spreads rapidly over some surface. The crack edge velocities are of the order of shear wave velocities in crustal rocks which are about 3 km/sec. Maximum particle velocities during slip are estimated to be of the order of a few meters/sec. The rate of loading of prestress is slow, thereby accounting for the long interval between successive large earthquakes; typical loading rates are of the order of several cm/year. In part, the energy radiated in an earthquake is a measure of the linear dimensions of the fracture: the larger the fracture, the larger the magnitude of the earthquake. The limits to the size of an earthquake fracture are strongly influenced by the geometrical fluctuations in the prestress and the fracture strengths; these fluctuations determine the size distribution of events.

To establish an intuitive understanding of the magnitudes of earthquakes, each increase of one magnitude corresponds to a 30-fold increase in energy. A magnitude 8 earthquake is an enormous event, capable of causing great death and property damage in populated areas; a magnitude 8 earthquake is a rare event locally on the scale of human lifetimes, and occurs only once every 100 years or so on the average in Southern California[9], but with large variability in interval times. A magnitude 3 earthquake is perceptible only very locally, causes no damage and occurs about once every two days in Southern California on average; magnitude 3 earthquakes occur about 2×10^4 more frequently than magnitude 8 events, but are felt by few people because of their small size and localization. Magnitude 8 events are huge earthquakes and are felt over wide regions.

The cumulative energy distribution in small earthquakes is

$$N(E) \sim E^{-B}. \tag{1}$$

(Seismologists prefer to use the cumulative distribution rather than the pdf.) The Gutenberg-Richter (G-R) power law is a universal and the exponent is 2/3 for tectonic earthquakes everywhere in the world with small fluctuations; there are heuristic arguments that give the exponent precisely. The relation (1) is known to hold for small earthquakes, since these are the only ones observable in sufficient quantity that their statistics can be identified. The catalog for Southern California earthquakes, which spans 60+ years, is the temporally longest local catalog in the world. In this interval the largest earthquake was a magnitude 7.5 event; the last earthquake in the magnitude 8 range took place in 1857 on the San Andreas Fault (SAF) which is the dominant fault of the Southern California system. Only 16 independent events with $M \geq 6.4$ have occurred in 65 years, and no M=8 earthquake has occurred in the interval. So we do not know the statistics of the energy distribution of large earthquakes since they are rare events within this relatively short, although longest of catalogs.

If the finite temporal sample of (1) is valid as a long-term average, the total energy radiated by earthquakes $\int E dN$ diverges if the upper limit to earthquake energy size is infinite, since $B < 1$. Since the earth's energy budget for generating earthquakes is finite, the power law cannot extend to infinity, and there must be a rolloff or cutoff in the distribution. This argues for the presence of at least two branches to the distribution: a low energy branch for which the power law is observed and is due to some universal process, and a high energy branch which does not fit the power law for small earthquakes and for which we have very little data. The high energy branch may or may not have its own power law distribution. I argue that the low energy branch, which we know is valid for frequent small earthquakes, is indeed due to a universal process (no surprise here), although the nature of the process may turn out to be something of a surprise. I shall also argue that the high energy branch is a response to physical processes that are strongly influenced by the geometry of earthquake faults locally. Since the geometry of fault maps varies from region to region, it will follow that the high energy branch must have a local character rather than a universal one, unless the major faults on which large earthquakes occur have some universal geometry of their own; this seems unlikely. I show below that the major part of the power law branch of the distribution involves organization from large scales to small ones, and not in the opposite direction.

The larger the region of potential energy storage before the earthquake, the larger the earthquake. If there is a rolloff or cutoff in the energy distribution, then there must be a corresponding characteristic length scale in the geometry of earthquakes. In Southern California, all small earthquakes are located in the uppermost 15 km of the earth's crust, which is only about 1/400 of the earth's radius; larger earthquakes may have fractures that extend as deeply as 18 or 19 km. Let us take 15 km to define the characteristic length. A critical depth is an appropriate measure of the crossover in the distribution since small fractures would be likely to grow as circles or ellipses without intersecting either the lower or upper boundaries of the seismogenic slab, while larger earthquakes would be constrained to grow horizontally, with a roughly rectangular geometry. Thus the statistics of large earthquakes might not be expected to follow the same rules as the small ones, since the fluctuations that stop growth would be expected to have different distributions on the two scales. In many other parts of the world, earthquakes occur at much greater depths, so the rolloff magnitude corresponding to the characteristic length scale is not expected to be a universal; instead it probably varies significantly from region to region.

The rupture dimensions of small earthquakes increase by about a half-order of magnitude for each increase in earthquake magnitude: a magnitude 3 earthquake corresponds to a fracture dimension of about 300 m, a magnitude 5 earthquake to a fracture of about 3 km, etc. An earthquake with linear dimensions of 15 km corresponds to a magnitude 6.4 earthquake approximately. Although this magnitude might be expected to to be the rolloff magnitude, the magnitude distribution for Southern California shows no kink at magnitude 6.4 nor at any other magnitude up to the largest ($M=7.5$) that has taken place in the last 65 years. Further, the net slip in all earthquakes of the last 65 years is less than that of the slow slip in the plates that drive the earthquakes; this has been interpreted as implying that we are currently in an episode of deficiency of strong earthquakes in the magnitude 7 range and above[10]; on this interpretation, the pdf should have a peak at its large energy end which is not observed; there should, of course, be a falloff at even larger energies. I argue below that the properties of aftershocks require that there be a rolloff in the distribution at energies around $M=6.4$, with relevance to the crustal thickness as above.

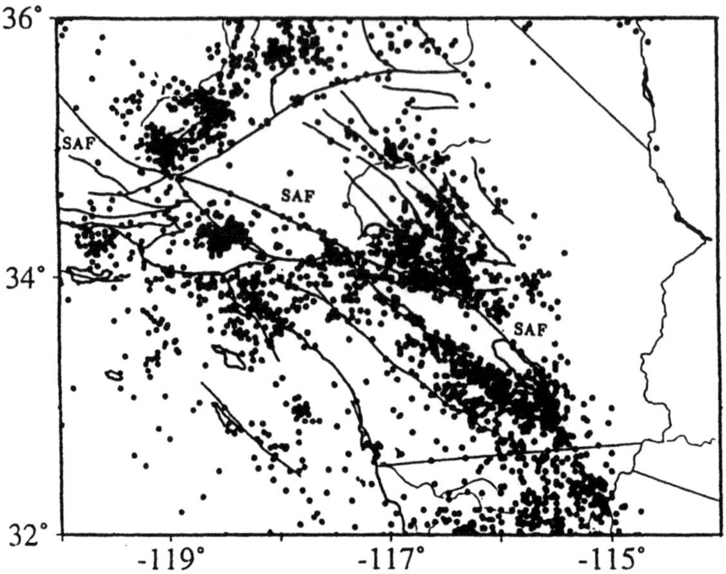

Fig. 1. — Epicenters of Southern California earthquakes with magnitudes greater than 3.5, 1935-1994. Most of the clusters of epicenters are due to aftershocks of large earthquakes.

2. AFTERSHOCKS

I turn to the incompatibility between the observation that power law behavior extends to the cutoff and the inference that there should have been a kink in the distribution at $M \approx 6.4$. A map of all earthquake epicenters in Southern California with a lower magnitude threshold on the low-energy branch of the distribution, is of course dominated by the overwhelming number of small earthquakes (Fig. 1). Rather remarkably, not only were there no large earthquakes on many of the sections of the mapped faults on which large earthquakes can be expected to occur, including the SAF, there was no seismicity on them whatsoever in this interval. This observation is completely at odds with expectations based on sandpile or other SOC models of self-organizing systems, which assert that any given section of fault should experience the full spectrum of sizes of events, with small events occurring more frequently than large ones.

Some faults have many small shocks on or near them. The densest clusters of these earthquakes can be identified as being the numerous small aftershocks of a few large earthquakes. Since these small earthquakes are the fallout of their predecessor parents, they are not self-organizing into a succession of events leading to future large earthquakes; indeed, these aftershock sequences terminate long before the occurrence of the next large event. Typically, large events occur without much preparation or build-up; hopes for earthquake prediction must rest on observations of fluctuations in what seismicity there is before a large earthquake. Some of the small earthquakes on the map are not aftershocks of known identifiable large earthquakes in the 65 year interval, but since some aftershock series may last for longer than 65 years (see below), some of these latter earthquakes may be aftershocks of large earthquakes that occurred before the start of the catalog.

Aftershocks are earthquakes that are highly localized spatially around fracture zones of large earthquakes and temporally clustered after the parent. Aftershocks have been much studied and have at least two universal properties. The rate of occurrence of aftershocks is a universal and is given by Omori's law[11]

$$\dot{n} \sim 1/t^p \tag{2}$$

. Observationally, the exponent on t is close to 1; it has been found to be $p = 1.11 \pm 0.25$ for 39 Southern California sequences[12]. Adjustments in the specific form of the time dependence have been suggested. We use (2) with $p = 1$ for these discussions. A second universal is the observation that the magnitudes of earthquakes in an individual aftershock series fit the G-R power law for small magnitude earthquakes with the expected exponent B [12-14]; since the aftershock magnitudes are less than those of the parents, their magnitude distribution has an upper cutoff. The durations of the individual sequences (2) can be long. In the longest well-documented aftershock series, which is that following the M=7.5 Kern County earthquake of 1952 in Southern California, aftershocks in the immediate neighborhood have a remarkably excellent fit to the Omori law, even after 43.5 years, a time that is a substantial fraction of the length of the complete catalog; for these earthquakes the fit to the power-law energy distribution is excellent: $1.5B = 1.03 \pm 0.04$. It is plausible to suppose that the law (2) holds for times longer than 65 years, at least for some very large parent earthquakes. On the other hand, detailed investigation of aftershock series for most smaller, strong earthquakes shows that, although they too have the property (2), aftershocks with a given magnitude do not occur after a finite time.

If the large earthquakes have the same self-similar character as do the smaller ones, as implied by the raw Gutenberg-Richter relation for Southern California, then the production of aftershocks should be also appropriately scaled, i.e. the number of aftershocks differing by ΔM from the magnitude of the main shock should be independent of the magnitude of the main shock. Then, under this assumption, if we were to strip off the identifiable aftershocks from a catalog, we should be left with a power law distribution for the residual events over the same magnitude range, but with a reduced exponent; the residual events would then be the parents that trigger aftershock series.

If we strip off the identifiable aftershocks from the catalog of Southern California earthquakes (there are several competing procedures[15-18], all of which depend on identification of bursts of activity in both space and time), we are left with a residual catalog having about 30% of the original total. The distribution of these events is not a power law over all magnitudes with reduced exponent, but is instead a power law over the magnitude range 4.0 to about 6.2 with the same exponent as before, with numbers uniformly reduced; at larger magnitudes, the distribution is a spike starting from about $M \approx 6.4$, admittedly of only about 16 or so earthquakes, that indicates that aftershocks of large earthquakes are about 70% of the smaller earthquakes, and the smaller independent earthquakes do not produce a proportionate percentage of aftershocks, by far. Since the exponent remains the same, the large earthquakes belong to a different class of events than the small ones, with a crossover at a magnitude of about 6.4, with corresponding fracture dimension. Although we do not know the pdf of the large earthquakes, the large earthquakes are not likely to be a product of the same *statistical* process as the smaller earthquakes.

3. STRESS REDISTRIBUTION

I leave the unfolding of the implications of earthquake phenomenology and turn to the constraints on the development of these patterns that are imposed by the physics of stress redistribution due to fractures and other relaxation processes. The discussion above has identified several targets for physical modeling:

- Why are there (at least) two classes of earthquakes, the small ones that fit the power-law distribution and the larger ones that do not have the same distribution, and may or may not fit a power-law distribution?
- Why are there two classes of small earthquakes, both of which fit what seems to be the same power law distribution, the one being the sequence of aftershock bursts and the other the residual catalog?
- Why do individual aftershock bursts have the universal properties of the power-law distribution in energy and the $1/t$ law of time delays? Despite the enormous disparity in scale sizes between the rupture dimensions of the parent, triggering event and the aftershocks, do the Omori/Utsu relations imply a local process that is a property of all sites around and near the large fracture, or is it a global process of self-organization into a state with long-range correlations, i.e. a sort of inverse SOC?
- How detailed must the physics of fracture and stress redistribution be specified to develop the above properties? Can we get away with invoking simplistic models of fracture and stress relaxation, or must we invoke a complex physics and/or geometry?

Earthquake systems are not SOC for geophysical reasons; they are not SOC for physical reasons as well. Because of the scaling of stress redistribution, large earthquakes must change the stress field significantly on the sites of occurrence of small earthquakes in the neighborhood, but the reverse is much less likely. Thus SOC is unlikely to develop on the small-energy branch of the distribution since any self-organization on the small energy branch is likely to be heavily punctuated by activity on the large-energy branch. It would be paradoxical if the universality of the small events were somehow connected with the non-universality of the larger ones.

How is the physics of fracture related to the distributions of the two branches? Because the earthquake catalogs do not indicate the relationship between the geometry of the fracture and the hypocenter, many authors have mistakenly treated the catalogs as describing a point process. The geometry of fracture is a vital influence on the geometry of the fluctuations in the stress and hence on the self-organization of seismicity.

The average slip $\langle u \rangle$ in a fracture is related to the average drop in stress $\langle \sigma \rangle$ from that in the prestressed state to that in the final state. From simple dimensional arguments based on Hookean elasticity, the scaling is

$$\langle u \rangle \sim \frac{\langle \sigma \rangle}{\mu} L \tag{3}$$

where μ is the elastic modulus of the material astride the crack and L is some critical length, presumably related to the geometry of the crack. Consider the application of (3) to the great ($M \approx 8$) San Francisco earthquake of 1906. The mean slip was about 4.5 m. The fracture length was about 450 km, which implies that there was a huge momentum exchange over the short time of the earthquake. (The energy released was of the order of 10^{24} erg, which is an enormous amount, and is of the order of the energy that would be released in a hypothetical 100 megaton explosion.) If we use typical values of stress drop and shear modulus, we get a scale size L for this earthquake that is at least one order of magnitude smaller than the length of the fracture. The coincidence of the order of magnitude of this dimension with the thickness of the seismogenic zone and with the crossover in the aftershock distribution is suggestive. The enormous discrepancy between the scale size and the fracture length implies that slip ceases when the fracture length is of the order of 15 km. Thus the fracture in this great earthquake must be described as a moving patch of rupture of size $\sim$ 15km; at the time the extremities of the fracture began to slip, the parts close to the point of initiation had healed a minute or two earlier; only $\sim$ 15km of the fracture slips at any instant. Thus the *dynamics* of healing is as important as the condition for initiation of fracture in the development of slip and hence in the self-organization of stresses for large earthquakes. The critical scale size for healing suggests that the smaller earthquakes may be approached by a more conventional model of extended fracture.

The drop in stress on a fracture is accompanied by a redistribution of stress in the neighborhood, for both static and dynamic cracks. Immediately to the side of the fracture, the stress is also lowered because of the continuity conditions on stress. Beyond the edge of the fracture and in its plane, the stress after fracture is increased. For classical cracks, the stress drop on the crack surface is counterbalanced by its increase outside it. Thus there is no way to relieve the inexorable buildup of stress from tectonic sources in this model, and ultimately this model must explode in a "non-earthquake" of size larger than the computational size of the system. These are not real earthquakes since they are not controlled by the same physics of healing. We can limit the sizes of earthquakes by redistributing stress onto noncoplanar faults and by introducing scaling through heterogeneity; these are subjects that are sketched in the next section.

The range of the stresses L outside the cracks is scaled by 15 km for the larger fractures or by the size of the fracture for the smaller fractures, $L = \min[15\text{km}, \ell]$, where ℓ is the length of the crack. The crossover in scale size is usually ignored in simple elastic fracture models of seismicity. In even simpler sandpile models, the scale size for the redistribution is uniformly nearest-neighbor, i.e. it is the lattice spacing. However this is a model of a dislocation and not of an extended fracture; in dislocation models healing, or restoration of strength, takes place before the neighboring lattice site can break, whereas on crack models healing does not take place until after the crack has reached the length L. Evidently the dislocation model is an adequate model for large earthquakes in view of the need to invoke rapid healing effects, but the dislocation models in the literature have targeted the power law instead of the large-earthquake branch of the distribution.

The self-organizing principle for single fault systems, is thus that the stress drops on the fractured segment and increases in its plane outside it. The slow increase of tectonic stress in the region slowly elevates the stress level on the faults over the entire region, causing a new segment to rupture, which in turn causes further stress redistribution. Thus the succession of fractures and the reloading by plate tectonics generates a system of stresses that fluctuate in time and space. Fractures are initiated when the prestress is brought to the level of the fracture strength. Fractures stop growing when they encounter a fluctuation in the difference between the local fracture strength and the prestress that is too great for the stress at the edge of the crack to deliver; the fluctuation is set up by the past history of fractures. From the Green's function for cracks with smooth edges, under either dynamic or static conditions of fracture, the stress enhancement at the edge of a crack is

$$\sigma_{edge} \sim \langle\sigma\rangle(L/d)^{1/2} \tag{4}$$

up to some numerical factors that take complex geometry into account, where σ and L are as before, where the average of the stress drop $\langle\sigma\rangle$ is taken over the active region of slip as measured by L, and where d is a distance of tapering at the edge of the crack that is introduced to avoid singularities; d is scaled by L but is less than it. Thus the larger cracks among those with $L < 15$km, have a greater likelihood that they will increase in size than smaller cracks; for the largest cracks, for which $L \approx 15$km, the likelihood of growth is independent of the present size of a fracture. The distance d is not a mere mathematical convenience: it can be shown that the edge of a crack must have a transition zone of dimension d, that lies between the regions of complete slip on the inside and the region of locking on the outside of the crack[19]; in this zone the material deforms nonelastically such as by inducing plastic deformation, slip-plane gliding, microfracturing, granulation, etc. In the case of earthquake faults, this requires that a zone of weaker material be developed around the fault laterally, either before or during the earthquake and whose strength rises again some time after the earthquake; thus the strength of this zone itself fluctuates over time. Weak zones have been identified as wave guides from observations of aftershocks of the very strong Landers earthquake (M=7.3, 1992) in Southern California[20].

4. THE PHYSICS OF THE CLUSTERING OF LARGE EARTHQUAKES

The broad features of much of the physics underlying the list of problems of seismicity outlined in Section 3 above is now known. Unfortunately, the space allocated for this contribution does not permit a full exposition; I can only give a skeletal presentation of the results in the space remaining.

Models of large earthquakes allow us to understand the qualitative properties of the self-organization. Since the largest events computed must correspond to the largest fractures observed, then numerical simulations perforce focus on the earthquakes with the largest energies in the distribution. A number of models of self-organization in earthquake systems have been constructed in which it is assumed that the power law is the modeling target. Since power laws imply self-similarity, it has been assumed that the earthquake histories are developed on the simplest of landscapes of the spatial distribution of fracture strengths, namely constant distributions. For spatially homogeneous threshold models of fracture, the power laws of the pdf's must have infinitely long tails. Thus for these models, there is a finite probability that a fracture will develop that extends across the entire model; at this point a "non-earthquake" or runaway develops that stops for reasons other than the internal fluctuations of difference between fracture strength and prestress; these finite-size effect events are of infinite length for periodic boundary conditions, or they encounter unbreakable barriers at ends (which are extraordinary inhomogeneities of strength), etc. In other words, the runaways interact with the computational boundaries and develop fractures that are not stopped for the same reasons as the smaller events; the largest earthquakes in nature are true earthquakes and are not finite size effects. After a runaway the subsequent evolutionary history is overwhelmed by and is a consequence of the reset of the stress conditions by the interaction with the boundaries, and no longer reflects a process that is characteristic of an unbounded region; in simple terms, after a fracture encounters the boundaries, we are concerned with surface rather than volume effects. The situation does not improve if one introduces dynamics and radiation damping into a finite planar fault model. It has been shown that evolutionary seismicity on a finite homogeneous lattice with dynamics and radiation and with rapid healing, develops a very long transient episode of activity that has the power law property, but even here the system ultimately organizes itself into an unacceptable lattice-wide event[21]. Indeed the power law in the transient phase does not describe an ensemble of random events; the events in the self-organizing phase are highly correlated: since the stress after any fracture is roughly uniform, all parts of it are likely to reach the threshold stress at about the same time, after a suitable elapsed time for reloading. Thus the next fracture on this segment has a very high probability of having about the same length as its predecessor. Thus the power law in these models reflects the numbers of clusters and the number of fractures in each cluster.

Since simulations of large events cannot make use of the power law branch of the distribution as a target, we must consider models for organization for which we have no statistical guidelines. Do inhomogeneous distributions of fracture strengths on planar faults restrict the occurrence of runaways? Recent work[22] has shown that there is a strong relationship between fracture strength and geometry of faulting, and that an irregular geometry implies an irregular fracture strength. This observation suggests that the long reaches of the SAF that are more or less parallel to the regional shear stress have relatively low shear strength, and are punctuated by occasional localized sites of great strength, a conclusion supported by observations of average shear stress on the SAF[23]. Simulations on a straight fault with a spiky distribution of breaking strengths shows a large number of weak slip events in the valleys between the peaks of strength. Occasionally, strong earthquakes develop at the spike asperities. However there are enormous instabilities in these patterns in the form of abrupt changes in rates and sizes of events[24,25]; these instabilities are associated with abrupt stress transfer from one part of the fault to another through broken asperities that act as gates. These patterns are extraordinarily different from those that are found to develop on models with large but smoothed fluctuations of breaking strengths, or with homogeneous thresholds. These instabilities of pattern from structure to structure, and instabilities of pattern evolution for a given structure, imply that the statistics of the large-energy branch of the distribution is strongly dependent on local fault geometry and cannot be expected to be a universal.

Two features of the observations suggest that models of evolutionary seismicity on single faults are inappropriate models for the large-energy branch of the distribution. First, all of the large earthquakes of the past 65 years in Southern California have occurred on their own individual faults, no two have occurred on the same fault, and at least one of these has an interval time of the order of 5000 years[26]; none of these has occurred on the SAF, which is the dominant structural element of the fault network of Southern California. Second, great earthquakes on the (presently dormant) SAF, although occurring with interval times of the order of 100 years, as we have noted, do not occur with the periodic regularity to have been expected if this part of the system were an isolated relaxation oscillator. The variability in repetition times for these latter earthquakes[9] is testimony to the importance of interactions with fractures on other faults of the system. Thus we must understand and model seismicity on a network of faults.

As a tutorial model to simulate earthquake pattern development and evolution on a network of faults, we have considered a simple case of stress interactions between two nearby, parallel faults with inhomogeneous distributions of fracture strengths[25]. We find a new instability of pattern: extended lacunae in the seismic activity are developed on individual faults, wherein activity on extended portions of one fault ceases for some time and activity is concentrated on the other fault; the cause is known. These instabilities are further testimony to the dependence of the large-energy branch on the local geometry of faulting on a two-dimensional network of faults capable of tearing in large earthquakes. The fact that these systems appear to be unstable due to a number of geometrical agents means that there is little likelihood at the present time of predicting interval times between events on the large earthquake branch of the distribution.

More generally, the time scales of earthquakes on purely elastic/brittle fracture models, even if dissipation due to dynamics is taken into account, is that of the recurrence times of the events. For these simulations of large-energy events, these times are long or very long compared to the human time scale. Thus these issues of clustering on the long-term are of academic interest at this time, if it is our purpose to generate some insights into the possibilities of earthquake prediction. With regard to the large energy branch of the distribution itself, this must certainly be dependent on the local geometry of large-scale faulting, and therefore must be different from region to region, unless the geometry of faulting and its excitation is a universal; that would seem to be doubtful.

5. THE PHYSICS OF SMALL EARTHQUAKES

We shift our discussion to the the universality of the Omori/Utsu law of aftershocks as a subproblem of the universality of the small energy branch of the distribution. From the discussion above, these events cannot be modeled by purely elastic/brittle fracture models. We must introduce some intermediate-term dissipation to account for the time delays of the aftershocks.

The spatial distribution of aftershocks provides a clue to the problem of the Gutenberg-Richter property of the aftershocks. The distribution of aftershocks after the Landers, California earthquake (M=7.3, 1992) was especially well documented[27]. Many, but not all of the aftershocks lie within 4 or 5 km of the rupture trace; more of the aftershocks are near the rupture trace than far from it. I discuss only the problem of the nearest aftershocks here.

There are two questions of concern: Why are aftershocks located in the neighborhood of the fracture and not solely on the fault (as some authors would prefer), and what is the cause of the time delays? I propose a modification of existing models[32,33]. We note first that the strength of the material a few km from the main fault is strong before the big earthquake, since the sites of the aftershocks did not tear before the main shock. Second, from continuity conditions, the rupture on the main shock lowered the stress on the aftershock sites, and probably by a large amount. Thus, on a standard model, the aftershock sites would be even less likely to tear at any time after the main shock. The solution to this dilemma is to generalize the observation that aftershocks of the Landers earthquake have been triggered at distances up to 1250km from the main rupture. We can suppose that the violence of the slip in the main shock shattered the rock nearby to a distance of several kilometers. The main fault itself comes to rest at irregularly located sites of topographic mismatch between the two walls. Thus the fragmented fault and nearby adjoining regions are in contact at points of geometrical irregularity; these points are interconnected by a spiderweb of cracks. Under a finite but small regional residual stress on the main fracture, weak cracks will have a large stress concentration at their edges and if these stresses are strong enough these asperities will disappear in delayed time by stress corrosion at their edges. But an isolated crack is unlikely to grow under its own stress concentration in long time, if it did not already do so in short time. However the stress at the edge of a crack is much intensified if the edge is an asperity bridge between two or more closely spaced cracks, and stress corrosion will act more rapidly in an accelerated fashion to weaken fracture strength if the stress is increased[28]. The rates are much enhanced by the presence of water because of the enhanced solubility (corrosion) of silicate bonds, and the possibility of developing granular rotations at high temperatures and confining pressures[29].

Thus aftershocks are a two-stage process. First, the large-slip event of the main shock shatters the near neighborhood over the time that it takes elastic waves to travel to the aftershock sites, which is of the order of a second or less. These fragments are prevented from further slipping by the geometrical irregularities of the shards. The radiation from these violent slips, as well as from the local shattering, generates seismograms of great complexity. In the second stage the geometrical contacts between the shards disappears in accelerated creep by stress corrosion processes under the influence of the residual shear stress on the main fault. The magnitude distribution of the aftershocks is thus dependent on the amount of energy stored in the geometrical contacts of the shards, and the time delays are regulated by the stress corrosion process. If there is a low density of small cracks are produced by the shattering for example at some distance from the main fault, then the stresses are not amplified significantly and no aftershocks occur in this region. Improved version of early simulations[30] have now reproduced most of this picture. The model generates universality of the aftershocks by virtue of the universality of the shattering process. The model is not dependent on the specific properties of large-scale fault geometry or on the specifics of slip on individual faults. Seismological tests of this model are under way at this time.

In this development, we have introduced several new components of the physics of fracture, beyond those of simple elastic loading and brittle fracture presented in the earlier parts of this lecture. I suggest that 1) the walls of the small cracks that participate in the development of small earthquakes remain open for some time after formation, in order that 2) their edges can deteriorate in time under non-elastic decay processes; 3) after the aftershock develops, the asperity reheals but the crack does not, i.e. the shattered region remains shattered. Thus we require that these fractures remain open for possibly decades. This is in contrast to the model of healing that we have developed for large earthquakes, in which healing is relatively rapid, and certainly shorter than the rupture time of the complete length of the fracture. How can this be? One possibility is to invoke the role of fluids in the fault zone and nearby: Fluids have been suggested as active agents in developing slip weakening prior to large-scale faulting, and in influencing the rupture dynamics. We suppose that there is rapid transfer of fluids from the large earthquake, thereby promoting rapid healing of it, to the small fractures in the shattered zone, thereby promoting crack opening and time-delayed fracture. Over long periods of time, the fluid can diffuse back into the main fault zone and reheal the shattered zone, thereby terminating the aftershock process, and preparing the region for the next great earthquake. There are several other indicators of the importance of fluids in fault zones and in fault networks to produce fluctuations in seismicity, but space limitations do not permit me to develop these ideas.

Further research along these lines must take us in the direction of understanding better the physics of granular materials: how they are locked together, and how they unlock in the presence of fluids and under local shear stress, produce main shocks, precursors and several post-seismic effects of large earthquakes.

6. DISCUSSION

I have proposed that the universals of the power-law distribution of energies for small earthquakes and the properties of aftershocks, may not be very helpful in allowing us to understand the self-organization of large earthquakes. The application of simplistic physics is not enough to solve the problems of earthquakes. The future will see a much greater effort on understanding better 1) the interaction between geometry and fracture, 2) the nature of deformation of granular materials and their behavior in fracture, 3) fluids as dynamic influences on stress in and strength of rocks, and 4) the healing process of rapid sliding. The problems of earthquakes are rich and well-deserving of serious attention, and there are many unsolved problems.

Acknowledgments

I thank M.S. Abinante, M.W. Lee and X.X. Ni for their kindness in acting as reflective and reflexive critics of some of these thoughts during the formative stages of this work.

References

[1] Bak P., Tang, C., *J. Geophys. Res.* **4** (1989) 15635-15637

[2] Sornette A., Sornette D., *Europhys. Lett.* **9** (1989) 197-202.

[3] Olami Z., Feder H.J.S., Christensen K., *Phys. Rev. Lett.* **68** (1992) 1244-1247.

[4] Olami Z., Christensen K., *Phys. Rev. A* **46** (1992) 1720-1723.

[5] Feder H.J.S., Feder J., *Phys. Rev. Lett.* **66** (1991) 2669-2672.

[6] Gutenberg B., *Q. J. Geol. Soc. Lond.* **112** (1956) 1-14.

[7] Gutenberg B., Richter C.F., *Bull. Seismol. Soc. Amer.* **34** (1944) 185-188.

[8] Gutenberg B., Richter C.F., *Ann. di Geofisica* **9** (1956) 1-15.

[9] Sieh K., Stuiver M., Brillinger D., *J. Geophys. Res.* **94** (1989) 603-623.

[10] Working Group on California Earthquake Probabilities, *Bull. Seismol. Soc. Amer.* **85** (1995) 379-439.

[11] Omori F., *J. Coll. of Science of Imper. Univ. of Japan* **7** (1894) 111-200.

[12] Kisslinger C., Jones L.M., *J. Geophys. Res.* **96** (1991) 11947-11958.

[13] Utsu T., *Geophys. Mag.* **30** (1961) 521-605.

[14] Utsu T., *J. Faculty Science Hokkaido Univ. Ser. VII* **3** (1969) 129-195.

[15] Gardner J.K., Knopoff L., *Bull. Seismol. Soc. Amer.* **64** (1974) 1363-1367.

[16] Prozorov A.G., *Dokl. Akad. Nauk SSSR* **227** (1984) 573-577.

[17] Reasenberg P., *J. Geophys. Res.* **90** (1985) 5479-5495.

[18] Kagan Y.Y., Knopoff L., *Science* **236** (1987) 1563-1567.

[19] Barenblatt G.I., *Adv. Appl. Mech.* **7** (1962) 55-129.

[20] Li Y.G., Vidale J.E., Aki K., and others, *Science* **265** (1994) 367-370.

[21] Xu H.J., Knopoff L., *Phys. Rev. E* **50** (1994) 3577-3581.

[22] Nielsen S., Knopoff L., submitted, *J. Geophys. Res.* (1997) .

[23] Zoback M.D., Zoback M.L., Mount V.S., and others, *Science* **238** (1987) 1105-1111.

[24] Knopoff L., *Proc. Amer. Philosoph. Soc.* **137** (1993) 339-349.

[25] Knopoff L., *Proc. Natl. Acad. Sci. USA* **93** (1996) 3830-3837.

[26] Sieh K., *Proc. Natl. Acad. Sci. USA* **93** (1996) 3764-3771.

[27] Hauksson E., Jones L.M., Hutton K. and other, *J. Geophys. Res.* **98** (1993) 19835-19858.

[28] Yamashita T., Knopoff L., *Geophys. J. Roy. Astron. Soc.* **91** (1987) 13-26.

[29] Griggs D.T., Handin J., *Geol. Soc. Amer. Memoir* **79** (1960) 347-364.

[30] Yamashita T., Knopoff L., *Geophys. J. Roy. Astron. Soc.* **96** (1989) 389-399.

[31] Yamashita T., Knopoff L., *J. Geophys. Res.* **97** (1992) 19873-19879.

[32] Dieterich J.H., *J. Geophys. Res.* **77** (1972) 3771-3781.

[33] Mikumo T., Miyatake T., *Geophys. J. Roy. Astron. Soc.* **59** (1979) 497-522.

[34] Hill D.P., Reasenberg P., Arabasz W.J., and others, *Science* **260** (1993) 1617-1623.

[35] Anderson J.G., Brune J.N., Louie J.N., and others, *Bull. Seismol. Soc. Amer.* **84** (1994) 863-891.

LECTURE 13

Models for Evolution and Extinction

Matteo Marsili

Institut de Physique Théorique, Université de Fribourg,
1700 Fribourg, Switzerland

1. INTRODUCTION

There has been considerable recent interest on large scale evolution biology. In the physics community this was triggered by the scenario of *punctuated equilibrium*, proposed by Gould and Eldredge [1], and by the work of Bak and co-workers stressing the similarities between this framework and the paradigm of Self Organized Criticality[2, 3]. Analyzing the fossil record[3, 4, 5], it has been realized that evolution might not take place smoothly, with small variations occurring at a more or less constant rate. Rather it occurred in bursts of very high activity, characterized by mass extinction, separated by periods of apparent stasis. More precisely, one can measure the number s_t of species (or of families) which went extinct in an interval dt (typically of the order of one million years) around time t. The signal s_t is intermittent, with quiet periods separated by large peaks on any size. The distribution $P(s)$ of s_t behaves as a power law $P(s) \sim s^{-\tau}$ with $\tau \approx 2$. As discussed at length in ref. [3, 4, 5, 6], the main issue is whether the ecosystem as a whole has to be regarded as being in a "normal" state, characterized by linear response to external perturbations, or whether it is in a "critical" state, where external perturbations can trigger fluctuations of any size. In the former case, one has to assume that mass extinction have been provoked by extreme external events, such as meteor impacts, volcano eruptions and the like. This assumption suggests that the statistics of mass extinction events should be related to that of extreme catastrophic events. The second scenario, on the other hand, describes ecology as a system which is poised in a critical state by its dynamics itself. Even small perturbations can propagate and amplify in such a state, and produce in the end a macroscopic change.

The practical implementation of such ideas led Bak and Sneppen [7] to propose a model which now goes under their name. In this model a species is characterized by a fitness and evolution operates by eliminating the less fit species. This "extinction" propagates in the ecosystem through the network of interactions among species. This simple model displays a quite rich behavior of self organization[8], with evolution happening in an intermittent way. Even though it lacks many ingredients of a real ecology, it has opened the way to more realistic models based on the same ideas[5, 9]. One of the shortcomings of these models lies in the abuse of the concept of fitness. In biology, the fitness is related to the number of offsprings an individual of a species can produce. But the rate of reproduction does not depend only on the intrinsic properties (genetic code) of a species, but also on its environment. For this reason the fitness of a species is not a well defined quantity in biology.

It would be of great interest if the avalanche evolution pattern could also arise in more realistic models. Here below we examine the classical Lotka-Volterra model, to see how to generalize it to describe large scale evolution and extinction events. The motivation for this is that Lotka-Volterra equations are a well established model for interacting populations widely used in the biology literature (see e.g. [10] and references therein). Our motivation for addressing biological evolution starting from Lotka-Volterra equations is, to some extent, the same motivation for which one would like to understand turbulence on the basis of Navier-Stokes equation, without invoking phenomenological models. In the next section we shall introduce and discuss the Lotka-Volterra equations. We also review recently proposed models of interacting species[15, 11]. In the last section we shall reexamine Lotka-Volterra equations and discuss how these can be extended to describe evolution of species. Our aim is mainly to understand under what minimal conditions can a scenario of self-organisation occur. For this reasons, we shall critically describe what are the ingredients in a model based on Lotka-Volterra equations, which produce punctuated equilibrium behavior and which are the features which do not. We shall present a model, which is a preliminary attempt in this direction, which reproduces a scenario of self-organized criticality. We shall however point out what are the features of a real ecosystem that this model fails to describe.

2. THE LOTKA-VOLTERRA EQUATIONS

The standard model for interacting populations are the Lotka-Volterra equations [12, 13, 10]

$$\partial_t n_i = n_i \left(f_i + \sum_j b_{i,j} n_j \right) . \tag{1}$$

$n_i(t)$ is the size of the population of the ith species. f_i is the Malthus term. It represents the "bare fitness" of a species, i.e. its growth rate in the absence of the other species. This term alone would cause exponential growth ($f_i > 0$) or decay ($f_i < 0$). The matrix $B = \{b_{i,j}\}$ represents the interaction among species. n_i represents the concentration of individuals of species i in the ecosystem. Therefore $n_i n_j$ is the probability that an individual of the ith species meets one of species j. If these species interact this will affect the population sizes of i and j. If species i interacts with j also j is affected by i. So $b_{i,j} \neq 0$ implies $b_{j,i} \neq 0$. Predator-prey interactions are described by $b_{i,j}$ and $b_{j,i}$ with opposite signs, whereas competition or collaboration are described by $b_{i,j}$ and $b_{j,i}$ with the same negative or positive sign respectively. The self interaction is usually non-positive $b_{i,i} \leq 0$. This term was first proposed by Verhulst, to avoid unbounded exponential growth in the population due to the Malthus term.

Lotka-Volterra equations assume that relevant interactions are mainly pairwise. Note that the equation is local in time. Therefore, loosely speaking, a cubic term $\Gamma_{i,j,k} n_i n_j n_k$ would describe processes where, for example, a member of a species can eat a member of another one only if it eats *at the same time* a member of a third specie. Such processes are rare in Nature and they are neglected in eq. (1).

These equations have been considered extensively in the literature since Lotka [12] and Volterra[13]. Their predictions have also been compared to the behavior of real ecosystems (see [10] and references therein) on time-scales of the order of years.

Biological evolution occurs, on the other hand, on much longer time-scales. In the next section we shall discuss how the Lotka-Volterra model can be extended to long time-scales. Before coming to this, it has to be said that network models for evolution of interacting populations have already been proposed. Solé and Manrubia [15] proposed a model where each species is represented by a spin variable s_i, representing the fact that the species is either extinct ($s_i = 0$), or not ($s_i = 1$). Each species interacts with any other species through a coupling $J_{i,j}$. The dynamics resembles that of neural networks. Mutations are included as random changes in the connections $J_{i,j}$. Finally speciation events occur: the "ecological niches" j of all extinct species $s_j = 0$ are filled by "duplicates" of one species. This model reproduces many of the statistical properties seen in the fossil record. Even though the interaction among species is introduced explicitly in this model, it still allows any species to interact with any other one, which is not very realistic.

The first attempt to use equations similar to (1) to describe large scale evolution has been made by Abramson [11]. He considered an ecosystem organized in a food chain, with species i interacting only with species $i-1$ and $i+1$. The interactions were predator-prey ones with species i feeding on species $i-1$ ($b_{i,i-1} > 0$ and $b_{i,i+1} < 0$). Species $i = 1$, which feeds all the chain has a fixed population n_1. There is no "bare" fitness f_i but only interactions among species. Also no Verhulst term $b_{i,i}$ was considered, and to avoid unbounded population growth a cubic term was introduced. Finally a discrete time evolution was considered. Explicitly the update equations between two time steps of the Abramson model reads

$$n_{i,t+1} = n_{i,t} + n_{i,t} \sum_j b_{i,j} n_{j,t} [1 - \theta(b_{i,j}) n_{i,t}] \tag{2}$$

where $\theta(x) = 1$ for $x > 0$ and $\theta(x) = 0$ otherwise. As shown in ref. [11], already for a simple uni-dimensional structure, the model shows interesting properties, such as a power law distribution of periods of stasis. The model can be generalized to a general random network, and it gives similar results. A detailed account of the numerical results will be given elsewhere [16].

An important feature of the dynamics of most models for evolution is that as soon as a species goes extinct, it is suddenly replaced by another one. This can be a new (totally random) species, as in the Bak Sneppen model [7] or in the Abramson[11] models, or a copy of another species in the ecosystem, as in Solé Manrubia model[15], which is produced by speciation. A second common feature lies in the discrete nature of the dynamics.

3. EVOLUTION AND EXTINCTION FROM LOTKA-VOLTERRA EQUATIONS

Let us come back to the Lotka-Volterra equations (1). As already mentioned the time occurring in eq. (1) varies on the time-scales of the lifetime of an individual. Using these equations directly on the time-scales of millions of years with a discrete time step, as in the Abramson model [11], seems a bit arbitrary. The long time limit of eq. (1) has been considered by many authors in the mathematical literature. It has been shown [14] that if the matrix $-b_{i,j}$ has all its principal minors positive (such matrices are called **P**-matrices), then there is a unique stable configuration $\{n_{i,\infty}\}$ such that $n_{i,t} \to n_{i,\infty}$ as $t \to \infty$. A simpler, sufficient condition for this result, is that the matrix $b_{i,j}$ be negative definite (1).

If the matrix $b_{i,j}$ satisfies this condition, over time-scales much longer than the life-time of individuals, the populations will be found in their equilibrium state, which satisfies $\partial_t n_{i,\infty} = 0$ or:

$$n_{i,\infty} \left(f_i + \sum_j b_{i,j} n_{j,\infty} \right) = 0 \tag{3}$$

$$n_{i,\infty} \geq 0, \quad \text{and} \quad f_i + \sum_j b_{i,j} n_{j,\infty} \leq 0. \tag{4}$$

(1) Let us recall that a principal minor of a matrix A is the determinant of the matrix obtained from A deleting certain columns and the same numbered rows. A matrix is positive (negative) definite if all of its eigenvalues are positive (negative).

The problem posed by this set of equations is well known in mathematical programming under the name of *Linear Complementarity Problem.* It has wide applications ranging from engineering to economics [17]. If (and only if) $-B = \{-b_{i,j}\}$ is a **P**-matrix, the problem has a unique solution which we shall denote by $\vec{n}_\infty = LCP(\vec{f}, B)$. This is a non-trivial result. At first sight, there can be as many as 2^N solutions to the first equation, which is the number of ways with which one can choose the indices i for which $n_i = 0$. Only one of the solutions will however satisfy the conditions (4), if $-B$ is a **P**-matrix.

Coming back to ecology, under the assumption that $-B$ is a **P**-matrix, n_i are given, for large times, by the solution of eqs. (3) and (4) (we shall omit the subscript $_\infty$ henceforth). In other words, *over long time-scales, we can regard n_i as a function of the interaction coefficients f_i and $b_{i,j}$*: $\vec{n} = LCP(\vec{f}, B)$.

On such long time-scales, mutation events occur which change the "phenotypes" coefficients f_i and/or $b_{i,j}$. Therefore the problem of evolution and extinction, in this framework, becomes simply that of studying the dynamics of $\vec{n} = LCP(\vec{f}, B)$ induced by random changes in $\vec{f}$ and B.

The separation of time-scales between a slow, driving process and a fast one, is one of the necessary conditions for the occurrence of self organized criticality[2]. In our case this separation arises naturally between the fast process of equilibration with fixed interactions $(\vec{f}, B)$, and the slow, driving process of mutations. The time-scale for equilibration is, in our case, given by the inverse of the real part of the minimum eigenvalue of the matrix $-B$.

Whether such a separation of time-scales exists in real ecosystems, is an issue which demands a separate discussion (Ref. [10] quotes characteristic times of the order of some tens of years). For our purposes, we shall assume this to be the case and consider the simplest non-trivial realization of a system with this characteristic. It is worth to stress that, because of the very wide applications the LCP has in various fields, the dynamics of the solutions of the LCP under random changes of the coefficients is an issue of interest in its own.

We shall restrict our attention to predator-prey systems, and in particular we shall only consider skew symmetric interactions: $b_{i,j} = -b_{j,i}$ for $i \neq j$. Furthermore we shall assume, for simplicity, a self interaction term $b_{i,i} = -u$ equal for each species. This choice ensures the negative definiteness of the matrix B (note indeed that all eigenvalues of B are of the form $\lambda_j = i\omega_j - u$) and futhermore it allows one to compute the characteristic time for decay to the equilibrium state (see the Appendix).

The solutions $\vec{n} = LCP(\vec{f}, B)$ are, generally[17], continuous functions of the parameters $\vec{f}$ and B. Therefore intermittence in the extinction events cannot occur under continuous random changes of $\vec{f}$ and B. We tested numerically that if e.g. $\vec{f}(t)$ performs a random walk extinction events also occur continuously in time. This means that in order to obtain an intermittent signal, with extinction of any size occurring, one needs discontinuous variation of the parameters. In some sense we have to focus on mutation events which change discontinuously the behavior of the specie in the ecosystem.

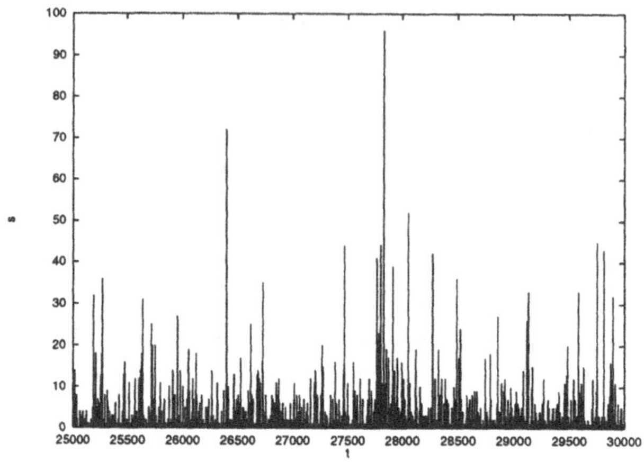

Fig. 1. — Typical realization of the signal s_t of extinction events.

In the model we have studied, the ecosystem starts from a set of non-interacting species: $b_{i,j} = -u\delta_{i,j}$ and f_i randomly drawn from a positive gaussian distribution. The dynamics consists of two steps:

Equilibration: given $\vec{f}$ and B, we find $\vec{n}$ solving the LCP (see ref.[17] for efficient iterative algorithms). If the solution yields extinct species $n_i = 0$, these are replaced by new, random species i. The number of interactions k_i is randomly chosen from a Poisson distribution of mean $\bar{k}$. The interacting species $j_1, \ldots, j_{k_i}$ are also randomly chosen and for each of them b_{i,j_l} (as well as f_i) is chosen from a gaussian of zero mean and unit variance. The replacement process goes on until $n_i > 0\ \forall i$.

Mutation: Two species i and j are randomly selected and a new interaction $b_{i,j} = -b_{j,i}$ is established between them (if an interaction already exists it is changed).

In this way the initial non-interacting set of species evolves into an ecosystem of interacting species. The connectivity of the network remains however small, with each species interacting with only a finite number of other species in the large N limit. Note that, as in most other models of biological evolution, the number of species is kept fixed suddenly replacing the extinct ones with new ones. During the first step in the time step t a number s_t of species can go extinct and be replaced by new ones[18].

A typical realization of the signal s_t is shown in figure 1 for a system of $N = 1000$ species with parameters $\bar{k} = 3$ and $u = 1$. The statistics $P(s) = \frac{1}{T}\sum_{t=1}^{T} \delta(s - s_t)$ of s_t is shown in figure 2. This displays a power law behavior with exponent close to $\tau \approx 2$, very similar to the one observed in the fossil record[3, 4, 5].

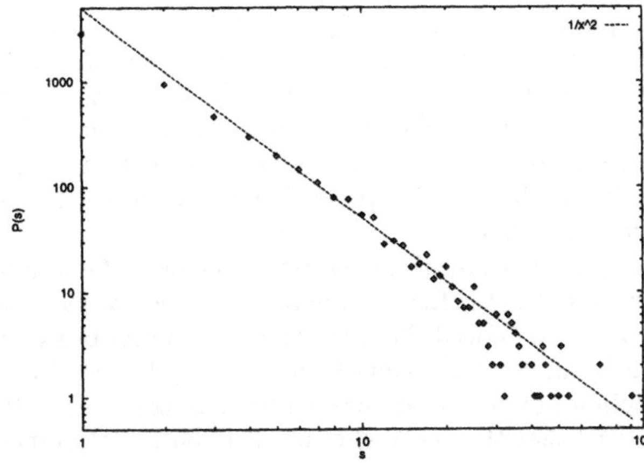

Fig. 2. — Distribution $P(s)$ of extinction event sizes s. The solid line is $\propto s^{-2}$.

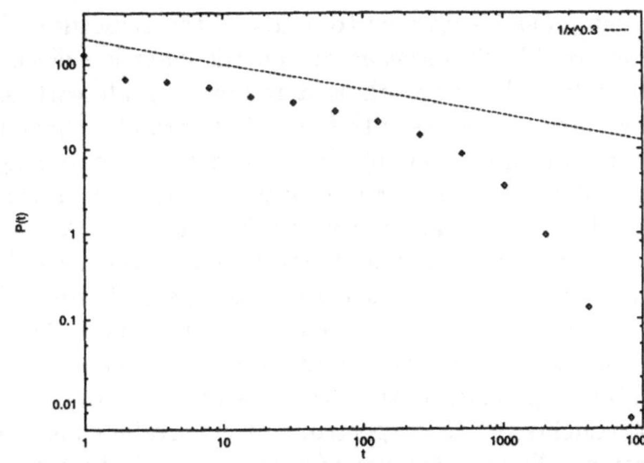

Fig. 3. — Distribution $P(t)$ of lifetimes t of species. The solid line is proportional to $t^{-0.3}$.

The distribution of life-times of "species" in our model also is characterized by a power law behavior as seen in figure 3. However the exponent in this case (≈ 0.3) is far from the one observed in ecosystems, which is also close to 2[3, 4, 5]. The cutoff time in the life-time distribution grows approximately linearly with N. Finally the distribution of periods of stasis is purely exponential with a characteristic time of ≈ 2.5 time steps between subsequent extinction events.

Generically as u decreases the system becomes more and more unstable.

However, investigating the cases $u = 5,\ 2,\ 1,\ 0.5$ and $u = 0.1$, we checked that the above results are very robust with respect to variations of this parameter. The results have a similar sensibility to the choice of the parameter $\bar{k}$: The larger $\bar{k}$ is the more unstable the system is, (i.e. the wilder the signal s_t is). The two parameters are in some sense related since the average connectivity also increases with u. However the power law behavior of $P(s)$ and of life time distribution does not change.

The most likely mechanism which creates avalanches of extinction of any size in this model is related to the fact that extinct species are suddenly replaced by new ones (with $n_i > 0$). Indeed the extinction of a species brings in the system a new species which can break the equilibrium among the remaining ones. In this sense the extinction process is similar to a branching process[19] (even though it has to be noticed that the avalanche size distribution of branching processes has $\tau = 3/2$[19]). The branching probability is driven to the critical value by the slow growth of connectivity in the web. All our attempts of models without sudden replacement of extinct species (i.e. allowing for a variable number of living species) failed to reproduce the intermittent signal of s_t. It therefore seems that the replacement condition plays a crucial role in the model.

One feature our model allows us to study is the structure of the network. One question one could ask about ecosystems is what is the dimension of the interaction network. This network is a random graph with species on the vertices and edges among species with $b_{i,j} \neq 0$. For random graphs, there is not a unique definition of dimension[20]. One can define a mass-length dimension counting the number $M(\ell)$ of vertices within a distance ℓ from any vertex (averaged over all vertices). The mass-length dimension D_m is then given by $M(\ell) \sim \ell^{D_m}$. In our case we found $M(\ell) \sim e^{b\ell}$ which implies an infinite dimension $D_m = \infty$. We also measured the spectral dimension[20] of the graph. This describes the dynamical behavior of the graph[20]. Operationally the spectral dimension can be extracted from the behavior of the density of eigenvalues of the Laplacian matrix of the graph[20] $\rho(\lambda) \sim \lambda^{D_s/2-1}$. Our results are not conclusive. The spectrum of eigenvalues has large sample to sample fluctuations. The low frequency part of it could be fitted with a power law behavior with $D_s \geq 3$. However we cannot exclude $D_s = \infty$.

With respect to the structure of the food web, one expects that in real ecology predation loops are very unlikely (species A eats B, B eats C and C eats A)[21]. Furthermore the web is organized in a hierarchy of trophic levels, with species on one level predating species of lower levels[21]. We found a considerable amount of predation loops and no hierarchic structure emerged from our model.

4. CONCLUSIONS

These results show that, even though the model reproduces a dynamics characterized by extinction of any size, it does not fully reproduces the statistical features observed in the fossil record [3, 4, 5]. Nor does it reproduces the structural properties of the food web[21]. This might be due to the over-simplification of the model (e.g. we didn't model speciation explicitly). Our model shows however that Lotka-Volterra equations offer a good starting point to study large scale evolution. In this sense, it can be regarded as a first step in the direction of deriving a more complete model. The advantage of this approach is that it allows one to test the results also with respect to the structure of the interaction network and its statistical nature. In a model based on Lotka-Volterra equations, the food web is "selected by evolution", i.e. it is self generated by the system itself and is not imposed *a priori*.

The model we have presented fails to reproduce a realistic food web structure. A natural extension of our results is to address the question of what is the simplest dynamics which "selects" realistic structures of the food web. Work in this direction is currently in progress.

Acknowledgments

The author wishes to thank Y.-C. Zhang for frequent discussions which stimulated his interest on this subject. He and R. Graber also contributed to the early stages of this work. I also gratefully acknowledges very stimulating discussions with S. C. Manrubia, B. Vidondo, G. Caldarelli and U. Bastolla which greatly contributed to his understanding of ecosystems and evolution.

Appendix A

Let the interaction matrix be $B = A - uI$, where A is skew symmetric and I is the identity matrix. We start by considering the case $u = 0$. Let us assume that there is a vector $\vec{q}$ such that $A\vec{q} = -\vec{f}$ and $q_i > 0\ \forall i$ (if this is not true one can always restrict the index set setting $q_j = 0$ for $j \in \alpha$ such that it holds for $i \notin \alpha$). Then, as first realized by Volterra[13], there is a constant of the motion $G = \sum_i [n_i - q_i - q_i \log(n_i/q_i)]$, of eqs. (1), i.e. $\partial_t G = 0$. This result is quite interesting since, given that a positive vector $\vec{q}$ exists, it implies that, starting from a positive initial condition $\vec{n}$, all n_i remain bounded for all times. Furthermore it is easy to show that the time average of $\vec{n}$ is $\vec{q}$ and that for $N \gg 1$, G measures the fluctuations of $\vec{n}$ around $\vec{q}$. Indeed for $n_i \simeq q_i$

$$G \simeq \frac{1}{2} \sum_i \frac{(n_i - q_i)^2}{q_i}. \tag{A1}$$

When the Verhulst term is included, $u > 0$, G is no more a constant of the motion $\partial_t G = -u\vec{n}(\vec{n} - \vec{q})$. Performing a time average, since $\langle n_i \rangle = q_i$, one finds

$$\partial_t \langle G \rangle = -u \langle (\vec{n} - \vec{q})^2 \rangle. \tag{A2}$$

Therefore the u term dumps the oscillations of $\vec{n}$. In order to quantify the timescale for this decay, let us assume that $(n_i - q_i)^2 \simeq (\vec{n} - \vec{q})^2/N$ does not fluctuate much. Then we can write eq. (A1) as $G \simeq (\vec{n} - \vec{q})^2/(2\bar{q})$ where

$$\bar{q}^{-1} = \frac{1}{N} \sum_i q_i^{-1}$$

Eq. (A2) can now be written as $\partial_t G = -2u\bar{q}G$ which implies an exponential decay of G within a characteristic time of order $T_{\text{eq}} = (2u\bar{q})^{-1}$.

References

[1] S. J. Gould and N. Eldredge, *Paleobiology* **3** (1977) 114.
[2] P. Bak, C. Tang and K. Wiesenfeld, *Phys. Rev. A* **38** (1988) 364.
[3] K. Sneppen, P. Bak, H. Flyvbjerg, and M. H. Jensen, *Proc. Nat. Acad. Sci.* **92** (1995) 5209.
[4] R. V. Solé, and J. Bascompte, *Proc. Roy. Soc. B* **263** (1996) 161;R. V. Solé, S. C. Manrubia, M. Benton and P. Bak, *Self-similarity in the fossil record*, Nature (to appear).
[5] M. E. J. Newman, *Proc. Roy. Soc. London B* **263** (1996) 1605.
[6] P. Bak and M. Paczuski, to appear in "Physics of Biological Systems" Lecture Notes in Physics (Springer-Verlag, Heidelberg , 1996).
[7] P. Bak and K. Sneppen, *Phys. Rev. Lett.* **59** (1993) 381.
[8] M. Paczuski, P. Bak and S. Maslov, *Phys. Rev. E* **53** (1996) 414.
[9] M. Marsili, G. Caldarelli and M. Vendruscolo, *Phys. Rev. E* **53** (1995) 13;M. Vendruscolo, P. De Los Rios, and L. Bonesi, *Phys. Rev. E* **54** (1996) 6053;D. A. Head and G. J. Rodgers, preprint adap-org/9611003 (1996).
[10] N. S. Goel, S. C. Maitra and E. W. Montroll, *Rev. Mod. Phys.* **42** (1971) 231.
[11] G. Abramson, *Phys. Rev. E* **55** (1997) 785.
[12] A. J. Lotka, *J. Phys. Chem.* **14** (1910) 271.*Proc. Nat. Acad. Sci.* **6** (1920) 410.
[13] V. Volterra, Leçon sur la Theorie Mathématique de la Lutte pour la Vie (Gautiers-Villars, Paris, 1931).
[14] R. Redheffer and Z. Zhiming, *Nonlinear Anal.-Th. Meth. & Appl.* **5** (1981) 1303.Y. Takeuchi and N. Adachi, *Ecol. Modelling* **32** (1986) 95.
[15] R. V. Solé and S. C. Manrubia, *Phys. Rev. E* **54** (1996) R42.*Phys. Rev. E* **55** (1997) 1.R. V. Solé, J. Bascompte and S. C. Manrubia, *Roy. Soc. London B* **263** (1996) 1407.
[16] R. Graber, M. Marsili, K. Sneppen and Y.-C. Zhang, in preparation.
[17] R. W. Cottle, J.-S. Pang and R. E. Stone, The Linear Complementarity Problem (Academic Press, London, 1992) p. 1.

[18] To be precise, s_t is the number of indices i where an extinction has taken place. We do not count unsuccessful species which did not survive at least one time step.
[19] S. Zapperi, K. B. Lauritsen and H. E. Stanley, *Phys. Rev. Lett.* **75** (1995) 4071.
[20] S. Alexander and R. Orbach, *Journ. de Phys. Lettres* **43** (1982) L625.K. Hattori, T. Hattori and H. Watanabe, *Progr. of Theor. Phys. Suppl.* **92** (1987) 108.
[21] J. E. Cohen, Community food webs : data and theory, (New York, Springer-Verlag, 1990). G. Caldarelli and U. Bastolla, private communication.

LECTURE 14

Scale Invariance in Economics and in Finance

Daniel Zajdenweber

Department of Economics, Paris-X Nanterre University, France

1. INTRODUCTION

"Scale invariance" is not a common expression in economics, and expressions like "self similarity" or "self affinity" are scarcely used. Even "stable non gaussian laws" (or "Lévy laws", or "Pareto-Lévy" laws) and "fractals", introduced in economics by B. Mandelbrot [1-2], remain relativily unheard outside financial markets and research centers devoted to finance. Actually, history of economics thought is full of scale invariant distributions, but under another name: Pareto. This famous sociologist and economist of the end of the XIXth century found numerous examples of power laws in income size-distributions [3]. Hence the common name "Pareto" for such a type of economic distributions. Nevertheless, income distributions are not the unique family of scale invariant economic distributions. Firm sizes and assets sizes [4-6], industrial insurance claim sizes [7-9], economic damages due to natural catastrophes, such as hurricanes or earthquakes [10-11], stock market prices fluctuations in the short run and in the long run [2, 6, 12-14] etc. display either one tail with a power law (for obvious reasons most economic size-distributions can only take a positive value) or two tails (all stock market prices, since they have positive and negative variations). In that case they have either a power law on both tails, a typical behavior of all stable non gaussian distributions fitting the whole distribution, or a gaussian law fitting the whole distribution. As regards stock market, scale invariance is not a new scientific discovery. Bachelier [12] in his now famous thesis on Paris "Bourse" settled in 1900 the founding stone of brownian motion, the prototype of gaussian scale invariance. In order to sort out the origins and implications of scale invariance in economics, we will restrict our analysis to one-tail economic phenomena, deferring the analysis of two tails scale invariant economic phenomena to another lecture.

2. EMPIRICAL EVIDENCE OF SINGLE-TAIL SCALE INVARIANT DISTRIBUTIONS IN ECONOMICS

Single tail means that only one tail is fully developed, generally the positive. The other is either inexistent or degenerate. As exposed in the introduction, numerous

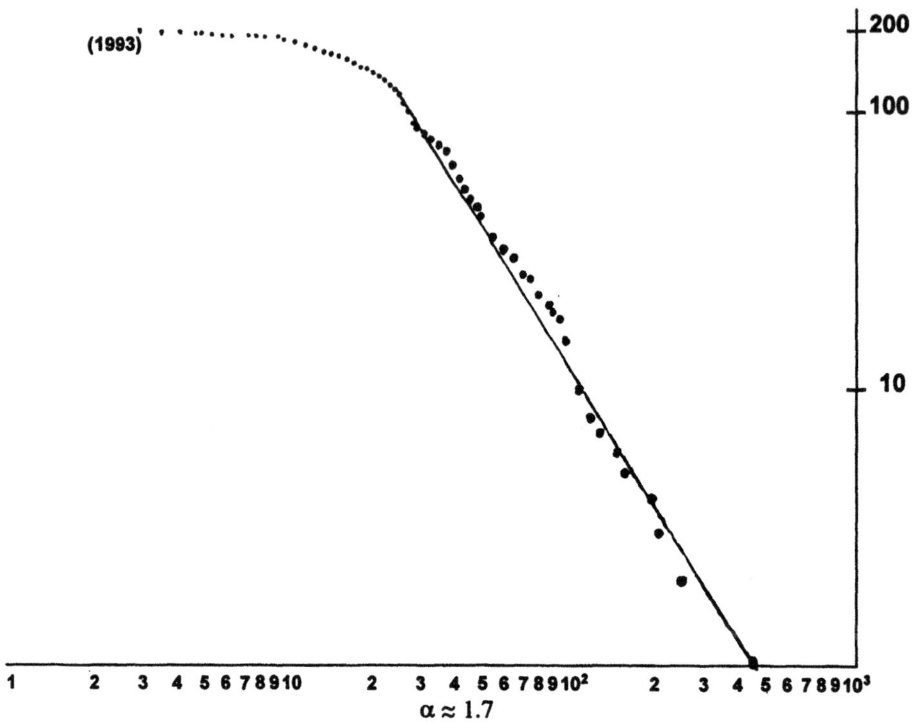

Figure 1: Log-Log plot of the upper tail of the cumulative size-distribution of incomes of CEO (data: FORTUNE Magazine 1993). Abscissa in $million.

significant size-distributions in economics have a tail with a power law (or Pareto) distribution. Let x an economic variable, income for instance, the cumulated distribution function of the tail is:

$$1 - (x/x_0)^{-\alpha} \qquad \text{with } x > 0$$

x_0 is a positive threshold below which the distribution is no more a power law, i.e.it tapers off. Caption I is a good illustration of that behavior. It shows the size-distribution of incomes of the largest paid CEO (chief-executice-officers) of the american companies. This income includes salaries, bonus and stock-options. (source: Fortune Magazine, 1993). The 120 largest yearly incomes run from $2.5 million up to a fancy $45 million! The straight-line, or Pareto line, is the typical behavior of a power law on a log-log plot. Its slope, -1.7, is the estimator of - α. This means that the second moment does not converge. Variance is infinite. For the record Pareto found = 1.5. Small undulations around the straight-line are due to a kind of accounting "window-dressing": people tend to estimate their revenues with rounding errors, especially around the even powers of ten. Below the threshold value $2.5 million the curve is bending. Two phenomena explains this behavior: incomes are less unequally distributed and statistics are less and less accurate. Some incomes could even remain unknown or so small that they are not reported. These

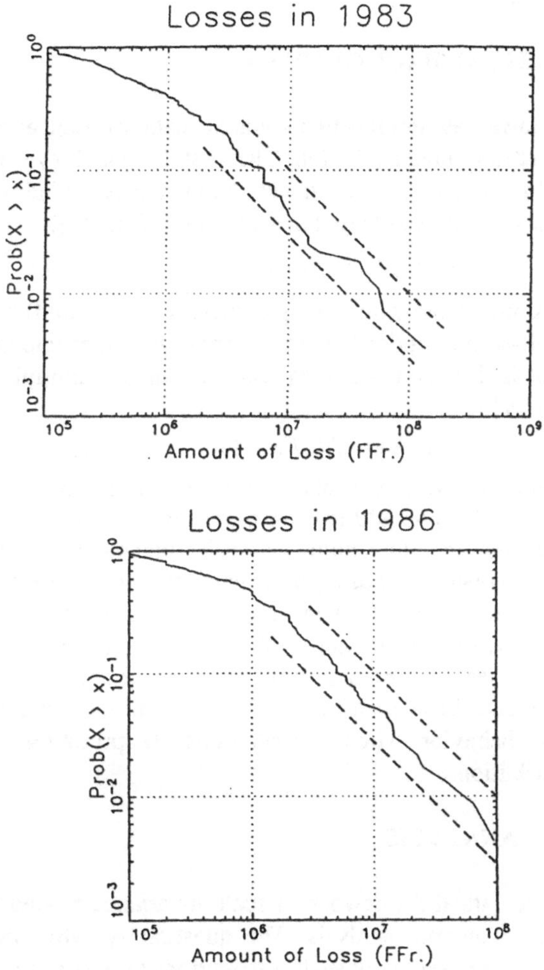

Figure 2: Log-Log plot of the cumulative size-distributions of business interruption losses in France (Years 1983 and 1986) from Zajdenweber [8]. The slope of the dot lines equals -1.

phenomena are not observed in income statistics only. Insurance claim size-distribution display the same statistical behavior: all the largest natural catastrophes damage have a power-law with $\alpha \approx 1$ (Barton and Nishenko [11]), and business interruption claim size-distributions have also a tail fitting a power law with index $\alpha = 1$ (see caption II reproduced from Zajdenweber [8]).

Nevertheless, small damages or claims do not fit a power law. In both examples this means that small claims are not fully reported or not insured either (due to insurance deductibles).

3. MATHEMATICAL BACKGROUND

A tail fitting a power law implies two kinds of scale invariance: scaling under lower truncation and scaling under addition. The first can be shown by conditioning the power law (or Pareto law) with characteristic exponent (or index) α. Let truncation at k' with $0 < x_0 < k' \leq x$, we obtain (as demonstrated in [15]):

$$P(X \geq x/X \geq k') = (x/k')^{-\alpha} \qquad x \geq k'$$

which is itself a power law with scale parameter (or threshold) k'. Thus the original power law is self-similar under lower truncation. An other important feature of that scaling behavior is that the conditional expectation is a multiple larger than one of the conditioning value k':

$$E(X) = k' [\alpha/(\alpha - 1)] \qquad (\alpha > 1)$$

The second scaling behavior was also introduced in economics by Mandelbrot [2]. He demonstrated that if one add two (or more than two) power laws with the same exponent, whatever the threshold or scale parameter x_0 the sum, suitably normalized, is also distributed as a power law with the same exponent. This scaling behavior can be generalized to a large class of distributions with an asymptotic "paretian" behavior, i.e. distributions with a tail fitting a power law with $\alpha < 2$ as $x \rightarrow \infty$. In those cases, they belong to the "domain of attraction" of a non-gaussian stable-Lévy or Pareto-Lévy distributions [2, 16] with the same exponent α and the same aymptotic behavior. Those distributions are prototypes of scale invariant probability distributions.

4. ECONOMIC ANALYSIS

Within the mathematical framework of scale invariance related to power laws, we now proceed to economic analysis. The question is: why such distributions in economics? More precisely why such distributions with $\alpha < 2$ or sometimes $\alpha \leq 1$? Three endogenous economic factors could give the answer[1].

1-Absence of characterisric scale in empirical and in theoretical economics.

[1]Many exogenous factors could explain the origin of power laws or Pareto distributions in economics, they are all related to size-distributions of natural ressources or hazards: oil, gas, mineral ore, water etc. [17, 18]. In this paper we focus on endogenous factors.

2-Non democratic sharing.
3-Economies of scale.

4.1. Absence of characteristic scale in empirical and in theoretical economics

In economics, contrary to other sciences, physics or biology, there are no constants or absolute sizes. In economics, absolute zero and light speed do not exist! Economic theory assumes that there are optimal sizes, for instance an optimum amount of investment, or an optimum amount of consumption, but those optima are related to a firm or a consumer or a nation without normative sizes. Economic theory always pretends that it is equally relevant for the large and for the small. There is no theoretical economic difference in managing a small company or a giant comglomerate, a small nation or a giant nation with hundred million citizen. Neo-classical theory of general economic equilibrium try to prove that this equilibrium exists, whatever the number of economic agents, provided it is large enough, whatever the number of markets, whatever the absolute value of the production etc. It always deals with relative prices, with proportions, with ratios. Even wealth has no norm. Poverty is measured on a relative scale, and financial rationality is only dealing with rates: interest rates, exchange rates, growth rate etc. In a dynamic framework the absence of constants or absolute sizes means there are no economic constraints in the accumulation of skill, of income, of wealth, of profits, of sales etc.

Of course empirical limits do appear: sizes of population, of production capacities could not be infinite, but empirical limits have no theoretical foundations. Scale invariance is thus a logical consequence of the absence of scale. Were the size-distributions scale dependent, then their tail would have been exponentially distributed. The complete size-distribution would have been either gaussian or lognormal, a skewed distribution with a very long tail, but a tail belonging to the "domain of attraction" of an exponential tail. For instance were the incomes in caption I lognormally distributed then the largest incomes would have been ten times smaller. The difference between extreme incomes and medium incomes would have been much less significant, making the existence of a "natural" average income more relevant. In the case of business interruption claims, the difference between a power law and a lognormal distribution is even more dramatic [9]. Were the claims lognormally distributed then the largest among them would have been nearly fifty times smaller. In the case of oil drilling the difference between a lognormal distribution and a power law has a very important economic meaning [15]. With a power (or Pareto) distribution the expected ultimate size of a reservoir is a multiple larger than one (it depends on α) of the reserves already proved. In a lognormal distribution, the ratio between expected ultimate size and actual reserves decreases with the size of the actual reserves. For an economist the difference is striking. The conditional probability of finding a giant oil field is always small, but in the lognormal case it is much smaller than in the power law case and as regards giant oil fields it could be considered as nil.

Absence of characteristic scale could explain the huge size of the empirical range between the smallest and the largest values of most economic variables. The two other economic factors explain why tails fitting power laws are so frequently observed.

4.2. Non democratic sharing

Non-democratic sharing is the academic translation of the popular maxim: "the winner takes all". Assume an amount of food, or territory to be shared among many people or nations. The share is the result of a random splitting device, that could be fair or unfair for instance the toss of a "perfect" (i.e. symmetric) coin or a skewed coin. The first toss splits food or territory in two halves that are either equal or unequal. But the second toss in a non-democratic sharing system is quite different. Instead of splitting each half (that coud be equal or unequal), one of the two equal halves or the largest is not split again and remains unchanged forever. The other half, or the smaller one is split again with the same procedure. If one keeps on splitting and splitting ad infinitum, the size-distribution of all the shares will converge to a power distribution, with an α that is a function of the skewness of the coin. The splitting device is a model of unequal power among people or nations competing to share a finite amount of food or land. The most powerful takes the lion's share, then the second most powerful takes the lion's share of what was not seized by the most powerful etc. This sharing system is very different from a democratic sharing system where each share is split again and again, from the largest down to the tiniest. In a democratic sharing system no one is protected against a split, in a non-democratic system the largest remains the largest forever (in a mathematical sense of course, because history has shown that "forever" means only: "for a relatively long time"). In a democratic system the size-distribution converges to a lognormal distribution, provided the random splitting device obeys to some norming constraints [19-20]. Size-distribution of the republics of the former USSR, measured in square kilometers, gives a stunning example of a non-democratic scheme. It is a power distribution with α=0.44, a very unequal distribution. What happened is quite simple to analyze. Russia, the most powerful republic and also the largest in former USSR grew very large at the expense of the remaining republics of USSR, the second largest (Kazakhstan) did the same thing etc. It is interesting to compare this country size-distribution with others. States in USA for instance, as nations of the world have a lognormal size-distribution [22-23]. They are still unequally distributed, democracy is not synonymous with equality. The main difference between those two size-distributions lies in the range.

The range between the largest countries (or states) and the smallest outside USSR is narrower than the range among republics from USSR. The way US states were formed was more democratic and in old countries history has eroded the largest nations by breaking the old empires unable to maintain their unity[2].

The first two factors could explain why most economic size-distributions are defined on non-compact supports, i.e. on the interval [0; $+\infty$], without scale

[2]Who knows? In the future, Russia may also be split in a democratic fashion, thus changing the power law (a well chosen name in that case) to a lognormal distribution.

dependent laws. They could explain also why some particular sharing devices make the relevant size-distributions belong to the "domain of attraction" of power laws. The third factor: "economies of scales" could reinforce the first two factors by explaining the ubiquity of such size-distributions and their persistency in many diffferent industries.

4.3. Economies of scale

This factor could be translated into a maxim: "The larger you are, the larger you will be or become". It is related to the preceding factor, but it is more dynamic. It could be relevant to economic variables in an open or nonstationary universe. The size of the pie to be shared is no more limited. It grows undefinitely.

There are many types of economies of scale in economics and the so-called "industrial economics", an important research field, describes numerous examples of economies of scale leading to oligopolistic competition or even to monopolies. The main idea is that large scale gives an advantage against competitors. For instance, large scale production induces costs reductions. This permits permanent price reduction, which reinforces the competitive power of the large scale producer, thus increasing its production, making the average costs even cheaper etc. Chips making in computer industry gives a famous example of such economies of scale induced by cost-price reductions. This phenomenon, however, is not restricted to industrial production. It appears in many social fields, such as city sizes (large cities attract many activities and services that attract more inhabitants)[3], box-office strong inequalities in entertainment industry (movie, popular music) due to fashions and fads effects (when a movie or a song is fashionable its reputation increases his success since everybody wants to see the movie or hear the song, thus increasing the success and so on..)[4], learning by doing [23] which explains why productivity grows with the number of hours already spent in the production process, hence giving a strong advantage to the first producer that can accumulate larger experience than its followers etc.

Stochastic models give two frameworks that help unifying those different economic and social fields: models of gravitational fields and theory of ruin asociated to gambling theory. In both cases one finds probability distributions with power laws fitting the tail(s).

4.4. Stochastic models of gravitational fields

Many economic phenomena could be modelled as "gravitational" models. The analogy between Newton's law of gravitation and economies of scale is obvious: the heavier a planet (or star) is, the more it attracts other celestial bodies. The famous

---

[3]City sizes is a famous example of Zipf's law [24].

[4]In 1993 the size-distribution of movie box-office is a power law with α=1.3. The most successful film was "Jurassic Park" with a total world box-office around $900 million! (Source: Variety). The tail of the power law fits about 80 movie box-offices that year. See also [25] for "superstardom" models in popular music.

law of "retail gravitation", the so-called Reilly's law in marketing, says that a commercial center attracts customers with a force that is a function of its size (or mass) and that is inversely related to the distance between the commercial center and potential customers [26]. This law is scale-free and what is remarkable, Feller [19] demonstrated with dimensions arguments only, that in a probabilistic framework, gravitational phenomena in the three dimensions space lead to non-gaussian stable-Lévy size-distributions with $\alpha = 1.5$. Its variance is infinite.

Those distributions are scale-invariant, and their tails fit a power law with the same α [5]. In the case of gravitational phenomena on a plane the same logic leads to stable-Lévy size-distributions with $\alpha = 1$, hence without expectation: the famous Cauchy distribution with two tails fitting a power law [6]. This law describes accurately the well-known concentration of economic activities in a country: most activities are tightly located, but some of them are extremely far from the center, farther than in a gaussian distribution.

The second probability framework, gambling theory and the classical ruin problem, gives new types of economies of scale, even if they are not labelled as such in probability theory. The main reference is Feller's textbook vol. I [19].

4.5. Theory of ruin and economies of scale

Classical results of gambling theory can easily be translated into economic language. Most investments are gambles. Investors bet that they will be profitable. Most of the time this the case, profits are (more or less) positive.

Nevertheless, nobody is absolutely sure that in the future profits will always be positive. On the contrary, economic and business history is full of "horror stories" depicting the sad fate of numerous investments without any profits and even worse, leading to the ruin of the investors.

Classical ruin theory deals with repeated identical trials or investments. The game is terminated when the gambler reaches the contemplated gain or is ruined, having lost all is capital endowment. The main results are:

In a fair game, which is the model of economic equilibrium and the basis of efficient financial markets [12-13], the wealthier the gambler, the smaller the probability of being ruined before reaching the gain terminating the game. With a large amount of capital a gambler can sustain long sequences of bad luck and avoid ruin. The probability of being ruined cannot be zero, nevertheless it can be sufficiently low to make investments profitable and end the game without being ruined.

In an unfair game, i.e. a game with negative expected gain, the probability of ruin depends even more strongly on the initial capital of the gambler. In a an unlimited sequence of trials the probability of ruin is obviously one. But in most real situations the economic problem is quite different. For instance, an investor

---

[5]The physicist Holtsmark obtained an equivalent answer by other methods. Its discovery remained for a long time the sole example of application of a stable non-gaussian distribution.

loses a certain amount of capital. Assume that the only way to recover it, thus avoiding bankruptcy, is to invest the remaining capital in a risky venture with two outcomes and a negative expectation. Either he wins and the gain is large enough to avoid ruin or he loses and he is ruined. If he does not gamble he is ruined also. In both cases the amount lost is the same, it is the amount of capital. The dilemma is obvious. The investor is compelled to gamble. Dubbins and Savage [27] demonstrated that in such risky gambles the only rational strategy is the bold strategy. Timid strategies are inefficient. Betting all the remaining capital may be the most efficient way of avoiding ruin. And the larger the remaining capital the larger the probability of avoiding ruin in a single trial [28].

Bankers know, at least intuitively, that the probability of ultimate survival is unequally distributed in fair games and in unfair games. Interest rates charged to borrowers are always higher for small firms than for large firms. Since the probability of a small firm to go bankrupt is higher, bankers ask for a higher risk premium, reinforcing the economies of scale effect. Hence, the larger a firm is, the more probable it will survive and keep on growing. This explains why large firms are extremely large and why power laws fitting the tail of the size-distributions have values below two and even in some industries below one.

CONCLUSION

Three endogenous factors, absence of natural or characteristic scale, non-democratic sharing and economies of scale explain why scale-invariant phenomena distributed as power laws are so common in economics. When other types of single-tail distributions are observed, for instance lognormal distributions or exponential distributions, those observations mean that economies of scale do not exist or are constrained by countervailing economic and political factors such as taxes, antitrust laws, new competition or new technologies. Those constraints can limit the length of the tails, they cannot give a natural or characteristic scale to economic variables. Scale-invariance exists in natural phenomena, nevertheless one of the main differences between natural sciences and economics is that in economics, scale-invariance is the result of the absence of scale and of a kind of permanent disequilibrium due to economies of scale.

References

[1] Mandelbrot B., The Pareto Levy Law and the Distribution of Income, *Inter. Econ. Review.* **1** (1960) 79-106.

[2] Mandelbrot B., The Variation of Certain Speculative Prices, *J. of Bus.* **34** (1963) 392-417.

[3] Pareto V., Cours d'Economie Politique, Lausanne (Paris, 1897).

[4] Gibrat R., Les Inégalités Economiques (Paris, 1930).

[5] Steindl J., Random Processes and the Growth of Firms, Griffin (London, 1965).

[6] Zajdenweber D., Hasard et Prévision, Economica (Paris, 1976).

[7] Zajdenweber D., Equité et Jeu de Saint-Petersbourg, *Revue Econ.* **45** (1994) 21-45.

[8] Zajdenweber D., Extreme Value in Business Interruption Insurance, *J of Risk and Insurance.* **63** (1996) 95-110.

[9] Zajdenweber D., Business Interruption Insurance, a Risky Business. A Study on Paretian Risk Phenomena. Fractals 3 (1995) 601-608.

[10] Leroy E. and Osouf S., Assurances Catastrophes, *Risques* **15** (1993) 129-136.

[11] Barton Ch. and Nishenko S., Natural Diasters-Forecasting Economic and Life Losses. USGS Marine and Coastal Geology Program, August (1994).

[12] Bachelier L., Théorie de la Spéculation, Annales de l'Ecole Normale Sup. Paris (1900).

[13] Fama E., The Behavior of Stock Market Prices, *J. of Bus.* **38** (1965) 34-105.

[14] Mandelbrot B., When can Price be Arbitraged Efficiently? A Limit to the Validity of the Random Walk and Martingale Models, *Rev. of Econ. and Stat.* **53** (1971) 225-236.

[15] Mandelbrot B., New Methods in Statistical Economics, *J. Pol. Econ.* **71** (1963) 421-440.

[16] Lévy P., Cours de Calcul des Probabilités (Gauthier-Villars, Paris, 1925).

[17] Barton Ch. and La Pointe P., Fractals in Petroleum Geology and Earth Processes (Plenum, New York, 1995).

[18] Mandelbrot B., 1962, The Statistics of Natural Resources and the Law of Pareto, IBM Res. Note, published in [17].

[19] Feller W., An Introduction to Probability Theory and its Applications, Vol. I 3rd ed Ed. Vol. II, 2nd Ed. (Wiley, New York, 1968, 1971).

[20] Frisch U., Turbulence. The Legacy of A.N. Kolmogorov (Cambridge U. Press, 1995).

[21] Reader's Digest Wide World Atlas (1979).

[22] Jaume S.C., Lamont-Doherty Geological Observatory (December 1991).

[23] Arrow K.J., The Economic Implications of Learning by Doing, *Rev. Ec. Stud.* (June 1962).

[24] Zipf G.K., Human Behavior and the Principle of Least Effort (Cambridge, Mass, 1949).

[25] Kee H. Chung and Raymond A.K. Cox., A Stochastic Model of Superstardom: An Application of the Yule Distribution, *Rev. of Econ. and Stat.* **76** (1994) 771-775.

[26] Isard W., Location and Space economy (Wiley, New York, 1956).

[27] Dubins L.E. and Savage L., How to gamble if you Must. Inequalities for Stochastic Processes (Mc Graw Hill, New York, 1965).

[28] Zajdenweber D., Risque et Rationalité en Période de Crise: Le cas des Intermédiaires Financiers, *Rev. Econ. Pol.* **105** (1995) 920-936.

LECTURE 15

Models of Artificial Foreign Exchange Markets

B. Chopard and R. Chatagny

Group of Parallel Computing, CUI, University of Geneva,
24 rue du Général Dufour, 1211 Geneva 4, Switzerland

1. INTRODUCTION

Economic processes are difficult to describe and understand because there is no theory explaining them in the same way as the more classical phenomena appearing in physical sciences. Yet, economic processes play a crucial role in our everyday life and there is a great deal of effort devoted to predict future trends and understand how the various parameters may affect the evolution.

The signature of an economical process is often given by a time series (for instance the exchange rate between two currencies as a function of time, or the value of some financial index in the stock market). The observed behavior looks rather chaotic and intricate but a more careful analysis reveals some underlying statistical laws describing the data distribution as well as some scaling properties. This indicates that some generic behaviors emerge out of the complicated interactions between the components of such a system. In particular, some scale invariance show up, indicating that the time resolution at which the system is observed is a crucial ingredient of the dynamics. The scaling form may reveal important information about the fundamental interactions that take place in an economical process. Thus, scaling may provide an unified framework to describe financial systems and relate them to other complex, dynamical phenomena.

In this paper, we focus on a specific financial system, the so-called Foreign Exchange Market. We first review its main rules and describe its most notable properties. Then, we propose a simple model (a market microsimulation) aimed at reproducing the behavior of each actor in such a market (traders, banks,...). Numerical simulations show that this model captures several of the qualitative features of the real market, as the results of the interaction of many actors. Finally, we give some indications on how we could improve our results. Our main purpose is not to make financial predictions but, rather, to understand how complexity builds up from the aggregation of many individual behaviors.

2. THE FOREIGN EXCHANGE MARKET

The Foreign Exchange (FX) market is slightly different from the other financial markets (like stock market, for instance) since one does not exchange a valuable against some money, but a currency for another currency. Every transaction price involves a pair of currencies, e.g. USD/DEM, USD/JPY or GBP/CHF([1]). The FX market is the most active financial market with a daily turnover of more than 1000 billions USD which is more than the total non-gold reserve of all industrial countries[1]. In fact these transactions concern only some major currencies (USD, DEM, JPY, GBP and CHF). The USD is evidently the main currency which is implied in 80% of the transactions. The DEM is the second major currency and 20% of the transactions are exchanges between USD and DEM. The other currencies are quite negligible in comparison.

Contrary to the other financial markets, the FX market is a 24 hours global market. Each pair of currencies can be traded theoretically in every financial places around the world and a trader can always find an open financial place somewhere. But, except for the four major currencies against the USD, the others are traded in their own geographical markets.

However, the FX market is not homogeneous in time and this causes difficulties when analyzing the data. The weekends are periods with a very low activity. Daily variations can also be observed, due to the continental components corresponding to the three major financial places: East Asia (Tokyo), Europe (London) and America (New York). The overlap of activities of these regions produces a regular pattern observed in the daily activity of the FX market.

Four types of actors are present in the FX market: the market makers, the brokers, the traders and the hedgers. Although we will mainly be interested in the behavior of market makers and traders, we first start with a brief description of all these agents.

2.1. The Market Makers and Brokers

The market makers decide on the selling and buying prices of the currencies and perform the transactions with the traders. Price determination is obviously

([1]) We use here the standard and obvious abbreviations of the International Organization for Standardization.

driven by the profit they expect. Its main source resides in the difference between the buying and selling price of a currency. By changing the exchange rate, market makers may favor one currency with respect to another and drive the market in some preferred direction. Of course, the precise strategy they use is not known publicly but we may assume generically that they try to avoid unbalanced stocks of currencies. The strategy we shall propose in our model goes along this line and turns out to be profitable. The role of market maker is played by the large banks. There are a limited number of them in the world, in the order of 100 to 200.

Brokers are similar to market makers except that they do not decide upon prices. They act as intermediary in the market, like smaller banks or provincial branches of large banks. With the advent of fully computerized transactions, their number is decreasing.

2.2. The Traders

A trader is anyone willing to exchange one currency for another. There are three main reasons for such a transaction: (i) the need to obtain a given foreign currency for some investment; (ii) pure speculation by which a person expect to sell at a higher price a currency he bought some time earlier; (iii) hedging by which a trader protects himself against the risk of exchange rates variation. Hedgers prefer to execute their operations with the current exchange rate rather than with the rate in effect at the time of the contract.

The hedgers represent now a very important part of the transactions, but their influence is mainly noticeable in the future market and less in the spot market which is the subject of our microsimulation.

In order for a trader to make a profit, it is important to forecast the exchange rate evolution. The trader's behavior is determined by its working time horizon. The duration of a transaction is defined as the time elapsed between the purchase and sale of a given currency. Short term traders performs intra-day transactions only, in a pure speculative mind. Middle term traders operate over a few days and up to a two weeks period. Finally, long term traders may keep their position open from a few months to one year. Since these different types of traders may have different strategies, the above three time scales may play a role in the global market behavior.

There are basically two methods for predicting the future market price: (i) The "fundamentalists" consider macro-economics factors like the interest rate, inflation and unemployment to infer the market evolution; (ii) The "chartists" rely on a technical analysis of past price history to determine the market trend. They may use more or less sophisticated mathematical tool for this purpose.

3. PROPERTIES OF THE FX MARKET

Before we present the known properties of the FX market let us defined precisely some of the basic quantities.

3.1. Exchange Rates and Market Price

Let us assume we have an amount s_1 of currency 1. This amount of money can be changed in another currency, say currency 2. The exchange rate λ_{12} from currency 1 to currency 2 is defined as $s_2 = s_1\lambda_{12}$, where s_2 is the amount of currency 2 one receives from such a transaction.

This amount s_2 can be changed back into currency 1 so as to give an amount $s_1' = s_2\lambda_{21} = s_1\lambda_{12}\lambda_{21}$, where λ_{21} is the exchange rate from currency 2 to currency 1.

Clearly one should have $s_1' < s_1$, otherwise one could produce money at no cost. This implies

$$\lambda_{21}\lambda_{12} < 1 \tag{1}$$

This relation expresses the market "friction" which allows market makers to make a profit.

Usually, the *bid* and *ask* prices (buying and selling prices) are used instead of λ_{12} or λ_{21}. They are defined as follows. First a currency of reference is chosen, say currency 1. Then we write

$$s_1 = p_{bid}s_2 \tag{2}$$

where $p_{bid} = \lambda_{21}$ is the buying price of currency 2, or the amount of currency 1 that the market maker gives for each unit of currency 2. Note that p_{bid} is sometimes called p_{buy}.

Similarly, one defines

$$s_2 = \frac{s_1}{p_{ask}} \tag{3}$$

where $p_{ask} = 1/\lambda_{12}$ is the selling price (also called p_{sell}), or the amount of currency 1 asked by the bank for each unit of currency 2.

From equation (1) one has $p_{bid} < p_{ask}$. If currency 2 is taken as a reference, the new bid and ask prices are $p'_{ask} = 1/p_{bid}$ and $p'_{bid} = 1/p_{ask}$. Usually, the USD is chosen as the reference in all transactions where it is present.

It would be natural to define the market price at time t, $x(t)$, as $[p_{ask}(t) + p_{bid}(t)]/2$. Here we prefer the following definition

$$x(t) = \frac{1}{2}[\log p_{ask}(t) + \log p_{bid}(t)] = \log\left[\frac{\lambda_{21}}{\lambda_{12}}\right]^{1/2} \tag{4}$$

This expression is largely used in statistical studies of the FX market. It has three advantages. First $x(t)$ changes sign when the reference currency changes. Second, x is dimensionless, except for a constant. Thus, two rates as different as USD/DEM and CHF/JPY can be compared, up to a constant. Finally, an increase or a decrease of the exchange rate of a given factor will result in a variation of x of the same amplitude, up to its sign.

3.2. Spread

The nominal spread denotes the friction $p_{ask} - p_{bid}$. Here, again, one prefers to express the spread S in its logarithm form

$$S = \log p_{ask} - \log p_{bid} = -\log \lambda_{21}\lambda_{12} \tag{5}$$

which has the same advantage as explained for the price x.

The spread is always positive and it quantifies the market maker profit when two traders are simultaneously buying and selling the same currency. Indeed, consider the case where a trader A sells to the bank an amount s_2 of currency 2 (for which he receives an amount s_1' of currency 1) and, at the same time, a trader B buy this same amount s_2, paying to the bank the amount s_1. Then, a simple calculation shows that the market maker's profit is

$$s_1 - s_1' = (p_{ask} - p_{bid})s_2 > 0 \tag{6}$$

Thus, in order to minimize risk, the market maker tries, in the shortest delay, to balance the transactions in both ways. If the time between two opposite transactions increases too much, there is a chance that a variation of the exchange rate screens out the effect of the spread and that the market maker lose money.

3.3. Change of Price

Consider a transaction $(s_1 \rightarrow s_2)$ at time $t - \Delta t$ followed by the inverse one $(s_2 \rightarrow s_1')$ at time t

$$s_1' = s_1 \frac{p_{bid}(t)}{p_{ask}(t - \Delta t)}$$

For the trader, the profit is $s_1[p_{bid}(t)/p_{ask}(t - \Delta t)] - s_1$ Thus, the quantity

$$\log[p_{bid}(t)/p_{ask}(t - \Delta t)] = \log p_{bid}(t) - \log p_{ask}(t - \Delta t)$$

indicates a gain or a loss, whether it is positive or negative.

From this observation, the *price change* or *return*, which is a measure of the market evolution over a time interval Δt, is defined as

$$r(t, \Delta t) = x(t) - x(t - \Delta t) \tag{7}$$

Usually, for $\Delta t < 10$ minutes, the return is smaller than the spread and no profit can be expected by a trader. Note that r is symmetrically distributed and believed to be a stationary process.

Studies have shown[1] that the time-autocorrelation $< r(t+\tau, \Delta t) r(t, \Delta t) >$ fluctuates around zero, except when the time lag Δt is shorter than four minutes, in which case it is significantly negative. Thus, on average, a positive price variation is followed by a negative one, which may reflect diverging price formation strategies among the market makers. For a larger Δt, price variation has no memory but, yet, the price is very slightly predictable in the sense that it is possible to define winning trading strategies.

3.4. Risk and Volatility

The risk describes the size of fluctuations of an economical process. Large risks means large fluctuations and large probability to guess wrong and lose money. The volatility $v(t, \Delta t)$ is a measure of the market fluctuations over a period T, where the price is considered with a time resolution Δt

$$v(t, \Delta t) = \frac{1}{T} \int_t^{t+T} |r(t', \Delta t)| \, dt'$$

Note that $|r|$ is sometimes preferred to r^2 because it better captures the behavior of the market[1]. This is due to the presence of fat tails when the distribution of returns is extrapolated to large values of r, making the fourth (and higher moments) of the distribution diverge. Using $|r|$ instead of the second moment put less weight to large and uncertain values.

It is found[1] that the volatility has some interesting properties. First, the volatility is not uniform: cluster of high volatility are followed by cluster of low volatility. This is known as the conditional heteroskedasticity and follows from the fact that an active market stimulates more transaction whereas a quiet market acts oppositely. Second, the volatility autocorrelation is positive and has long memory (3 weeks). Third, daily volatility is a good predictor of hourly volatility, probably because important news first affect middle term traders before they propagate to short term traders.

For what we are concerned, the most important properties of the volatility is the so-called *scaling law*[1]. It has been observed that

$$v(t, \Delta t) \sim \left(\frac{\Delta t}{\Delta T} \right)^{1/E} \tag{8}$$

for Δt ranging from 10 minutes to two months (i.e. over four decades). Here ΔT is some amplitude dependent on the FX rate. The sampling period T is several years and, for the USD/DEM market, we get

$$\frac{1}{E} = 0.582 \pm 0.006$$

which is called the drift exponent. For a pure Gaussian random walk model, one has $1/E = 1/2$. It is also observed that $1/E$ varies according to the currencies and may change over time although it is remarkably stable for

USD/DEM. However, the drift exponent can be smaller than 1/2 in some cases (DEM/FRF). This may be due to anti-correlations (or anti-persistences) in the return distribution.

3.5. Return Distribution

In order to extract the market properties, it is natural to measure the probability distribution (histogram) of the return r. The main results[1] are: (i) the distribution is not Gaussian, unless $\Delta t > 2$ months; (ii) the distribution is leptokurtic (there is a more pronounced peak than a Gaussian of same variance); (iii) there are fat tails; (iv) the central part is well described by a Levy or Pareto distribution.

The dependence of the distribution upon the time resolution Δt is an important question in order to verify whether or not scaling holds. First, one notices that

$$r(t, 2\Delta t) = r(t, \Delta t) + r(t - \Delta t, \Delta t)$$

which obviously follows from definition (7) of r. The process of changing the time scale by summing up the contributions at the smaller scale is called *time aggregation.*

If $r(t, \Delta t)$ and $r(t - \Delta t, \Delta t)$ are independent random variables, one expects that r scales with $\Delta t^{1/2}$. On the other hand, if $r(t, \Delta t)$ and $r(t - \Delta t, \Delta t)$ are deterministically related, one expect a scaling like $r \sim \Delta t$.

From this remark, one may assume that the probability distribution $P(r, \Delta t)$ is stable, i.e it can be written as

$$P(r, \Delta t) = \Delta t^{-\alpha} F\left(\frac{r}{\Delta t^{\alpha}}\right) \tag{9}$$

where F is a universal distribution function (see [2] for the stock market case).

For compatibility with the scaling law (8), one should have $\alpha = 1/E = 0.58$. However, the above scaling does not work ([O.Pictet, private communication]) when $\Delta t > 1$ day. This can be explained by the different strategies used by traders working with a different time horizon.

Power laws fat tails have been observed in the FX market[1], namely

$$P(r, \Delta t) \sim r^{-(\beta+1)} \tag{10}$$

for $r >> 1$. This relation is stable with time aggregation. From this expression, it is clear that moments of order k exist if $k < \beta$. Measurements on real data give $\beta = 3.5 \pm 0.5$ for many currencies against USD and $\beta = 2.7 \pm 0.5$ for many currencies against DEM. Therefore the second moment of the return distribution exists, but not the fourth.

4. SIMULATION OF THE FX MARKET

Modeling the behavior of a financial system is an important issue. Stochastic processes like GARCH (Generalized Auto-Regressive Conditional Heteroskedastic) have been considered for this purpose and can account for the conditional heteroskedasticity [7]. Among the limitations of this approach, there is no clear connections between the parameters of the market and the GARCH mathematical expressions.

In this section we propose a model implementing, at the level of the market makers and traders, the rule of the FX market we described above. Such an approach has been already attempted in the case of stock market[3, 4, 5] but, to our knowledge, no such model exists for the FX market.

Although we restrict ourselves to a simple situation, we try to make the model as realistic as possible. We consider only one pair of currencies and the actors of the simulation have a deterministic, non-evolving behavior. However, each individual has a different profile, that is a different set of parameters (time horizon, risk profile and so on) for interacting in the market. The two main steps involved in each time step of the model are: (i) price formation, (ii) transactions.

The price formation rule (for market makers) is based on the stock balance of the two currencies, S_1 and S_2. Let $\Delta = S_1 - S_2\lambda_{21}$ be the stock excess. The new price is determined by the relation $p_{ask}(t+\tau) = p_{ask}(t) + f(\Delta)$ where f is a truncated linear function (10^{-4} per 50 million USD). It is truncated so that the price does not move out of the spread range.

Traders are characterized by two thresholds ϵ_1 and ϵ_2 ($\epsilon_1 < \epsilon_2$) defining their trading strategy. They all act as chartists and analyze the past data to predict the future. Other parameters define their risk profile (how much they expect to win or are ready to loose) and their time horizon.

Their decision to sell, buy or stay neutral is obtained by comparing the so-called momentum $m(t) = x(t) - \text{ema}_T(t)$ with ϵ_1 and ϵ_2. Here, the quantity $\text{ema}_T(t)$ is an exponential moving average giving the average past price over a time scale T, as defined for instance in [6]

$$\text{ema}_T(t) \equiv \frac{1}{T}\int_{-\infty}^{t} x(t')\, e^{-\frac{t-t'}{T}}\, dt' \tag{11}$$

The momentum $m(t) \neq 0$ indicates a market trends because the current price departs from its average. The parameter T can be used to specify the time horizon and may vary for each trader.

When the momentum raises above ϵ_1 (or ϵ_2), the corresponding trader enters the market by buying currency 1 (or currency 2). He leaves it as soon as: (i) he reaches the maximum affordable loss; or (ii) he reaches the minimal expected profit; or (iii) its time is over (intra-day trader must complete all transactions within the day). The thresholds $-\epsilon_1$ and $-\epsilon_2$ are used similarly but with the reverse order of the currencies.

Usually, a trader may realize a transaction with only one or a few market makers where he has a deposit of at least 10% of the desired operation. As a result of the transactions, price will change according to the price formation rule described above and the cycle is repeated.

5. RESULTS OF OUR MICROSIMULATION

We have initialized our simulation with the following GARCH process so as to produce a history of price necessary for the traders to predict the future trend.

$$\begin{aligned} r(t) &= v(t)\xi(t) \\ v^2(t) &= 2.39 \cdot 10^{-8} + 0.1247r^2(t-1) = 0.8446v^2(t-1) \end{aligned}$$

where $\xi(t)$ is a gaussian noise, r the price change and v the volatility.

Figure 1 shows the time evolution of several quantities in our artificial market simulation.

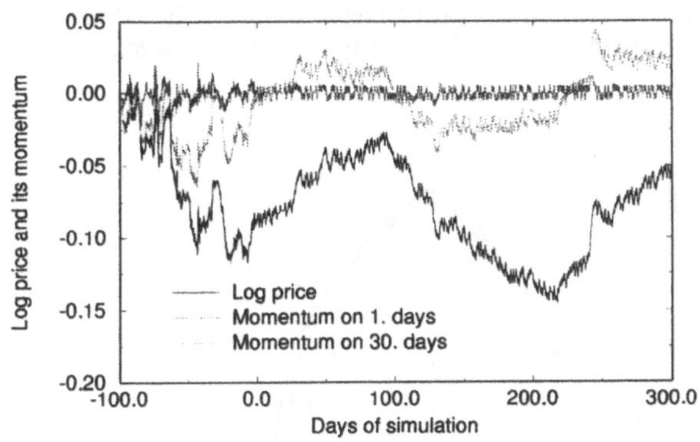

Fig. 1. — Behavior of our simulated artificial market. The lowest curve indicates the logarithm of the average market price as a function of time. The upper curves correspond to moving averages of the price series and are used by the dealers to take their trading decision.

In figure 2 we compute the volatility scaling law for our artificial market and other simple processes. We observe a drift exponent $1/E = 1$, which is not the expected 0.58 result. This value may reflect that there is too much trend in our model (our dynamics is deterministic) and probably not enough randomness.

The scaling form (9) can be checked by the data collapse shown in figure 3. We observe a satisfactory scaling for the central part of the distribution which, in addition, clearly departs from a Gaussian. However, the tails of the distribution are not satisfactory since they are practically non existent. This may be due to the lack of collected data to capture the rare events, or to the fact that our price variation is bounded (as suggested by figure 1), thus introducing a cutoff in the probability distribution.

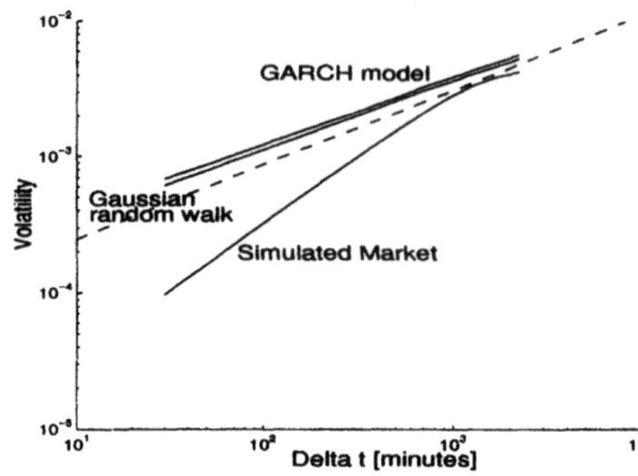

Fig. 2. — Scaling law for the volatility versus the time resolution Δt for three models (solid lines): our simulation, a GARCH process and a Gaussian random walk. The drift exponents are 1 and 1/2, respectively. The dashed line represents the plot obtained from the USD/JPY rate [1] and has a slope of 0.58.

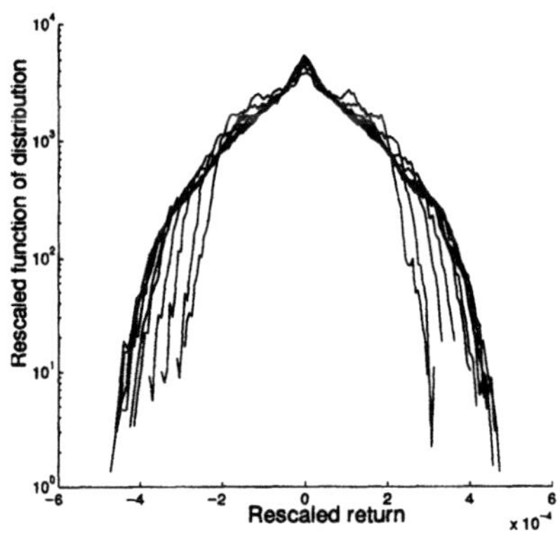

Fig. 3. — Data collapse illustrating the validity of the scaling form $\Delta t^{-1}F(r/\Delta t)$ for the return probability distribution.

6. CONCLUSION

These preliminary results are promising since our model captures many of the qualitative features of the real FX market. Of course some improvements are necessary to obtain the correct exponents and produce fat tails. Adding more randomness in the actors' behavior (to mimic the effect of external news) or including different strategies should help. Another interesting step would be to add a *crowd effect*[8] by which undecided traders would follow the behavior of the majority.

Acknowledgments

The authors wish to thank O. Pictet from Olsen and Associates for helpful discussions about the mechanisms of the FX market.

References

[1] Domimique M. Guillaume, Michel M. Dacorogna, Rakhal R. Davé, Ulrich A. Müller, Richard B. Olsen, and Olivier V. Pictet. From the bird's eye to the microscope: A survey of new stylized facts of the intra-daily foreign exchange markets. Discussion paper DMG.1994-04-06, Olsen & Associates, Seefeldstrasse 233, CH-8008 Zürich, 1994.

[2] Rosario N. Mantegna and H. Eugene Stanley. Scaling behaviour in the dynamics of an economic index. *Nature*, 376:46–49, July 1995.

[3] W. Brian Arthur. Complexity in economics and financial markets. *Complexity*, 1:20–25, 1995.

[4] Blake LeBaron. Experiments in evolutionary finance. Technical report, University of Wisconsin, 1995.

[5] M. Levy, H. Levy, and S. Solomon. Microscopic simulation of the stock market. *J. de Phys. I (France)*, 5:1087, 1995.

[6] O.V. Pictet, M.M Dacorogna, U.A. Mueller, R.B. Olsen and J.R. Ward. Real-time trading models for foreign exchange rate. *Neural Network World*, 6:713–744, 1992.

[7] Ulrich A. Müller, Michel M. Dacorogna, Rakhal D. Davé, Richard B. Olsen, Olivier V. Pictet, and Jakob E. von Weizsäcker. Volatilities of different time resolutions: Analysing the dynamics of market components. Preprint UAM.1995-01-12, Olsen & Associates, Seefeldstrasse 233, CH-8008 Zürich, 1995.

[8] P. Bak, M. Paczuski, and M. Shubik. Price variations in a stock market with many agents. Technical report, Brookhaven National Laboratory, Upton NY 11973, 1996.

BEYOND SCALE INVARIANCE

From Scale Invariance to Scale Covariance

Scale Symmetry Breaking

LECTURE 16

From Scale-Invariance to Scale-Covariance in Out-of-Equilibrium Systems

A. Pocheau

IRPHE, Université Aix-Marseille I & II, CNRS, Centre de Saint-Jérôme, Case 252, 13397 Marseille, France

1. INTRODUCTION

In principle, the concept of scale-invariance enables relevant information to be gained on physical systems by different means. For example, minimization of the set of independent variables from dimensional analysis, selection of self-similar laws from evolution equations, selection of power laws for functions of a single scaling variable and characterization of geometry by a fractal dimension. In practice however, most of these approaches hardly apply. Various reasons for this include the fact that selection of power laws is invalid for multivariable functions, selection of other self-similar laws requires analytic modelling of systems, physical cut-offs (e.g. finite size effects) break scale-invariance (e.g. scale-similarity) and forbid fractal.

How can the concept of scale-invariance be applied to finite size systems, multivariable functions and phenomena whose evolution equations are unknown ? To what extent is it still valid and powerful in such situations ? These questions are addressed here in the context of two out-of-equilibrium systems: turbulent front propagation [1-3] and fully developed turbulence [4]. In the former system, we legitimize the selection of scale-covariant laws, even in non-scale-invariant states, by emphasizing the difference between the scale-invariance of a process and the scale-invariance of a state. In the latter system, we restrict our attention to a scale-invariant state of a turbulent flow and select its possible statistics by invoking the scale-covariance of the relation which describes them. In both cases, scale-invariance with respect to both uniform and stretched changes of scale are considered.

2. DEFINITIONS AND CONCEPTS

Let us first clarify the definitions and concepts pertaining to the present problem. By "scale", we mean the magnitude of a variable, not necessarily related to space. By "scale-invariance", we mean that scales are physically equivalent so that the same physical conclusions might be drawn from any scale. We stress that this invariance may address either a physical process, a physical state, or both. For instance, as the Newtonian gravitational interaction involves no characteristic spatial scale, it behaves in the same way at the planetary or the laboratory scales and is thus a scale-invariant *phenomenon*. On the other hand, at a critical point in a phase transition, fluctuations do not rely on a characteristic scale so that observing the fluid at one or the other scale leads the same statistical picture : this is a scale-invariant *state*.

We emphasize that these invariances may be satisfied for scale changes that may or may not preserve scale ratios [1]. The former case simply traces back to a change of scale unit according to which all scales (e.g. L_i) are multiplied by the same prefactor: $L_i \rightarrow L_{i'} = \lambda L_i$ with $\nabla\lambda = 0$, $L_i/L_j \rightarrow L_{i'}/L_{j'} = L_i/L_j$ (fig.1a). We call such invariance by a global change of scales a 'global scale-invariance'. It refers to the usual meaning of scale-invariance as contained in dimensional analysis. The second case of scale changes is more subtle. It yields the scale ratios, and thus the size of scale ranges, to be modified. This may be achieved by changing the scale unit so as to make it vary with scale: $L_i \rightarrow L_{i'} = \lambda(L_i) L_i$ with $\nabla\lambda \neq 0$, $L_i/L_j \rightarrow L_{i'}/L_{j'} \neq L_i/L_j$ (fig.1b). We call such invariance by a local change of scales a 'local scale-invariance'. This property, which is stronger than global scale-invariance, goes beyond dimensional analysis.

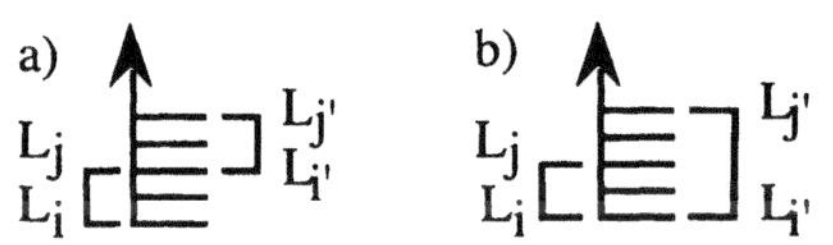

Figure 1: Sketch of scale change in scale space : a) global change, b) local change.

Global (resp. local) scale-invariance means that all scales (resp. all scale ranges) are physically equivalent. Examples of such situations are provided by fully developed turbulence at infinite Reynolds number. Within this limit, the dissipation scale vanishes, so that the system a priori involves no characteristic *scale*. In such a state, it thus satisfies a *global* scale-invariance. On the other hand, as it is usually considered that the integral scale L_I of turbulence bears no absolute meaning, any of its subscales $L_j < L_I$ might play a similar role. Then the same physics should also be encountered in any sub-range $[L_i, L_j]$ of the inertial range, so that all sub-ranges should be physically equivalent. In other terms, the system should involve no characteristic *scale range* and thus also satisfy a *local* scale-invariance.

An operational consequence of the above invariances can be obtained by considering, for a given observer, what actually specifies the physics. It is not the values of variables which, being dependent on units, appear somewhat arbitrary. Instead, it is the relation between them since its mathematical structure is preserved by unit changes, up to coefficient variations. It then corresponds to a law $\mathcal{L}(.)$ either governing the system (e.g. Navier-Stokes equations) or characterizing its state (e.g. the Kolmogorov 4/5th law). Usually, this law bears no particular property regarding scale changes. This is, however, no longer valid when scale-invariance is satisfied, since changing scales should not modify the physics. Despite the scale changes, the law should then keep the same structure and the same coefficients and thus be "scale-covariant" : $\mathcal{L}(.) \rightarrow \mathcal{L}'(.) = \mathcal{L}(.)$ for $L_i \rightarrow L_{i'}$.

As covariance is a quite specific mathematical property, it constrains the structure of laws and selects few possible forms for them. This property is well-known in mechanics and in field theories and has also been used to unify theories developed at different scales [5]. Here, our purpose is to use it in scale-invariant out-of-equilibrium systems to select statistical laws by scale-covariance.

3. TURBULENT PROPAGATIVE FRONTS

We consider here a reaction front propagating in a metastable turbulent medium. This pertains for instance to epidemic propagation or turbulent combustion (fig.2a). The turbulent medium is characterized by a turbulence intensity u' and the reaction by the normal velocity U_N of the front with respect to the medium (fig.2b). Both u' and U_N will be considered as constant in the following.

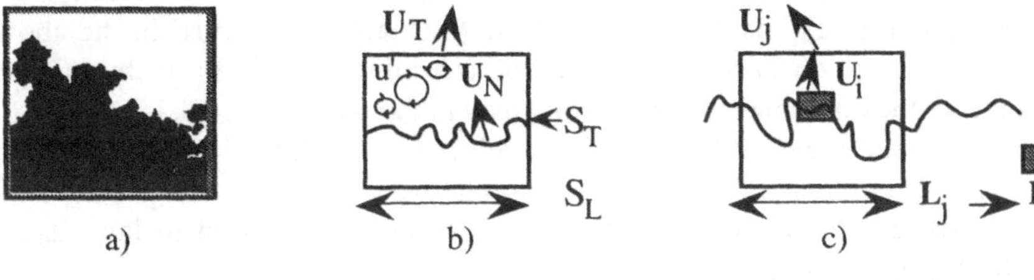

Figure 2: a) Tomographic cut of a turbulent flame front. b) Sketch of turbulent propagation. c) Generalization of b) to any scale range.

3.1 Scale-invariant regime

The system involves various spatial and time scales: the front thickness δ and the transit time $\tau=\delta/U_N$ of the medium through the front ; the scales $L_i=a^i L_0$ of the turbulent vortices (in between the Kolmogorov scale L_K and the integral scale L_I) and their turn-over times τ_i.

At a scale L_i, the nature of the interaction process between the vortices and the front depends on the scale ratios δ/L_i and τ/τ_i. Hereafter, we restrict our attention to a regime where, for any vortex in the flow scale range $[L_K, L_I]$, these ratios are small: $\delta/L_i \ll 1$ and $\tau/\tau_i \ll 1$. Then, for *all* vortices, the front behaves in the *same* way, as an infinitely thin interface with infinitely fast kinetics. It thus provides the *same* interaction process, *independently* of the vortex scales. In this regime, neither the front geometry nor the flow state are scale-invariant because of the cut-off scales L_K and L_I. Nevertheless, the interaction process between them is scale-invariant.

3.2 Scale-covariance ; Scale-covariant laws

In order to formalize the above property, we address turbulent front propagation in the most general situation regarding scales [1]. For this, we explicitly introduce two scales required for observing turbulent fronts: a field scale L_j equal to the size of the observation window and a resolution scale L_i equal to the lowest observed scale (fig.2c). Viewed this way, the front displays a mean velocity U_j and a mean normal velocity U_i.

The mean velocity U_j corresponds to that of a window of scale L_j comoving with the front. It thus does not depend on the resolution scale of observation L_i and may therefore indeed be labelled U_j. Of course, it varies with the scale L_j since the larger L_j, the more vortices advect the front and improve combustion.

The normal velocity U_i corresponds to the velocity of sub-windows of size L_i comoving with the front. It is thus analogous to the mean velocity U_j but stands at a smaller scale L_i. The fact that the mean and normal velocity are defined in a similar way will allow us to implement a renormalization procedure where the same velocity U_k will be alternatively considered as a mean velocity or as a normal velocity. For this reason, it will be called in both cases turbulent front velocity.

We finally define $u'_{i,j}$ as the r.m.s. of flow fluctuations observed in the above window, i.e. with an observation scale L_j and a resolution scale L_i. It thus simply corresponds to the turbulence intensity contained in the scale range $[L_i,L_j]$.

Since the enhancement of the mean front velocity from U_i to U_j follows from its interaction with turbulence, we seek a relation between U_i, U_j and $u'_{i,j}$. As there is no characteristic velocity in the system, we restrict our attention to laws $\mathcal{L}_{i,j}(.)$ defined by $U_j/U_i = \mathcal{L}_{i,j}(u'_{i,j}/U_i)$.

The indices i and j recall that the law $\mathcal{L}_{i,j}(.)$ refers to observations in the scale range $[L_i,L_j]$. Both are actually relevant since laws usually vary with the scale range in which they are considered. This is no longer the case in a scale-invariant regime since, according to scale-covariance, laws do not vary with scales: $\mathcal{L}_{i,j}(.)=\mathcal{L}(.)\ \forall i\ \forall j$. We stress that this property includes not only global covariance (i.e. $\mathcal{L}_{i,j}(.)=\mathcal{L}_{i+n,j+n}(.)\ \forall n$ (fig.1a)), but also local covariance (e.g. $\mathcal{L}_{i,j}(.)=\mathcal{L}_{i,n*j}(.)\ \forall n$ (fig.1b)).

Given the turbulence intensities $(u'_{i,j})$, the laws $\mathcal{L}_{i,j}(.)$ provide a link between the turbulent velocities U_j and U_i. Combining two of them, $\mathcal{L}_{i,j}(.)$ and $\mathcal{L}_{j,k}(.)$, yields a relation between U_i and U_k which, by definition, must be the law $\mathcal{L}_{i,k}(.)$. In the case of covariance, such a combination must yield an invariance and thus provides a criterion sufficiently strong for selecting a few possible kinds of law [1]. In the opposite case, we nevertheless notice that the law $\mathcal{L}_{0,I}(.)$ obtained by combination of the laws $\mathcal{L}_{i,i+1}(.)$, for $0 \le i < I$, corresponds to their renormalization by a change of resolution in scale space ($L_{i+1}/L_i=a \rightarrow L_I/L_0=a^I$) ; by iteration, this should yield a fixed point of renormalization, i.e. a scale-covariant law too.

These remarks show that scale-covariant laws may be selected by two equivalent methods: solving the criterion for scale-covariance or iterating the renormalization of a given law. As they address laws instead of their variables, both of them differ from renormalization methods based on elimination of small scale variables [6].

We provide below an example of selection of laws by iterative renormalization. We first consider an approximate relation proposed by Clavin and Williams at low turbulence intensity ($u'/U_N<<1$) [7]: $U_T=U_N+\beta u'^2/U_N$. We apply it in any consecutive scale range $[L_i,L_{i+1}]$ by replacing U_N, U_T, u' by U_i, U_{i+1}, $u'_{i,i+1}$ respectively. This makes the laws $\mathcal{L}_{i,i+1}(.)$ all equal to the Clavin-Williams law :

$$U_{i+1} = U_i + \beta\, u'^2_{i,i+1} / U_i \qquad \forall i \qquad (1)$$

The law $\mathcal{L}_{0,I}(.)$ relevant to the large scale range $[L_0,L_I]$ is obtained by integrating relations (1) from i=0 to i=I. For a sufficiently small resolution $a=L_{i+1}/L_i$ in scale space, the difference between U_{i+1} and U_i is small enough for being approximated by a first order variation dU_i. The relations (1) can then be rewritten :

$$U_i\, dU_i = \beta\, u'^2_{i,i+1} \qquad \forall i \qquad (2)$$

To integrate the right hand side, we use the fact that, since the flow is turbulent, its fluctuations at different scales are uncorrelated. This yields the same property as for incoherent waves in optics : the square of the sum of flow fluctuations reduces to the sum of their squares. As the turbulence intensity $u'_{i,j}$ corresponds to the r.m.s of velocity fluctuations in the scale range $[L_i,L_j]$, this implies the additivity of its squares with respect to scale ranges :

$$u'^2_{0,I} = \sum_{k=0}^{k=I-1} u'^2_{k,k+1} \qquad (3)$$

Integration of the left hand side from i=0 to I-1 gives $(U_I^2 - U_0^2)/2$. This, together with (3) yields the integral form of (2) :

$$U_I^2 = U_0^2 + 2\,\beta\, u'^2_{0,I} \qquad (4)$$

Whereas the relation (4) is quadratic, the Clavin-Williams relation (1) is not. The latter relation has thus changed by renormalization : it is not scale-covariant.

Let us now turn to the quadratic law (4) itself and implement the same renormalization procedure. Applied to consecutive scale ranges, relation (4) gives:

$$U_{i+1}^2 = U_i^2 + 2\,\beta\, u'^2_{i,i+1} \qquad \forall i \qquad (5)$$

Integration of relation (5) from i=0 to I-1 yields, by using (3) again, the quadratic relation (5) back ! This shows the invariance by renormalization of the quadratic law (5) and proves its scale-covariance. Owing to this property, this law stands as the *exact* non-linear form of the Clavin-Williams relation in a scale-invariant regime.

The specificity of scale-covariant laws is emphasized in figure 3 by considering, for different laws, the effect of a scale change. In figure 3a, the data relevant to a turbulent front are generated in large scale ranges $[L_0,L_j]$ according to a given law. They give rise in figure 3b to the data pertaining to the analysis of the same front in consecutive scale ranges. If covariance is satisfied, the new data should be related at *all* scales by the *same* law and thus yield the *same* curve. Conversely, any change of curve would evidence a breaking of scale-covariance.

Motivated by the quadratic form of scale-covariant laws, we find it convenient to consider the squares $R_{i,j}^2$ of the ratio $R_{i,j}=U_j/U_i$ and those $m_{i,j}^2$ of the ratio $m_{i,j}=u'_{i,j}/U_i$. Then the quadratic law simply corresponds to a line in these coordinates $(m_{i,j}^2, R_{i,j}^2)$. Given a relation $R_{0,j}^2$ vs $m_{0,j}^2$ in large scale ranges, its form in consecutive scale ranges is deduced by the following algebra, derived from the definition of variables and the additivity of turbulence intensities squared (3) :

$$R_{i,i+1}^2 = R_{0,i+1}^2 / R_{0,i}^2 \quad ; \quad m_{i,i+1}^2 = (m_{0,i+1}^2 - m_{0,i}^2) / R_{0,i}^2 \qquad (6)$$

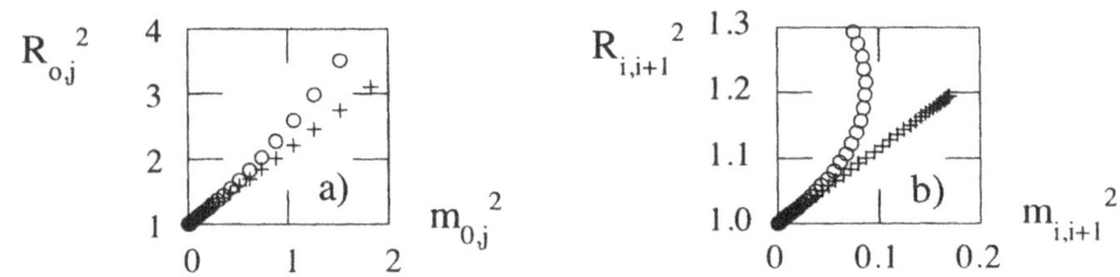

Figure 3: Clavin-Williams law (circles) and quadratic law (crosses) in increasing scale ranges $[L_0, L_j]$ (a) and in constant scale ranges but increasing resolution $[L_i, L_{i+1}]$ (b).

Figure 3 shows that the quadratic law maintains the same form in either large or small scale ranges but that the Clavin-Williams law explodes when viewed in small scale ranges. In the latter case, such variation would easily enable an observer to discriminate scale ranges according to their size. It would therefore correspond to an objective breaking of local scale-invariance.

Figure 3 also shows a pathological behaviour of the Clavin-Williams relation viewed in small scale ranges in terms of a turning point. Especially, whereas fronts yield a one to one relation in large scale ranges (fig.3a), they do not in small scale ones (fig.3b). This property is by no way paradoxical and simply indicates that, to provide the Clavin-Williams relation in large scale ranges, the wrinkling process viewed in consecutive scale ranges must be more and more efficient above the scale L_t pertaining to the turning point. Then, for the same ratios $m_{h,h+1}=m_{k,k+1}$ between turbulence intensity and normal velocity, the front roughness is larger at large scales, $L_k>L_t$, than at small scales, $L_h<L_t$: $R_{k,k+1}>R_{h,h+1}$. This, again, evidences of a breaking of scale-covariance.

3.3 Experimental evidence

We now turn to the experimental evidence for scale-covariance in a scale-invariant regime of turbulent front propagation [2]. For this, a combustion front has been generated in a closed chamber filled with turbulent premixed gases. A grid turbulence was generated so as to satisfy the criteria for a scale-invariant regime. The required variables were then computed in various scale ranges in the following way :

As the flow scales are more dissipative than inertial, turbulence intensities do not follow the Kolmogorov scaling ($u'_{i,i+1} \alpha L_i^{1/3}$), but rather a close corrected law [8]. Fitting it by a power law $u'_{i,i+1} \alpha L_i^K$, we then find a larger exponent K=1/2.

Consider the volume burnt per unit time by the front (fig.2b). On one hand, it is given by the product $U_N S_T$ since the front propagates at speed U_N all along its surface S_T. On the other hand, it is equal to $U_T S_L$ since the front propagates at mean velocity U_T in a window of size S_L. This gives $U_T/U_N=S_T/S_L=R$, where R labels the relative surface increase, i.e. the front roughness. More generally, the same analysis performed in any scale range $[L_i,L_j]$ shows that the ratio $R_{i,j}=U_j/U_i$ corresponds to the averaged front roughness within windows of size L_j and resolution scales L_i (fig.2c). They may thus easily be determined from image processing. Moreover, taking the smallest scale L_0 to be smaller than the first wrinkling scale implies that U_0 is the normal velocity of a planar front: $U_0=U_N$. Then, the turbulent front velocity at scale L_i simply follows from $U_i=U_0R_{0,i}$.

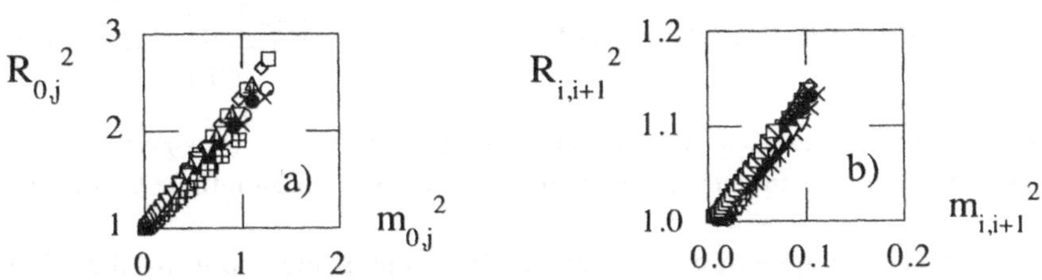

Figure 4: Experimental analysis of flame fronts in : a) scale ranges $[L_0,L_j]$ involving different scale ratios L_j/L_0 but a constant resolution scale L_0. b) scale ranges $[L_i,L_{i+1}]$ involving constant scale ratios $L_{i+1}/L_i=a$ but a varying resolution scale L_i.

From the above procedure, the mean values of roughnesses $R_{i,j}=U_j/U_i$ and of so-called "mixing variables" $m_{i,j}=u'_{i,j}/U_i$ can be obtained for any scale ranges. The relation between them, for various flames obtained at different net turbulence intensities $u'=u'_{0,I}$, is shown in figure 4. Figure 4a refers to scale ranges $[L_0,L_j]$ involving the same resolution scale L_0. Whatever the observation scale L_j, the same relation, a line in the quadratic coordinates ($m_{0,j}^2$, $R_{0,j}^2$), is obtained. The corresponding laws $\mathcal{L}_{0,j}(.)$ are thus all *equal* to the quadratic law (4) with I → j. As the scale ratio L_j/L_0 varies with j, this covariance corresponds to a *local* scale-covariance. Figure 4b refers to scale ranges $[L_i,L_{i+1}]$. Whatever the scale L_i, the same relation, a line in the quadratic coordinates ($m_{i,i+1}^2$, $R_{i,i+1}^2$), is obtained. The corresponding laws $\mathcal{L}_{i,i+1}(.)$ are thus all *equal* to the quadratic law (5). As the scale ratio L_{i+1}/L_i is fixed, this covariance corresponds to a *global* scale-covariance.

Taken together, figures 4a and 4b show that, regardless of the scale unit and scale ratio of observation, the same relation between front roughness $R_{i,j}$ and mixing variables $m_{i,j}$ is obtained. Under these conditions, the physics of turbulent front propagation is thus unable to distinguish between scales or between scale ratios. This shows the scale-covariance of the wrinkling process of a front by turbulence in the scale-invariant regime pertaining to this experiment.

3.4 Fractal construction in scale space

We emphasize that, although the wrinkling process of fronts by turbulence is scale-invariant, neither the front geometry nor the flow structure are scale-invariant themselves, owing to cut-off scales. In particular, the non-scale similar nature of front geometry is shown in figure 5a by the bent section of the curve relating the front roughness $R_{i,I}$ to the scale ratio L_i/L_0 in log-log coordinates. However, a straight section is nevertheless displayed at large scales in this figure. Does this imply that this front has eventually reached a fractal geometry at large scales ?

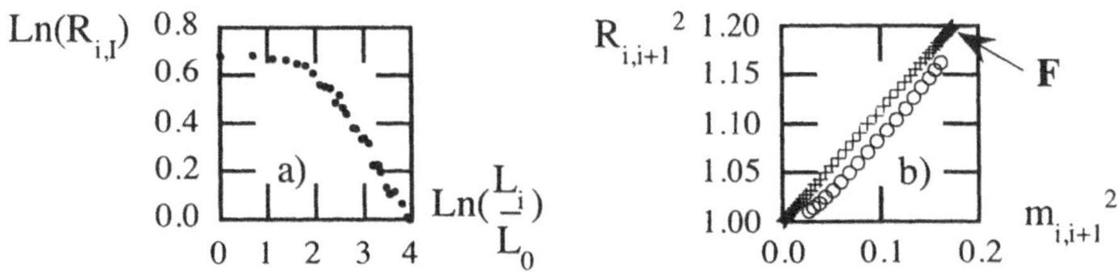

Figure 5: Classical scale analysis (a) and "causal" analysis (b) of a given front. In b, circles refer to experiment and crosses to the covariant law for an infinite set of scales.

To answer this question, we can analyse the same front in the alternate representation relevant to the wrinkling process within consecutive scale ranges $[L_i,L_{i+1}]$ (fig.5b) [3]. Here, a fractal state corresponds, beyond a scale L_f, to a value of $R_{i,i+1}$ *constant* with the scale L_i : $\forall i>f$, $R_{i,i+1}=C$. Indeed, as in deterministic fractals, any additional scale L_{i+1} with i>f then increases the front roughness by the same factor $R_{i,i+1}=C$, so that the front is a random fractal with a fractal dimension $D = 2 + dLn(R_{i,i+1})/dLn(L_{i+1}/L_i) = 2 + Ln(C)/Ln(a)$. But a constant $R_{i,i+1}$ means a constant $m_{i,i+1}$ since both are linked by the turbulent conbustion law : a fractal state also corresponds to a fixed point in the representation of figure 5b.

For the scaling of turbulence intensities satisfied in this experiment, $u'_{i,i+1} \alpha L_i^K$, the front of figure 5a gives rise, in figure 5b, to a series of points $(m_{i,i+1}^2, R_{i,i+1}^2)$ displayed by circles. In agreement with the scale-covariance of the wrinkling process, they describe an almost straight line. In addition, their distance to the origin (0,1) grows with the scale L_i, meaning that turbulence intensities $u'_{i,i+1}$ increase faster with scale L_i than turbulent velocities U_i. Above $R_{i,i+1}=1.10$, the relative distance between consecutive circles decreases with L_i (i.e. with $R_{i,i+1}$) but *no* fixed point is displayed : in contrast with the impression left by figure 5a, the front is thus definitely *not* in a fractal state.

The fact that the front would be abusively taken for a fractal according to figure 5a stems from the choice of variables. In this classical representation, a fractal state corresponds to a straight line. Then, the slow drift of roughness $R_{i,i+1}$ with scale L_i corresponds to an exponentially weak curvature in figure 5a, which, indeed, can hardly be noticed on the graph. Interpreting this weakly curved line as an actual straight line, usually yields to improperly attribute a fractal dimension D_a to such an object: $D_a=2+Ln(R_{I-1,I})/Ln(a)$. However, this number is explicitly linked to the largest scale of observation L_I since, even in the vicinity of L_I, the elementary roughnesses $R_{i,i+1}$ still evolve with scale L_i (fig.5b). It is then not solely relevant to the physical process (the covariant evolution shown in figure 5b) and might therefore more appropriately be called "apparent" fractal dimension to remind us of its dependence on the finite range of the observation scales.

To remove finite size effects in this study, we can extrapolate the theoretical covariant evolution (5) to scales extending to *infinity*. Labelling, for simplifying the notations $\rho_i=R_{i,i+1}^2$, and using the scaling $u'_{i,i+1}=(L_i/L_0)^K=a^{iK}$, one obtains from (5) a deterministic evolution of mean roughness in scale space, parametrized by an initial condition ρ_0 and by $\rho_\infty=a^{2K}$:

$$(\rho_{i+1} - 1) / (\rho_i - 1) = \rho_\infty / \rho_i \qquad (7)$$

Here, the link between squared roughness $\rho_i=R_{i,i+1}^2$ is conveyed by turbulent velocities : U_{i+1} results from the roughness $R_{i,i+1}$ but also drives the roughness $R_{i+1,i+2}$. For $\rho_0<\rho_\infty$, the series (7) provides the increase of elementary roughness $R_{i,i+1}$ with scale L_i but also shows evidence of an accumulation point F of the roughness evolution : $\rho_i \rightarrow \rho_\infty$ for $i \rightarrow \infty$ (fig. 5b). In the very vicinity of F, any scale provides the same additional roughness, $R_{i,i+1} \approx \rho_\infty^{1/2}$, so that the front has indeed reached a fractal state of dimension $D=2+\mathrm{Ln}(\rho_\infty^{1/2})/\mathrm{Ln}(a)=2+K$.

This analysis shows that scale-covariance actually yields the construction of a fractal front at an *infinitely* large scale, but with a *definite* single fractal dimension representative of the wrinkling process itself. In comparison, the apparent fractal dimensions D_a are many; they depend on both the integral scale L_I and the ratio u'/U_N: $D_a \equiv D_a(u'/U_N, L_I)$. Accordingly, without knowing the law of fractal construction (5) (7), the fractal dimension D of the process and the distance to the fractal regime, it would be quite difficult, from the set of apparent fractal dimensions only, to extract the intrinsic physical informations pertaining to this system.

3.5 Scale-covariance versus scale-similarity

The above study points to a comparison between scale-covariance and scale-similarity. Scale-similarity addresses a fractal state of the front. It is sensitive to finite size effects which interrupt the approach to a pure fractal regime, and thus calls for either large scale ranges or large mixing regimes (e.g. $u'/U_N>>1$ here). Furthermore, it treats the euclidean part of the front, the apparently fractal part (the straight section of figure 5a) and the part in-between (the bent section of figure 5a) as physically different, and usually ignores the latter. In comparison, scale-covariance is insensitive to system size, treats the euclidean, the fractal or the intermediate geometries as physically equivalent (compare figures 5a and 5b), and extracts more information from the weak mixing regime than from the large one.

These differences between scale-covariance and scale-similarity stem from the nature of their representation: causal for scale-covariance, since the effect (the front roughness $R_{i,j}$) is considered with respect to its cause (the ratio $m_{i,j}$ of turbulence intensity to normal velocity), and descriptive for scale-similarity, since front roughnesses ($\mathrm{LnR}_{i,l}$) are considered with respect to scales ($\mathrm{Ln}(L_i/L_0)$) which in no way represent their cause.

4. INTERMITTENCY IN FULLY DEVELOPED TURBULENCE

We address the statistical features of fully developed turbulence. We thus consider a fluid in the limit of infinite Reynolds number and infinite extent and assume, following the discussion of section 2, that it stands in a state satisfying both global and local scale-invariances. Our goal is to extract from these symmetries informations regarding the laws characterizing this statistical state.

We thus consider the fluctuations of the one-dimensional increments $u(l,t)=[\mathbf{v}(\mathbf{r}+\mathbf{l},t)-\mathbf{v}(\mathbf{r},t)].\mathbf{l}/l$ of velocity $\mathbf{v}(\mathbf{r},t)$ across a distance l and focus attention on its structure functions $<u^i>(l)$, $i\in\Re$, the brackets $<.>$ standing for a time average. As recalled in section 4.1, global scale-invariance selects them as scale power laws, $<u^i>(l)\sim l^{\zeta(i)}$, but does not give further information on the exponents $\zeta(i)$. Our purpose will then consist in looking whether local scale-invariance constrains the possible relation between them, i.e. the structure of the function $\zeta(.)$, hereafter called intermittency function. This question has been addressed by Dubrulle and Graner within the framework of a relativistic group structure [9] and will be considered here within the framework of gauge invariance [4].

4.1 Scale-invariance as a gauge-invariance

Although this is usually implicitly implied, we stress that, in physics, data always result from measurements. They thus refer to some definite measurement procedure and to some definite choice of standards. For instance, stating explicitly the velocity standard u_s and the length standard l_s used to measure u and l, the structure functions $<(u/u_s)^i>(l)\sim(l/l_s)^{\zeta(i)}$ appear explictly standard-dependent. The following analysis aims at clarifying the nature of this dependence in scale-invariant regimes.

Global and local scale invariances state that changing the measures of scales, $l/l_s\rightarrow\lambda l/l_s$ with $\nabla\lambda=0$, or of scale ratios, $l/l_s\rightarrow\lambda l/l_s$ with $\nabla\lambda\neq0$, should not necessarily change the physics. Such changes can be equally obtained either by keeping the same scales but changing the length standard, i.e. $l\rightarrow l$ and $l_s\rightarrow l_s/\lambda$ or by changing scales and keeping the same standard, i.e. $l\rightarrow\lambda l$ and $l_s\rightarrow l_s$ [4,9]. We chose the former interpretation in the following. It turns out interpreting scale-invariance as an invariance by change of standards or, following Weyl [10], by change of gauge. It thus yields us to consider scale-invariance as a gauge-invariance.

Of course, the concept of scale-invariance used here is exactly the same as that implemented in turbulent combustion (section 3). However, in this case, length scales were *implicit* variables of the relation between front velocities and turbulence intensities so that stating explicitly the standard changes was not necessary. This is no longer the case for structure functions since length scales, i.e. l/l_s, are *explicit* variables of them. As illustrated in section 4.2 below for single variable functions, invariance of functions by change of length standard then requires also changing the standards pertaining to the other variables, e.g. the velocity.

4.2 Scale-covariance of single scale functions

We consider a quantity u depending on a single scale l : $u \equiv u(l)$. When measured with the standards u_s and l_s, the variables u and l provide data, u/u_s, l/l_s, which are related by some function $\varphi(.)$: $u/u_s=\varphi(l/l_s)$. Changing length standards $l_s \to l_s'=l_s/q$ $\nabla q=0$, and velocity standards, $u_s \to u_s'=u_s/p$ $\nabla p=0$, in a global way yields $u/u_s'=\varphi'(l/l_s')$ with $\varphi'(qx)=p\varphi(x)$. Global scale invariance implies that for any q there exists p(q) such that $\varphi'(.)=\varphi(.)$. As is well known, this selects power laws : $\varphi(x)=\varphi(1)\, x^e$.

The same procedure may also be applied to implement local scale invariance. Then the possibility of preserving physics is extended to standards change modifying scale ratios. Doing so without using explicit scales implies working with new standards $u_s'(u,l)$, $l_s'(u,l)$ related to the old ones u_s, l_s by power laws :

$$u_s \to u_s' = u_s\,(u/u_s)^{1-\alpha}\,(l/l_s)^{-\beta} \quad ; \quad l_s \to l_s' = l_s\,(u/u_s)^{-\gamma}\,(l/l_s)^{1-\lambda} \tag{8}$$

Measured with the new standards, the variables u and l give the new data : $u/u_s'=(u/u_s)^{\alpha}(l/l_s)^{\beta}$, $l/l_s'=(u/u_s)^{\gamma}(l/l_s)^{\lambda}$. In case of the former power law, $u/u_s=(l/l_s)^e$, they yield the following relation : $u/u_s'=(l/l_s')^{e'}$ with $e'=(\alpha e+\beta)/(\gamma e+\lambda)$. Covariance by local standard changes turns out satisfying the criterion $e'=e$. This can be achieved for any change of scale ratio, i.e. any λ, by suitably choosing the other gauge parameters α, β and γ : power laws are also scale-covariant in a local sense.

4.3 Observers, measurement procedure and gauge covariance

We apply the above procedure to structure functions. As they are functions of a single variable, the scale l, global scale-invariance selects them as scale power laws:

$$\langle (u/u_s)^i \rangle (l) \quad \alpha \quad (l/l_s)^{e_i} \qquad \forall i \tag{9}$$

No further constraints on the exponents could however be obtained by applying global or local scale covariance to each structure function *separately*. The family of structure functions $\{\langle (u/u_s)^i \rangle (l),\ i \in \Re\}$ has thus to be considered as a *whole* for improving the selection by local scale-covariance. This turns out focusing attention of the intermittency function $\zeta(.)$ relating the family of scaling exponents e_i to their moment orders, hereafter labelled quite generally o_i : $e_i=\zeta(o_i)$.

Implementing local scale-covariance on the intermittency function requires several steps : i) identifying the measurement procedure involved to determine $\zeta(.)$, ii) identifying the standards required by this procedure, iii) changing the standards so as to simulate a change of the ratio of variables iv) requiring, for *different* values of the gauge parameters but a *fixed* measurement procedure, the possibility of covariance.

Keeping the same measurement procedure while changing standards is actually essential to make the comparison between observers relevant to scale-covariance. In particular, if the procedure was allowed to vary, a breaking of covariance could simply arise from the fact that different things have been measured and would thus not be conclusive. Here, the measurement procedure pertaining to the intermittency function $\zeta(.)$ addresses the measure of scaling exponents e_i on one hand and of moment orders o_i on the other hand.

Regarding scaling exponents, we emphasize that their determination follows from two kinds of measurements only: a measure of velocity increments u(l,t) for a given distance l between probes and a measure of this distance l. Then, the value of any i-th structure function at distance l is obtained by raising the data $u/u_s(l,t)$ of velocity increments at the i-th power and averaging : $<(u/u_s)^i>(l)$. Moreover, for *any* i, these values are attributed to the *same* value of distance: l/l_s. Although, this remark might seem superfluous, it points to an essential feature of the measurement procedure : for any observer, the value of length attributed to a structure function only depends on the distance l between probes and *not* on the moment order ; it is thus the *same* for *all* structure functions provided that they refer to the *same* distance l. We call this property "locality" of the procedure.

Regarding moment orders, we stress that the series of data $\{m(o_i)=(u/u_s)^i(l,t), \forall l, \forall t, i\in\Re\}$ provides for any observer a multiplicative algebra: $\forall i, \forall j, \exists k$; $m(o_i)*m(o_j)=m(o_k)$. Although different parametrizations of this algebra could follow from different choices of moment orders, it is implicitly considered in the definition of the intermittency function $\zeta(.)$ that any observer keeps track of the underlying multiplicative structure by choosing moment orders o_i so that: $o_i+o_j=o_k$. Then the parametrization o_i is linear, $o_i=\varepsilon i$, and $m(n)=m(1)^n, \forall n$. Although the latter relation might seem natural, we emphasize that it corresponds to a convention regarding the measurement procedure of moment orders. If it was not implicitly adopted by all workers in this field [11], then comparison between the intermittency functions obtained by each of them would be meaningless, since possible differences might simply trace back to a non-linear parametrization of moment orders. This remark, of course, also addresses the different observers considered here. Regardless of their different standards, they shall therefore subscribe to this convention.

Another convention, more explicit, is also widely used. It consists in choosing moment orders so that $\zeta(0)=0$. This turns out to state that, for $o_i=0$, the structure function $<(u/u_s)^i>$ is uniform. Here, this is of course satisfied for i=0, so that the linear parametrization $o_i=\varepsilon i$ is relevant. However, we shall meet in section 4.4.2 a different measurement procedure according to which the scaling exponent for i=0 no longer vanishes. To nevertheless satisfy the above initial condition for $\zeta(.)$ while keeping track of the multiplicative algebra for moments, a translation of moment orders is required. Following it, the most general parametrization of moment orders will be affine : $o_i=\varepsilon i+\omega$ and $[m(n)/m(0)] = [m(1)/m(0)]^n \; \forall n$.

The latter relation shows that two standards, m(0), m(1), are required for measuring moment orders. In addition to the velocity and length standards u_s, l_s, they fully specify an observer O for our purpose: $O=(u_s, l_s, m(0), m(1))$. Change of observers, $O\rightarrow O'=(u_s', l_s', m(0)', m(1)')$ will be obtained from relations (8) and (10):

$$m(0) \rightarrow m(0)' \quad ; \quad m(1) \rightarrow m(1)' \tag{10}$$

Regarding the change of length standards, we note that the parameter γ allows the length measure to depend on the velocity field. As, according to the locality of the measurement procedure, the length data are however the same for all moments, this degree if freedom will be ineffective here. For simplicity, we shall skip it out in the following : $\gamma = 0$.

4.4 Scale-covariant intermittency function

We select scale covariant intermittency functions by changing standards according to (8)(10) while requiring covariance. This is done first by ignoring another fundamental symmetry, statistical homogeneity, and then by invoking it explicitly.

4.4.1 Statistical homogeneity ignored

We consider two observers O and O' measuring $\zeta(.)$ according to the definition (9). The changes of standards from O to O' induce the following change of data :

$$e_i \rightarrow e_i' = (e_{\alpha i} + \beta i) / \delta \quad ; \quad o_i \rightarrow o_i' = \varepsilon o_i + \omega \qquad (11)$$

We chose the observer O so that $o_i=i$ and $e_i=\zeta(i)$. By comparison, the observer O' determines the intermittency function as $\zeta'(.)$ such that $e_i'=\zeta'(o_i')$ (fig.6a). Scale-covariance requires that, despite the change of data, both observers conclude at the same law : $\zeta(.) = \zeta'(.)$.

According to (11), this criterion implies: $\zeta(\varepsilon i+\omega)=[\zeta(\alpha i)+\beta i] / \lambda \ \forall i$. Here, $G=(\alpha, \beta, \lambda, \varepsilon, \omega)$ are the gauge parameters making difference between the two observers, i.e. $G=(1, 0, 1, 1, 0)$ for O, and $G'\neq G$ for O'. In such case, the gauge change (11) makes the graph of the function $\zeta(.)$ glide on itself and thus drives a reparametrization invariance of $\zeta(.)$ (fig.6b).

The corresponding constraint on the function $\zeta(.)$ may be expressed by considering an elementary change of parametrization, $\delta G\equiv(\delta\alpha, \delta\beta, \delta\gamma, \delta\varepsilon, \delta\nu)$. It induces a coupled change of e_i and o_i, i.e. δe_i, δo_i, which must be directed along the tangent of the curve to make it glide on itself: $\delta e_i= d\zeta/do_i \ \delta o_i$. Applied in the vicinity of G, this necessary criterion for covariance gives an ordinary differential equation for $\zeta(.)$ parametrized by δG : $(Ai+B)\, d\zeta/di+C\zeta(i)+Di =0$. Its solutions involve two classes :

i) Log-Poisson statistics : $\zeta(i) = a\, i + b\,(1- \exp\text{-}ci)$

This kind of intermittency function has also been obtained by Dubrulle and Graner [9]. It includes the relation proposed by She and Levesque from phenomenological arguments [12]. Its link with a log-Poisson statistics has been identified in [13].

ii) Log-normal / truncated log-Levy statistics: $\zeta(i) = a\, i + d\,[\,|b|^c - |i+b|^c\,]$

The log-normal statistics corresponds to c=2. Log-Levy statistics [14] may be linked to other cases [15].

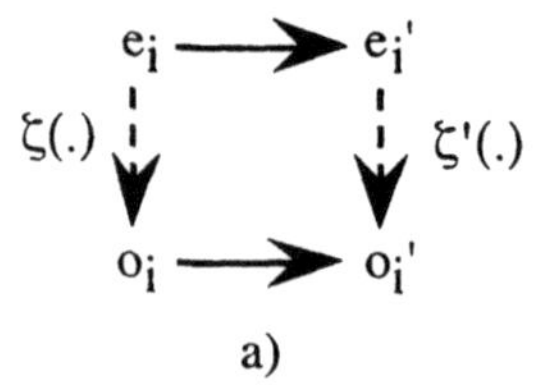

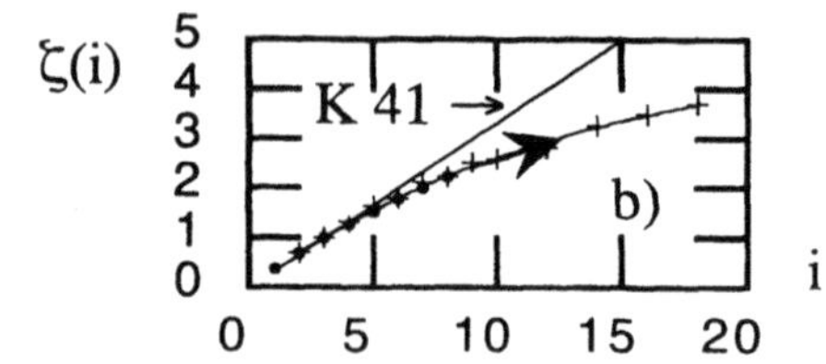

Figure 6: a) Sketch of the transformation induced by gauge change. b) Scale change (arrow) makes the covariant curve glide on itself. Crosses (resp. circles) refer to experimental points [16] (resp. [17]). Both data are compatible with one or the other class of solution for $\zeta(.)$ with no linear term, i.e. a=0.

4.4.2 Statistical homogeneous state

We consider an observer O according to which the statistics of the fluid is homogeneous. Then the mean flow $\mathbf{V}(\mathbf{r})=\langle \mathbf{v}(\mathbf{r},t)/u_s \rangle u_s$ is independent of $\mathbf{r}$ so that its averaged amplitude U(l) over a distance l is a constant. To express the constraints imposed by this additional symmetry, we slightly modify the measurement procedure by considering the scaling exponents of the i-th structure function of velocity increments divided by the i-1-th structure function of the field U(l) :

$$\frac{\langle (u/u_s)^i \rangle (l)}{\langle (U/u_s)^{i-1} \rangle (l)} \quad \alpha \quad \left(\frac{l}{l_s}\right)^{e_i} \qquad \forall i \tag{13}$$

For the observer O, this change of procedure does not make difference since U(l) is a constant. This is no longer the case for other observers generated by the gauge transformations (8)(10). Compared to section (4.2.1), the term βi is replaced by β, Di by a constant D and the value of a vanishes for the two classes of solutions: a=0 [4]. Such a log-Poisson (resp. log-Levy) solution has been considered by Chen and Cao with a=0, b=9/2 and c=ln(9/7)/3 [18] (resp. Nelkin [19] with a=0 and c=1/2).

Few remarks are in order. The modification of intermittency functions according to statistical homogeneity shows that the constraints imposed by covariance depend not only on gauge changes but also on the measurement procedure and on various fundamental symmetries. On the other hand, the intermittency functions found in both cases include the linear scaling of Kolmogorov as a specific case (e.g. $c=\infty$ or c=1) but also display non-linear scalings. This shows that intermittency deviations [11] are compatible with scale-invariance. Here, we shall leave opened the interesting question of the class of intermittency functions possibly relevant to fully developed turbulence. Instead, we stress that the present solutions are obtained from a necessary condition for scale-covariance. Further analyses show that the solutions relying on statistical homogeneity are actually scale-covariant but that those found without invoking it may or may not be so depending on the coefficients a,b,c,d.

5. CONCLUSION

The scale-invariance of a physical situation implies the scale-covariance of the corresponding physical law. Here, we have shown that this concept can be successfully applied in out-of-equilibrium systems, in both a global and a local sense and either to physical processes or to physical states. As this approach works even in finite sized systems, it enables one to go beyond scale-similarity and fractal analysis. In particular, in turbulent front propagation, it elucidates the asymptotic construction of fractal and unifies the description of euclidean, fractal or intermediate geometries. As it also works within a family of functions, it can address the set of structure functions of fully developed turbulence. It then selects scaling exponents showing intermittency. Owing to its universal basis, namely scale-invariance, this approach should, in principle, apply to many other out-of-equilibrium systems.

References

[1] Pocheau A., *Phys.Rev.E* **49** (1994) 1109-1122
[2] Pocheau A. and Queiros-Condé D.,*Phys.Rev.Lett.* **76** (1996) 3352-3355
[3] Pocheau A. and Queiros-Condé D., *Europhys..Lett.* **35** (1996) 439-444
[4] Pocheau A., *Europhys.Lett.* **35** (1996) 183-188
[5] Nottale L., Fractal space-time and microphysics, World Scientific (1993)
[6] Sivashinsky G.I., *Combust.Sci.Technol.* **62** (1988) 77-96
Yakhot V., *Combust.Sci.Technol.* **60** (1988) 191-214
[7] Clavin P. and Williams F.A., *J.Fluid Mech.* **90** (1979) 589-604
[8] Villermaux E. and Gagne Y., *Phys.Rev.Lett.* **73** (1994) 252-255
[9] Dubrulle B. and Graner F., *J.Physique II France* **6** (1996) 797-816
Dubrulle B. and Graner F., *J.Physique II France* **6** (1996) 817-824
Dubrulle B. and Graner F., Submitted to *Phys.Rev.E.* (1996)
[10] Weyl H., *Naturwissenschaften* **19** (1931) 49
[11] Frisch U., Turbulence, Cambridge University Press (1995)
[12] She Z-S. and Levesque E., *Phys.Rev.Lett.* **72** (1994) 336
[13] Dubrulle B., *Phys. Rev. Lett.* **73** (1994) 959
She Z-S. and Waymire E.C., *Phys.Rev.Lett.* **74** (1995) 262
[14] Cont R., Potters M. and Bouchaud J.P., Lecture 5, these proceedings
Mantegna R.N. and Stanley H.E., *Phys.Rev.Lett.* **73** (1994) 2946
Koponen I., *Phys.Rev.E* **52** (1995) 1197
[15] Dubrulle B., private communication
[16] Anselmet F., Gagne Y., Hopfinger E.J. and Antonia R.A., *J.Fluid Mech.* **140** (1984) 63-89
[17] Benzi R., Ciliberto S., Baudet C. and Ruiz Chavarria G., Physica **D80** (1995) 385-398
[18] Chen S. and Cao N., *Phys.Rev.E* **52** (1995) 5757
[19] Nelkin M., *Phys.Rev.E* **52** (1995) 4610

LECTURE 17

Turbulence: Statistical Approach

B. Castaing

Centre de Recherches sur les Très Basses Températures,
Associé à l'Université Joseph Fourier, CNRS, B.P. 166,
38042 Grenoble cedex 9, France

1. INTRODUCTION

The statistics of velocity differences in turbulence gives an example of both the limitations of scale invariance and the generic character of the corrections it implies. In this lecture, as an introduction I shall present another case where such limitations appear, by successive refinements of a dimensional analysis approach: the turbulent velocity profile close to the wall of a channel. Then I shall discuss the velocity intermittency phenomenon in homogeneous, isotropic turbulent flows. I shall finish with a new Markovian approach for this problem, mainly stressing the connection with the previous one.

Thus all this lecture will be concerned with turbulence, with which everybody in this school is not familiar. Without entering in details let us precise that we are concerned with incompressible turbulent flows, governed by the Navier-Stokes equation. Out of geometrical parameters, defining the particular flow, dynamics depends on one non-dimensional quantity, the Reynolds number

$$\mathrm{Re} = \frac{UL}{\nu},$$

where U and L are characteristic velocity and size and ν the kinematic viscosity. Turbulence concerns high Reynolds numbers, where one could think ν to be of little importance. However only viscosity can dissipate the kinetic energy, mainly present in the large scale motion. The main transfer process from large scale to small one [1] is by stretching which allows smaller scale vortical structure to grow spontaneously, creating new local stretching and so on.

2. THE TURBULENT VELOCITY PROFILE

2.1. Dimensional Analysis

Let us consider a developed turbulent flow in a channel. The determination of the velocity profile close to the wall is a celebrated example of the possibilities of dimensional analysis in University courses [2]. The dimension perpendicular to the wall is called y, and we are interested in the y values much smaller than the total transverse size of the channel L, but sufficiently large to neglect the viscosity. The problem is fully determined by giving the average momentum flux (or the stress) from the wall, which is conserved across the fluid layers. Let us call it $\rho v_\star^2$, where ρ is the fluid density, defining a characteristic velocity $v_\star$. If $U(y)$ is the average velocity parallel to the wall at a distance y of it we can write, due to the above assumptions:

$$\frac{U(y)}{v_\star} = f(y, \rho, v_\star).$$

Following Buckinkham theorem [3], this non-dimensional quantity can only depend on non-dimensional parameters made from the variables. There is none!

$$U(y) = \text{cst } v_\star, \tag{1}$$

thus results from the "zeroth order" application of scale invariance versus y.

2.2. A First Refinement

Eq. 1 is not so bad a result on a practical point of view. Turbulent profiles are known to be much "flatter" than viscous ones. However, a momentum flux only can result from velocity differences between the various fluid layers. Considering two different values y_1 and y_2 we can again apply dimensional analysis:

$$U(y_1) - U(y_2) = v_\star g(\frac{y_1}{y_2}). \tag{2}$$

The function g verifies the obvious property that: $g(a) + g(b) = g(ab)$, which determines it as a logarithm. Thus:

$$U(y) = K v_\star \ln(\frac{y}{y_0}). \tag{3}$$

y_0 is the distance where scale invariance ceases because of the viscosity. The only length which can be obtained from $v_\star$ and the kinematic viscosity ν is $\frac{\nu}{v_\star}$ and we obtain the von Karman [2] law:

$$U(y) = K v_\star \ln \frac{y v_\star}{\nu}. \tag{4}$$

This can be seen as a first refinement on the brute previous application of scale invariance. It is generally obtained by considering the average gradient of viscosity $\frac{\partial U}{\partial y}$ and its non-dimensional counter-part $\frac{y}{v_\star}\frac{\partial U}{\partial y}$ which must be constant if scale invariance applies

$$\frac{y}{v_\star}\frac{\partial U}{\partial y} = K. \tag{5}$$

2.3. Second Refinement

Let us go now towards a second refinement by explicitly taking into account the two limitations of scale invariance: $K = K(y, v_\star, \rho, \nu, L)$. We have now two non-dimensional parameters:

$$K = K\left(\frac{yv_\star}{\nu}, \frac{Lv_\star}{\nu}\right). \tag{6}$$

Both are Reynolds numbers, the local one and the global one. The interesting range of $\ln\frac{yv_\star}{\nu}$ goes from zero to $\ln\frac{Lv_\star}{\nu}$. Being far from these limits means limiting ourselves to a small interval of $\ln\frac{yv_\star}{\nu}$ in the middle of the total one. On this small interval, K should be considered as constant, following the previous analysis. To be specific, defining $\theta\ln\frac{Lv_\star}{\nu}$ as the length of the small interval, the variation of $\ln K$ on this interval:

$$\frac{\partial \ln K}{\partial \ln y}\theta \ln\frac{Lv_\star}{\nu},$$

must be small if θ is small. When the global Reynolds number $\mathrm{Re} = \frac{Lv_\star}{\nu}$ takes larger and larger values, the product

$$\frac{\partial \ln K}{\partial \ln y}\ln \mathrm{Re}$$

remains of order 1. A power law correction to scaling gives:

$$K = A\left(\frac{yv_\star}{\nu}\right)^\beta \Rightarrow U(y) = \frac{A}{\beta}v_\star\left(\frac{yv_\star}{\nu}\right)^\beta, \tag{7}$$

as proposed long ago by Barenblatt [4] with the assymptotic behavior of β [5]:

$$\beta = \frac{\beta_0}{\ln \mathrm{Re}}. \tag{8}$$

which both gives the global Reynolds dependence and restores the previous law at the $\mathrm{Re} \to \infty$ limit.

3. VELOCITY INTERMITTENCY IN TURBULENCE

3.1. The Problem

Let us first say a few words on the way experimental data are taken. In order to avoid interference between probes, a single probe is often used which measures the fluctuations of the x component of the velocity, x being the direction of the average velocity U. Information about scales is obtained through the Taylor hypothesis [6], assuming that velocity fluctuations are frozen and carried out with the velocity U. Thus

$$\delta v = v(t) - v(t+\tau) = v(x+U\tau) - v(x), \tag{9}$$

and δv is assumed to probe the velocity field at the scale $r = U\tau$. This is a longitudinal velocity difference as being its component on the axis joining the two virtual sampled points. For such longitudinal differences, Kolmogorov [6] derived a relation based on the assumed isotropy of the velocity field:

$$\langle \delta v^3 \rangle = -\frac{4}{5}\epsilon r + 6\nu \frac{\partial}{\partial r}\langle \delta v^2 \rangle. \tag{10}$$

ϵ is the average dissipated power per unit mass. Being proportional to the viscosity the last term should be negligible for large enough scale r, i.e:

$$L > r > \eta = (\nu^3/\epsilon)^{1/4}, \tag{11}$$

where L is of order of the size of the flow.

Generalizing this simplified relation assumes that any moment $\langle |\delta v|^p \rangle$ only depends on ϵ and r. Then by dimensional analysis (zeroth order !):

$$\langle |\delta v|^p \rangle = K_p \epsilon^{p/3} r^{p/3}. \tag{12}$$

The consequence would be a universal distribution, independent of r, for the reduced variable $(\delta v/\epsilon^{1/3} r^{1/3})$. This is wrong. The shape of the δv distribution evolves from a gaussian shape at large scale r to a stretched wings one at small scales [7] (Fig. 1). This is the intermittency phenomenon.

3.2. The Cascade of Multipliers

The proposed "explanation" of this phenomenon comes through the hypothesis that, given two scales r_0, r_1 ($r_1 < r_0$) in the range of interest, the probability distribution of δv at scale r_1, $P_{r_1}(\delta v)$, can be seen as uniquely dependent on that at scale r_0 [8]. Then, due to the possibility of multiplying all the velocities in the flow by the same factor without changing the Reynolds number (by multiplying ν by the same factor) thus without changing the relation between distributions, this relation must have the following shape [9]:

$$P_{r_1}(\delta v) = \int G_{r_1 r_0}(\ln\alpha)\frac{1}{\alpha}P_{r_0}(\frac{\delta v}{\alpha})d\ln\alpha. \tag{13}$$

Again this must be taken as a functional relation between P_{r_1} and P_{r_0}. If we are able to change P_{r_0} without changing the Reynolds number the relation between the new P_{r_1} and P_{r_0} will be the same one, with the same distribution

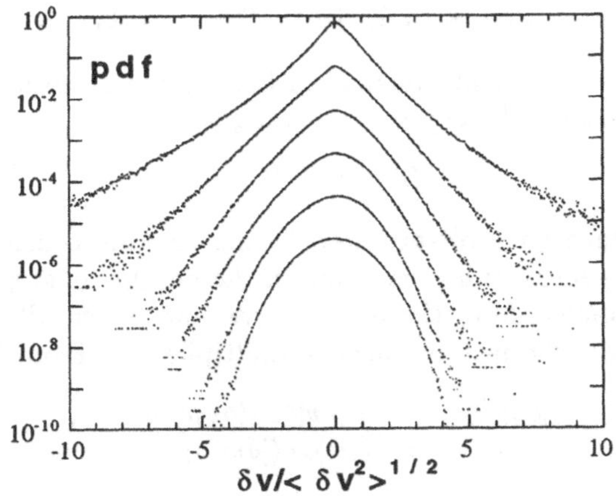

Fig. 1. — Evolution of the probability density functions (pdf) of velocity differences in a low temperature helium gas turbulent jet (ref. 7). Abscissas are normalized to the root mean square of δv for each pdf, in order to stress the shape evolution. Vertical shifts of one decade are applied for clarity.

$G_{r_1 r_0}$. Calling $\bar{P}$ the distribution of $\ln|\delta v|$, we note that the above relation can be viewed as a convolution ($\ln|\frac{\delta v}{\alpha}| = \ln|\delta v| - \ln\alpha$):

$$\bar{P}_{r_1}(\ln|\delta v|) = G_{r_1 r_0} \otimes \bar{P}_{r_0}. \tag{14}$$

Introducing a new scale $r_2 < r_1$ we can write

$$\bar{P}_{r_2} = G_{r_2 r_1} \otimes \bar{P}_{r_1} = G_{r_2 r_1} \otimes G_{r_1 r_0} \otimes \bar{P}_{r_0}. \tag{15}$$

As this is a functional relation between $\bar{P}_{r_2}$ and $\bar{P}_{r_0}$ we can identify:

$$G_{r_2 r_0} = G_{r_2 r_1} \otimes G_{r_1 r_0}, \tag{16}$$

and generalize to an arbitrary cascade of scales:

$$G_{r_n r_0} = G_{r_n r_{n-1}} \otimes \ldots \otimes G_{r_1 r_0}. \tag{17}$$

Now, a natural question arises. Are we able to choose the r_i in such a way that all $G_{r_i r_{i-1}}$ are equal to the same distribution ?

In case of strict scale invariance, the answer is clearly yes: G_{ab} only can depend on the ratio a/b. We thus have to choose all the ratio r_i/r_{i-1} equal and for any n we can choose H such that [9, 10]:

$$G_{r,r_0} = H^{n\otimes}, \tag{18}$$

for any pair of scales r, r_0. Eq. 18 means that G_{rr_0} is an infinitely divisible distribution [11, 12]. Is it the case for real, finite Reynolds experiments ?

3.3. Experimental Check of Infinite Divisibility

Let us change our view point. Instead of looking for H with given n at fixed r, let us keep H fixed and change r for changing n:

$$G_{rL} = H^{n(r)\otimes}. \tag{19}$$

We thus introduce an arbitrary step in the cascade considering H as the propagator which generates all the G and n the number of steps. The quantity which can be extracted from the experiments and which allows to check the infinite divisibility is the p-th moment of multipliers α. From Eq. 13 we infer:

$$\begin{aligned} \langle|\delta v|^p\rangle_r &= \int |\delta v|^p P_r(\delta v) d\delta v \\ &= \langle\alpha^p\rangle_{rL}\langle|\delta v|^p\rangle_L. \end{aligned} \tag{20}$$

Here L must be seen as an arbitrary reference scale and the discussed properties will not depend on its actual choice. In the following we thus often drop the index L. The p-th moment of α is thus accessible from experiments by:

$$\langle\alpha^p\rangle_r = \frac{\langle|\delta v|^p\rangle_r}{\langle|\delta v|^p\rangle_L}. \tag{21}$$

On the other hand, this moment is the Laplace transform of G_{rL}. Using Eq. 19 [13]:

$$\langle\alpha^p\rangle_r = \int e^{px} G_{rL}(x) dx = \left[\int e^{py} H(y) dy\right]^{n(r)}. \tag{22}$$

Defining the function $F_r(p)$ which can be measured from experiments [14]:

$$\begin{aligned} F_r(p) &= \tfrac{1}{p} \ln \tfrac{\langle|\delta v|^p\rangle_r}{\langle|\delta v|^p\rangle_L} \\ &= n(r)\left(C_1 + C_2 \tfrac{p}{2!} + \ldots\right), \end{aligned} \tag{23}$$

where the cumulants C_i are defined by the development:

$$\int e^{py} H(y) dy = \exp\left(C_1 \frac{p}{1!} + C_2 \frac{p^2}{2!} + \ldots\right). \tag{24}$$

The shape of $F_r(p)$ is shown on Fig. 2a for a jet experiment at Reynolds number 2×10^4. It is nearly linear which corresponds to a gaussian shape for G_{rL} (for a gaussian shape, all cumulants, except the first two, are zero). However a small systematic downward curvature is observed.

From Eq. 23 the p and r dependences of $F_r(p)$ are separable [13]:

$$F_r(p) = n(r)\xi(p). \tag{25}$$

This property is equivalent to the infinite divisibility. It is the point we can experimentally check. For instance it would infer that:

$$\ln \frac{\langle|\delta v|^p\rangle_r}{\langle|\delta v|^p\rangle_L} = \frac{p}{q}\frac{\xi(p)}{\xi(q)} \ln \frac{\langle|\delta v|^q\rangle_r}{\langle|\delta v|^q\rangle_L}, \tag{26}$$

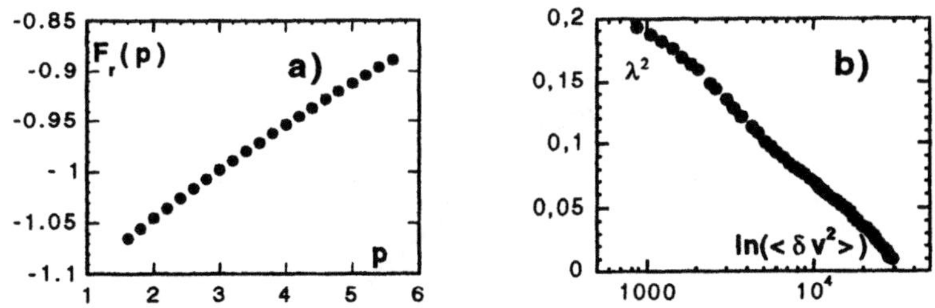

Fig. 2. — a) For calculating the above characteristic function $F_r(p)$ (see text) we choose L close to the integral scale ($\simeq 400\eta$) and $r = 23\eta$. The linearity would be equivalent to log-normality. Note the small downward curvature and the small range of the vertical scale. b) Up to a translation, $\ln\langle \delta v^2 \rangle$ is $2F_r(2)$ and λ^2 is twice the slope of $F_r(p)$ at $p = 0$. Their linear relation shows that all $F_r(p)$ are proportional, which is equivalent to the infinite divisibility of G_{rL} distributions.

showing that all moments of $|\delta v|$ behave as power law from each other, even in the absence of pure scaling of these moments with r which would correspond to $n(r)$ being linear with $\ln \frac{L}{r}$. This is the *Extended Self-Similarity* [15].

A more stringent test is to check the linearity of the second cumulant of G_{rL} ($\lambda^2 = \langle (\delta \ln \alpha)^2 \rangle$ which is twice the slope of $F_r(p)$) with $\ln\langle |\delta v|^2 \rangle_r$ Fig. 2b. The infinite divisibility of the G_{rL} distributions can be considered as reasonably verified on an experimental point of view [16].

3.4. Behavior of $n(r)$; First and Second Refinements

In discussing the $n(r)$ behavior we find the same steps than for the turbulent boundary layer velocity profile. Zeroth order is equivalent to Kolmogorov 41 theory, that is a constant shape for the δv distributions, which can be seen as a constant value for $n(r)$. With the first refinement, assuming complete scale invariance, we remark that the difference: $n(r_1) - n(r_2)$ must only depend on r_1/r_2, which infers a linear dependence of $n(r)$ with $\ln(\frac{L}{r})$:

$$n(r) = \mu \ln \frac{L}{r}. \tag{27}$$

It corresponds to the Kolmogorov-Obukhov 62 theory [6] and to all scale invariant theories as the multifractal one [17].

In the second refinement we take into account the limits of the scale invariant range, i.e. $L > r > \eta$. From the general argument presented in section 2, we expect a power law behavior

$$\frac{\partial n}{\partial \ln r} \propto r^{-\beta} \implies n(r) = n_0 \left(\left(\frac{L}{r} \right)^{\beta} - 1 \right), \tag{28}$$

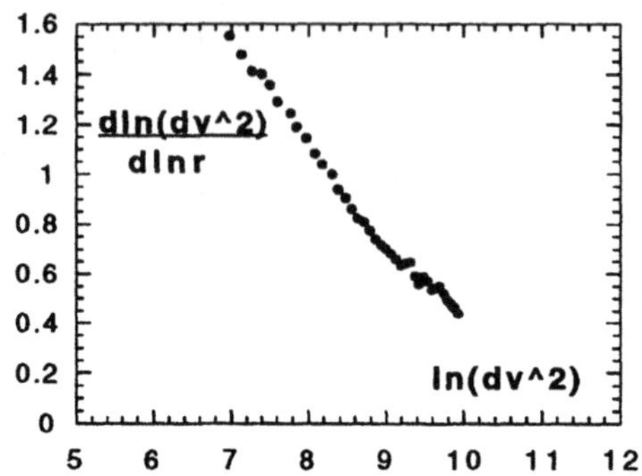

Fig. 3. — The absence of characteristic scale would imply a power law behaviour for $\langle \delta v^2 \rangle$ versus r. The above shown variations of $\partial \ln\langle \delta v^2 \rangle / \partial \ln r$ versus $\ln\langle \delta v^2 \rangle$ imply better agreement with Eq. 28.

or a linear dependence of $\frac{\partial n}{\partial \ln r}$ versus n, with the slope $-\beta$. Fig. 3 seems to support this conclusion. The measurements are also in good agreement with the asymptotic law

$$\beta = \beta_0 / \ln(\frac{L}{\eta}), \tag{29}$$

which was also predicted on various grounds [18, 19, 20].

4. MARKOVIAN APPROACHES AND CONCLUSION

I shall end this lecture by evocating a recent interesting work [21] in which the evolution of δv statistics is seen as a Markovian process along the scales. The authors first verify the validity of the Chapman-Kolmogorov relation:

$$P(\delta v_3|\delta v_1) = \int P(\delta v_3|\delta v_2) P(\delta v_2|\delta v_1) d\delta v_2, \tag{30}$$

where $P(u_1|u_2)$ is the probability density of u_1 for a given u_2. The δv_i:

$$\delta v_i = v(x + \frac{r_i}{2}) - v(x - \frac{r_i}{2}),$$

are longitudinal velocity differences at scale r_i ($r_1 > r_2 > r_3$) for the same point x. The authors find good experimental agreement between the two sides of Eq. 30. As they remark, this is not a proof, but a good indication for a markovian behaviour which is also an essential assumption behind Eq. 13.

Pushing further, they show that the probability density function of δv is verifying a Fokker-Planck equation where the time is replaced by the logarithm of the scale. In this paragraph, we simply want to signal the connection with what preceeds (which is natural for specialists of random walk).

Limiting ourselves to log-normal G_{rL}, its Laplace transform is:

$$\tilde{G}_{rL}(p) = \int e^{px} G_{rL}(x)dx = \exp\left\{n(r)(C_1 p + C_2 \tfrac{p^2}{2})\right\}$$
$$\tfrac{\partial \tilde{G}_{rL}}{\partial n} = (C_1 p + C_2 \tfrac{p^2}{2})\tilde{G}_{rL},$$

which is equivalent to the following equation in the direct $(\ln \alpha)$ space:

$$\frac{\partial G_{rL}}{\partial n} = C_1 \frac{\partial G_{rL}}{\partial \ln \alpha} + \frac{C_2}{2} \frac{\partial^2 G_{rL}}{(\partial \ln \alpha)^2}. \tag{31}$$

Using Eq. 13 and the fact that G_{rL} contains all the dependence in the scale:

$$\frac{\partial P_r(\delta v)}{\partial n} = \int \frac{\partial G_{rL}}{\partial n}(\ln \alpha)\frac{1}{\alpha} P_L(\frac{\delta v}{\alpha}) d \ln \alpha. \tag{32}$$

Substituting the expression Eq. 31 for $\partial G/\partial n$ and integrating in parts gives a Fokker-Planck equation for P_r:

$$\frac{\partial P_r}{\partial n} = -\frac{\partial}{\partial \delta v}(D_1 P_r) + \frac{\partial^2}{(\partial \delta v)^2}(D_2 P_r),$$

where:

$$D_1(\delta v) = (C_1 + \frac{C_2}{2})\delta v; \; D_2(\delta v) = \frac{C_2}{2}\delta v^2.$$

The results of reference [21] are indeed coherent with these dependencies within uncertainties, the main difference being an additional constant in D_2, which could be attributed to a noise.

In the two examples shown in this lecture, we have seen that scale invariance is not limited to the total absence of characteristic scale. It can be extended to the case where the extreme scale have marginal influence. The generic deviations to usual laws it implies can be seen as finite size effects of the scale invariant range [16, 19, 20].

They are essential in determining the proper extrapolation to the infinite range.

As for the intermittency problem, we have shown the equivalence of various approaches, giving a complete and cross checked description of the velocity difference statistics.

What next ? The passive scalar case gives an interesting extension where two cascade processes are present [22, 23, 24]. The parameters introduced in this approach give a powerful characterization of the observed phenomenon: for instance β gives an estimation of the Reynolds number, i.e. the scaling range, even when this scaling range is not wholly accessible. The approach can be extended to many other multiplicative processes.

The stricking point is just the generality of this approach. Out of the putative exact determination of the parameters, Navier-Stokes equation only enters through the scaling properties of Eq. 10 and the Reynolds number. No sign of the often invoked coherent structures. If they have some importance, these structures should reveal in some statistical measurements. An interesting goal would be to discover how.

References

[1] Tennekes H., Lumley J.L., *A First Course in Turbulence* (M.I.T. Press, Cambridge, 1994).
[2] Landau L.D., Lifshitz E.M., *Fluid Mechanics* (Pergamon Press, New York, 1959).
[3] Buckingham D., *Phys. Rev.* **14** (1914) 345.
[4] Barenblatt G.I., *Similarity, Self-similarity and Intermediate Asymptotics* (Consultant Bureau, Washington, 1979).
[5] Barenblatt G.I., Goldenfeld N., *Phys. Fluids* **7** (1995) 3078.
[6] Monin A.S., Yaglom A.M., *Statistical Fluid Mechanics, 1 and 2* (M.I.T. Press, Cambridge, 1975).
[7] Chabaud B., Naert A, Peinke J., Chillà F., Castaing B. and Hébral B., *Phys. Rev. Lett.* **73** (1994) 3227.
[8] Castaing B., *J. Phys. II (France)* **6** (1996) 105.
[9] Castaing B., Gagne Y., *Turbulence in Spatially Extended Systems*, R. Benzi, S. Ciliberto, C. Basdevant Eds. (Les Houches, 1993).
[10] Saito Y., *J. Phys. Soc. Jpn* **61** (1992) 403.
[11] Novikov E.A., *Phys. Rev. E Rapid Com.* **50** (1994) R3303; Feller W., *An introduction to Probability Theory and its Applications* (Wiley, New York, 1971, Vol. 2.
[12] She Z.S. and Waymire E.C., *Phys. Rev. Lett.* **74** (1995) 262.
[13] Castaing B., Dubrulle B., *J. Phys. II (France)* **5** (1995) 895.
[14] Naert A., Chabaud B., Hébral B., Castaing B., *Advances in Turbulence VI*, S. Gavrilakis *et al.* eds (Kluwer Academic Publishers, 1996) p. 251.
[15] Benzi B., Ciliberto S., Baudet C., Tripiccione R., Massaioli F. and Succi S., *Phys. Rev. E* **48** (1993) R29.
[16] Roux S., thesis University Aix-Marseille II (1996) unpublished; Arnéodo A., Muzy J.F., Roux S., *J. Physique II France* **7** (1997) 363.
[17] Parisi G. and Frisch U., *Turbulence and Predictability*, Varenna, Summer School, M. Ghol, R. Benzi, G. Parisi Eds (North-Holland, Amsterdam, 1985) p. 84.
[18] Castaing B., *J. Phys. France* **50** (1989) 147.
[19] Castaing B., Gagne Y., Hopfinger E., *Physica D* **46** (1990) 177.
[20] Dubrulle B., *J. Phys. (France) II* **6** (1996) 1825. See also the B. Dubrullee lesson in these proceedings.
[21] Friedrich R., Peinke J., *J. Phys. Rev. Lett.* **78** (1997) 863.
[22] Castaing B., *Physica D* **73** (1994) 31.
[23] He G., Dubrulle B., *J. Physique II France* **7** (1997) 793.
[24] Dubrulle B., Graner F., *J. Physique II France* **6** (1996) 797.

LECTURE 18

Discrete Scale Invariance

Didier Sornette

Laboratoire de Physique de la Matière Condensée,
CNRS and Université de Nice - Sophia Antipolis,
Parc Valrose, 06108 Nice, France
and
Department of Earth and Space Sciences
and Institute of Geophysics and Planetary Physics,
University of California, Los Angeles, California 90095-1567, U.S.A.

1. WHAT IS DISCRETE SCALE INVARIANCE (DSI)?

In a nutshell, continuous scale invariance means reproducing itself on different time or space scales. More precisely, an observable $\mathcal{O}$ which depends on a "control" parameter x is scale invariant under the arbitrary change $x \to \lambda x$ ([1]), if there is a number $\mu(\lambda)$ such that

$$\mathcal{O}(x) = \mu \mathcal{O}(\lambda x) \tag{1}$$

Eq.(1) defines a homogeneous function and is encountered in the theory of critical phenomena, in turbulence, etc. Its solution is simply a power law $\mathcal{O}(x) = Cx^{\alpha}$, with $\alpha = -\frac{\log \mu}{\log \lambda}$, which can be verified directly by insertion. Power laws are the hallmark of scale invariance (for a function of a single variable) as the ratio $\frac{\mathcal{O}(\lambda x)}{\mathcal{O}(x)} = \lambda^{\alpha}$ does not depend on x, i.e. the relative value of the observable at two different scales only depend on the *ratio* of the two scales. This is the fundamental property that associates power laws to scale

([1]) Here, we implicitely assume that a change of scale leads to a change of control parameter as in the renormalization group formalism. More directly, x can itself be a scale.

invariance, self-similarity ([2]) and criticality ([3]).

Discrete scale invariance (DSI) is a weaker kind of scale invariance according to which the system or the observable obeys scale invariance as defined above only for specific choices of λ (and therefore μ), which form in general an infinite but countable set of values $\lambda_1, \lambda_2, \ldots$ that can be written as $\lambda_n = \lambda^n$. λ is the fundamental scaling ratio. This property can be qualitatively seen to encode a *lacunarity* of the fractal structure [1]. In addition, there can be several prefered ratios corresponding to several log-periodicities that are superimposed. This can lead to a richer behavior such as log-quasi-periodicity.

Note that, since $x \to \lambda x$ and $\mathcal{O}(x) \to \mu \mathcal{O}(\lambda x)$ is equivalent to $y = \log x \to y + \log \lambda$ and $\log \mathcal{O}(y) \to \log \mathcal{O}(y + \log \lambda) + \log \mu$, a scale transformation is simply a translation of $\log x$ leading to a translation of $\mathcal{O}$. Continuous scale invariance is thus the same as continuous translational invariance expressed on the logarithms of the variables. DSI is then seen as the restriction of the continuous translational invariance to a *discrete* translational invariance: $\log \mathcal{O}$ is simply translated when translating y by a multiple of a fundamental "unit" size $\log \lambda$. Going from continuous scale invariance to DSI can thus be compared with (in logarithmic scales) going from the fluid state to the solid state in condensed matter physics! In other words, the symmetry group is no more the full set of translations but only those which are multiple of a fundamental discrete generator.

2. WHAT ARE THE SIGNATURES OF DSI?

We have seen that the hallmark of scale invariance is the existence of power laws (when the "observable" depends on a single parameter). The signature of DSI is the presence of power law behaviors with *complex* exponents α which manifests itself in data by log-periodic corrections to scaling. To see this, consider the triadic Cantor set. This fractal is built by a recursive process as follows. The first step consists in dividing the unit interval into three equal intervals of length $\frac{1}{3}$ and in deleting the central one. In the second step, the two remaining intervals of length $\frac{1}{3}$ are themselves divided into three equal intervals of length $\frac{1}{9}$ and their central intervals are deleted, thus keeping 4 intervals of length $\frac{1}{9}$. The process is then iterated ad infinitum. It is usually stated that this triadic Cantor set has the fractal dimension $D_0 = \frac{\log 2}{\log 3}$, as the number of intervals grows as 2^n while their length shrinks as 3^{-n} at the n-th iteration.

([2]) Self-similarity is the same notion as scale invariance but is expressed in the geometrical domain, with application to fractals.

([3]) Criticality refers to the state of a system which has scale invariant properties. The critical state is usually reached by tuning a control parameter as in liquid-gas and paramagnetic-ferromagnetic phase transitions. Many driven extended out-of-equilibrium systems seem also to exhibit a kind of dynamical criticality, that has been coined "self-organized criticality".

It is obvious to see that, by construction, this triadic Cantor set is geometrically identical to itself *only* under magnification or coarse-graining by factors $\lambda_p = 3^p$ which are arbitrary powers of 3. If you take another magnification factor, say 1.5, you will not be able to superimpose the magnified part on the initial Cantor set. We must thus conclude that the triadic Cantor set does not possess the property of continuous scale invariance but only that of DSI under the fundamental scaling ratio 3.

This can be quantified as follows [2]. Call $N_x(n)$ the number of intervals found at the n-th iteration of the construction. Call x the magnification factor. The original unit interval corresponds to magnification 1 by definition. Obviously, when the magnification increases by a factor 3, the number $N_x(n)$ increases by a factor 2 independent of the particular index of the iteration. The fractal dimension is defined as

$$D = \lim_{x \to \infty} \frac{\ln N_x(n)}{\ln x} = \lim_{x \to 0} \frac{\ln N_x(n)}{\ln x} = \frac{\ln 2}{\ln 3} \approx 0.63 \tag{2}$$

However nothing requires us to use magnification $x = 3^p$ only. If we increase the magnification continuously from say $x = 3^p$ to $x = 3^{p+1}$, the numbers of intervals in all classes jump by a factor of 2 at $x = 3^p$, but then remains unchanged until $x = 3^{p+1}$, at which point they jump again by an additional factor of 2. For $3^p < x < 3^{p+1}$, $N_x(n)$ does not change while x increases, so the measured fractal dimension $D(x) = \frac{\ln N_x(n)}{\ln x}$ decreases. The fractal dimension $D = 0.63$ is obtained only when x is a positive or negative power of three. For continuous values of x one has

$$N_x(n) = N_1(n) x^D P\left(\frac{\ln x}{\ln 3}\right), \tag{3}$$

where P is a function of period unity. Now, since P is a periodic function, we can expand it as a Fourier series

$$P\left(\frac{\ln x}{\ln 3}\right) = \sum_{n=-\infty}^{\infty} c_n \exp\left(2n\pi i \frac{\ln x}{\ln 3}\right) \tag{4}$$

Plugging this expansion back into (3), it appears that D is replaced by an infinity of complex values

$$D_n = D + ni\frac{2\pi}{\ln 3} \tag{5}$$

We now see, that a more proper characterization of the fractal is given by this set of *complex dimensions* which quantifies not only the asymptotic behaviour of the number of fragments at a given magnification, but also its modulations at intermediate magnifications. The imaginary part of the complex dimension is directly controlled by the prefered ratio 3 under which the triadic Cantor set is exactly self-similar.

If we keep only the first term in the Fourier series in (4) and insert in (3), we get

$$N_x(n) = N_1(n) x^D \left(1 + 2\frac{c_1}{c_0} \cos(2n\pi \frac{\ln x}{\ln 3})\right) \quad , \tag{6}$$

where we have used that $c_{-1} = c_1$ to ensure that $N_x(n)$ is real. Expression (6) shows that the imaginary part of the fractal dimension translates itself into a log-periodic modulation decorating the leading power law behavior.

For phase transitions, the existence of prefered scales appears in contradiction with the notion that a critical system exhibiting scale invariance has an infinite correlation length, hence only the microscopic ultraviolet cut-off and the large scale infra-red cut-off (for instance the size of the system) appear as distinguishable length scales. The mathematical existence of complex exponents and their related log-periodic corrections has been discussed quite early in renormalization group solutions for the statistical mechanics of critical phase transitions [3, 4, 5]. However, these solutions were rejected for translationally invariant systems, on the not totally correct (see [6] for a fuller discussion) basis that a period (even in a logarithmic scale) implies the existence of one or several characteristic scales, which is forbidden in these ergodic systems in the critical regime. Exponents are real if the renormalization group is a gradient flow [7], a rather common situation, but, as we shall see, not the only one by far. Examples when this is not the case can be found especially in random systems, out-of-equilibrium situations and irreversible growth problems.

3. SCENARIOS LEADING TO DSI

The list of mechanisms below is by no mean exhaustive and other unknown (at least to the author) mechanisms may exist. It is essential to notice that all the mechanisms involve the existence of a characteristic scale (an upper and/or lower cut-off) from which the DSI can develop and cascade.

3.1. Built-in geometrical hierarchy

The most obvious situation occurs when some physical system is put on a pre-existing discrete hierarchical system, such as the Cantor set described above, the Bethe lattice, or a fractal tree. Since the hierarchical system contains by construction a discrete hierarchy of scales occurring according to a geometrical series, one expects and does find complex exponents and their associated log-periodic structures. Examples are fractal dimensions of Cantor sets [8, 9, 10], percolation on fractal lattices[11], ultrametric structures [12], wave propagation in fractal systems [13], magnetic and resistive effects on a system of wires connected along the Sierpinski gasket[14], Ising and Potts models [15], fiber bundle rupture [16, 2], sandpiles [17], all put on hierarchical lattices.

3.2. Diffusion in anisotropic quenched random lattices

In this scenario, the DSI hierarchy is constructed dynamically in a random walk process due to intermittent encounters with slow regions [18]. Consider a random walker jumping from site to site. Bonds between sites are of two types: (i) directed ones on which the walker surely goes from his site to the next on his right ("diode" situation); (ii) two-way bonds characterized by a rate u (resp. v) to jump to the neighboring site on his right (resp. left). The fraction of two-way bonds is $1-p$ and the fraction of directed bonds is p. We construct a *frozen* random lattice by choosing a given configuration of randomly distributed mixtures of the two bond species according to their respective average concentration p and $1-p$. The exact solution of this problem has been given in [18] and shows very clearly nice log-periodic oscillations in the dependence of $\langle x^2 \rangle$ as a function of time.

We now present a simple scaling argument [2] which recovers the exact results. To do so, we assume $\frac{u}{v} << 1-p << 1$. We are thus in a situation where most bonds are directed and dilute clusters of two-way bonds are present. In addition, the two bonds are strongly impeding the progress of the walker as the forward rate to the right is much smaller than the backward rate to the left.

In this situation, the random walker progresses at constant velocity to the right as long as it encounters only diode bonds and gets partially trapped when it encounters two-way bonds. To see how DSI is spontaneously generated, we estimate the typical number of jumps τ_k needed for the random walker to pass k adjacent two-way bonds, i.e. a connected cluster of k two-way bonds. In the limit $\frac{u}{v} << 1$, $\tau_k \sim \left(\frac{v}{u}\right)^k$ ([4]). Using the fact that the average separation between k-tuples of two-way bonds is approximately $(1-p)^{-k}$, if $\frac{u}{v} << (1-p)$, the typical number of jumps needed for the random walker to go beyond the first k-tuple of consecutive two-way values is completely dominated by τ_k. One thus expects the rate as a function of the number of jumps to exhibit local minima at $\tau = \tau_k$: these are the jump-numbers time scales. The second part of the scaling argument consists in recognizing that the random walker has to cover a typical distance from the origin to encounter the first k-tuple of consecutive 'two-way' values, of the order of $(1-p)^{-k}$. In the limit where $\frac{u}{v} << (1-p) << 1$, we can thus write the approximate renormalization group equation

$$x(\tau) \approx (1-p)x(\lambda\tau) + g(\tau) \quad , \tag{7}$$

where we have set $\lambda \equiv \frac{v}{u}$ and $g(\tau)$ is some regular function taking into account various local effects that correct the main scaling. Because this renormalization group equation can be written only at scales which are powers of λ, we are back to the situation discussed in the first sections. We see in particular that $x \sim \tau^\alpha$ with $\alpha = \frac{\log(1-p)}{\log(u/v)} + i\frac{2\pi n}{\log(u/v)}$. It turns out that the range of parameters over which this holds is much larger than suggested by this intuitive argument.

([4]) and not $\tau_k \sim k\frac{v}{u}$. This is due to the fact that the walker goes back many times before escaping from the cluster of size k.

More precisely, as soon as $\frac{u}{v} < (1-p)$, one finds $x(\tau) \sim \tau^{\nu} P(\frac{\log \tau}{\log(v/u)})$, where $\nu = \frac{\log(1-p)}{\log(u/v)}$ and P is a periodic function of unit period. This prediction is remarkably well-confirmed by numerical simulations and recovers the exact calculation of [18].

3.3. Cascade of ultra-violet instabilities: growth processes and rupture

Numerical analysis of large diffusion-limited-aggregates have uncovered a *discrete* scaling invariance in their inner structure, which can be quantified by the introduction of a set of *complex* fractal dimensions [19]. The values of the complex fractal dimensions can be predicted *quantitatively* from a renormalization group approach using the quasi-periodic mapping found in [20].

A theoretical investigation of a simplified model of DLA, the needle problem, which is also of direct application to crack growths has been done to identify the underlying physical mechanism [21]. Based on perturbative analysis and some exact results from the hodograph method in the 2D conformal plane, we find that the two basic ingredients leading to DSI are the short wavelength Mullins-Sekerka instability ([5])and the strong screening of competing needles. The basic simple picture that emerges is that *non-linear* interactions between the unstable modes of the set of needles lead to a succession of period doubling, the next sub-harmonic catching up and eventually screening the leading unstable mode. The succession of these period doubling explains the existence of discrete scale invariance in these systems. We thus think that short wavelength instabilities of the Mullins-Sekerka type supplemented by a strong screening effect provides a general scenario for the *spontaneous* formation of log-periodic structures in growth processes. This scenario suggests, in addition, an explanation of the ubiquitous observation of a prefered scaling ratio close to 2.

Numerical simulations of the needle problem, using various growth rules (DLA, angle screening, η-model, crack approximation) on systems containing up to 5000 needles confirm clearly the proposed scenario. The density of needles as a function of the distance to the base presents clear evidence of log-periodic modulations of the leading algebraic decay. Geological data on joints competing in their growth in a similar fashion also exhibit approximately the log-periodic structure [21, 22].

A growing body of evidences indicate that the log-periodic oscillations appear in the time dependence of the energy release on the approach of impending rupture in laboratory experiments [23], numerical simulations [24, 25] and earthquakes [26, 27, 2, 28, 29, 30, 31]. It is thought that a similar type of inverse cascade, from progressive damage at small scale to coalescence and

([5]) The Mullins-Sekerka instability results from the "lightning rod effect" well-known in electrostatics, according to which large curvature concentrate the gradient of the potential field. Here, the growth velocity is proportional to the gradient of the concentration field and convex perturbations will grow faster.

unstable growth, is controlling the appearence of log-periodicity. The typical time-to-failure formula used in these works is

$$E \sim (t_r - t)^m \left[1 + C cos\left(2\pi \frac{\log(t_r - t)}{\log \lambda} + \Psi\right)\right] \quad , \tag{8}$$

where E is the energy released or some other variable quantifying the on-going damage, t_r is the time of rupture, m is a critical exponent, and Ψ is a phase in the cosine that can be get rid of by a change of time units. It has been found that the log-periodic oscillations enable a much better performance of the prediction due to "lock-in" of the fit on the oscillating structure. Physically, the oscillations contain information on t_r and thus help significantly in its determination. A link, which is not obvious in general, between log-periodicity in space and in time is given in [32].

This hierarchy of scales is reflecting, in the time-to-failure analysis given by (8), a hierarchy of local positive maxima of the "observable" E. They are given by

$$t_r - t_n = \tau \lambda^{\frac{n}{2}} \quad , \tag{9}$$

where $\tau \propto \exp(-\frac{\log\lambda}{2\pi} tan^{-1}\frac{2\pi}{m\log\lambda})$. The spacing between successive values of t_n approaches zero as n becomes large and t_n converges to t_r. This hierarchy of scales $t_r - t_n$ are not universal but depend upon the specific geometry and structure of the system. What is expected to be universal are the ratios $\frac{t_r - t_{n+1}}{t_r - t_n} = \lambda^{\frac{1}{2}}$. From three successive observed values of t_n, say t_n, t_{n+1} and t_{n+2}, we have

$$t_r = \frac{t_{n+1}^2 - t_{n+2}t_n}{2t_{n+1} - t_n - t_{n+2}} \quad . \tag{10}$$

This relation is invariant with respect to an arbitrary translation in time. In addition, the next time t_{n+3} is predicted from the first three ones by

$$t_{n+3} = \frac{t_{n+1}^2 + t_{n+2}^2 - t_n t_{n+2} - t_{n+1}t_{n+2}}{t_{n+1} - t_n} \quad . \tag{11}$$

These relations have been used in [16, 28, 29].

For prediction purposes, it is much more constrained and thus reliable to fit a part of an oscillating data than a monotonous power law especially in the presence of noise. This remark has been used and is vigorously investigated in several applied domains, such as earthquakes [26, 27, 2, 28], rupture prediction [23] and financial crashes [33, 34, 35].

3.4. Cascades of sub-harmonic bifurcations in the transition to chaos

An area where log-periodic structures should be expected is low-dimensional dynamical systems exhibiting the Feigenbaum sequence of subharmonic bifurcations to chaos [36]. Indeed, this route to chaos can be understood from an asymptotically exact *discrete* renormalization group with a universal scaling factor. The existence of this prefered scaling ratio should thus lead to complex exponents and log-periodic oscillations around the main scaling as the dynamics converges to the invariant Cantor set measure at criticality. It was noticed quite early [37] that the length of the stable period diverges as a power law with log-periodic modulations as the control parameter approaches the transition to chaos. Argoul et al. [38] have studied the transitions to chaos in the presence of an external periodic field and show figures exhibiting very clearly that the Lyapunov exponent has a power dependence with log-periodic oscillations as a function of the amplitude of the external field. Similarly, the topological entropy at the onset of pruning in generalized Baker transformations is a power law function of the distance to the onset of pruning with log-periodic oscillations [39]. The oscillations are due to the self-similar structure of the Cantor set forming the attractor [37].

3.5. Non-local geometry and non-unitary field theory

We have noticed [6] that, in contrast to common lore, complex critical exponents should generally be expected in the field theories that describe geometrical systems, because the latter are *non unitary*. In particular, evidence of complex exponents in lattice animals, a simple geometrical generalization of percolation has been presented [6]. The model of lattice animals is the most natural generalization of the percolation model, which itself is the prototype of disordered systems. The animal problem is the statistics of connected clusters on a lattice [40] and thus also describes unrooted branched polymers. Using transfer matrix techniques, the number of unrooted branched polymers of size N is found to exhibit a correction to the main scaling with a complex exponent.

3.6. Quenched disordered systems

Renormalization group analysis of a variety of spin problems with long-ranged quenched interactions have found complex critical exponents [41, 42, 43, 44, 45]. However, these authors have in general remained cautious as to the interpretation of their results. Indeed, one could argue that uncontrolled approximations (present in all these works) rather than physics could be the cause of the complex exponents.

On the other hand, with the qualitative understanding of the ultrametric structure of the energy landscape of spin glasses [46, 47] in the mean field approximation, these above results could be interpreted as the observable signature of the hierarchical structure of energy states in frozen random systems. It is generally believed that such topology occurs more generally in other complex systems with highly degenerate, locally stable states. However, much works remain to be done to clarify this problem as the additional presence of long-range interactions complicate the matter.

4. OTHER SYSTEMS

It has been pointed out that the morphology of the bronchial airway of the mammalian lung is roughly hierarchical leading to a log-log plot of the average diameter of a branch of a mammalian lung as a function of the branch order which exhibits a full S-oscillation (log-periodic) decorating an average linear (power law) dependence. This fractal structure has been argued to allow the organ to be more stable with respect to disturbance [48, 49] but the physical mechanism underlying its appearence is not understood.

Probably the first theoretical suggestion of the relevance of log-periodic oscillations to physics has been put forward by Novikov to describe the influence of intermittency in turbulent flows [50]. The idea is that the DSI could stem from the existence of a prefered ratio in the cascade from large eddies to small ones. The existence of log-periodic oscillations has not been convincingly demonstrated as they seem quite elusive and sensitive to the global geometry of the flow and recirculation [51, 52].

Dubrulle and Graner [53] (see F. Graner in these Proceedings) have noticed that the Titius-Bode law of planets distance to the sun $r_n = r_0 K^n$ with $K \approx 1.7$ can be seen a discrete scale invariant law. They show that it may result from any mechanism which presents a scale invariant instability coupled to a breakdown of angular symmetry.

5. OPEN PROBLEMS

We have kept the analysis of the renormalization group at the level of a linear expansion of the flow map. Taking into account the nonlinear structure of the flow map may lead to an infinite set of singularities accumulating at the main critical point or even to the whole axis of the control parameter being critical (in the chaos regime of the flow map [54]). The question of the relevance of these regimes to nature is still open (see [27] for an application to earthquakes). Generalization to several control parameters and multicritical points would be useful.

We have seen that DSI embodies the concept of lacunarity. The set of complex exponents or singularities has been called *multilacunarity* spectrum [9]. Generalizations with several different incommensurate log-frequencies would be of great interest and seem to appear for instance in the DLA problem [19]. Complex multifractal dimension spectrum in the presence of disorder can be handled using probabilistic versions of the renormalization group [55] and their development and impact are just emerging.

A very important practical question is the effect of disorder and the process of averaging. Disorder is expected to scramble the phases of the log-periodic oscillations and it is a priori not clear whether the log-periodic oscillations are robust. It turns out that small fluctuations around the log-periodic structure do not seem to spoil DSI, as found in many examples quoted above. For instance, in the needle DLA problem, intervals between needles were taken to fluctuate by a few percent without altering significantly the log-periodic structure of the growth process [21]. DLA clusters themselves are formed under a very strong annealed noise, corresponding to the random walk motion of the sticking particles. Nevertheless, clear evidence of log-periodicity in the mass as a function of radius has been found [19] giving confidence in their robustness with respect to disorder. This has been further substantiated by explicit calculations [6] showing generally that the complex exponents are robust.

However, we must stress that disorder introduces a sensitive dependence of the phase in the cos log formula : different realizations have a different phase and averaging will produce a "destructive interference" that makes vanish the log-periodic oscillations. Indeed, notwithstanding the fact that the phase corresponds to a characteristic scale, the finite size leads to fluctuations from sample to sample with fluctuating values of the characteristic scale. It is thus important to carry out analysis on each sample realization separately, *without* averaging. For instance, in the DLA case, 350 clusters of 10^6 particles have each been analyzed one by one and an histogram of the main log-frequencies has been constructed. Theoretically, preventing averaging is a problem as one is usually able only to calculate quantities averaged over the different realizations of the disorder. However, it must be stressed that the fact that log-periodic oscillations are mainly present before averaging tells us that they are *specific* fingerprints of the specific system one is looking at. This is obviously a desirable property for prediction purposes in engineering and other practical applications. An open problem however is to devise optimal tools to decipher the log-periodic structures in highly noisy data, as is usually the case due to limitation of sizes for instance. We note also that going to very large systems will in general progressively destroy the log-periodic structures as they are often correction to scaling ([6]).

From the point of view of non-unitary field theory, we should expect generically the existence of complex exponents. However, there is no known recipe to tell us what are the relevant observables that will have complex dimensions. Practically, this means that it is not a priori obvious what measure must be made to identify log-periodicity. In other words, it is important to look carefully at available data in all imaginable angles to extract the useful information. Of course, one must always been aware of statistical traps that noise can be taken for log-periodicity. Analysis must thus be carried out with synthetic tests used as tests for the null hypothesis, bootstrap approach, etc (see for instance [28]).

([6]) This is often due to the disorder which can be shown to renormalize the real part of the complex exponents so that they correspond to sub-leading correction to the main scaling behavior [6, 19].

Acknowledgments

A special mention should be given to A. Johansen and H. Saleur.

References

[1] B.B. Mandelbrot, The fractal geometry of nature (San Francisco: W.H. Freeman, 1982).
[2] H. Saleur, C.G. Sammis and D. Sornette, J.Geophys.Res. 101, 17661-17677 (1996).
[3] Jona-Lasinio G., Nuovo Cimento 26B, 99 (1975)
[4] Nauenberg M., J.Phys. A 8, 925 (1975).
[5] Niemeijer Th. and J.M.J. van Leeuwen, in Phase transitions and critical phenomena, Vol.6, C. Domb and M.S. Green, eds. (Academic Press, London, 1976), p.425.
[6] H. Saleur and D. Sornette, J.Phys.I France 6, 327-355 (1996).
[7] D. J. Wallace, R. K. P. Zia, Phys. Lett. 48A (1974) 325.
[8] D. Bessis, J.-D. Fournier, G. Servizi, G. Tourchetti and S. Vaienti, Phys. Rev. A 36, 920 (1987)
[9] J.-D. Fournier, G. Tourchetti and S. Vaienti, Phys.lett.A 140, 331 (1989)
[10] E. Orlandini, M.C. Tesi and G. Turchetti, Europhysics Lett. 21, 719-722 (1993).
[11] A. Kapitulnik, A. Aharony, G. Deutscher and D. Stauffer, J. Phys. A 16, L269-L274 (1983)
[12] M.O. Vlad and M.C. Mackey, Physica Scripta 50, 615-623 (1994)
[13] Bessis D., J.S. Geronimo and P. Moussa, J.Physique-Lett. 44, L977-L982 (1983)
[14] Douçot B., W. Wang, J. Chaussy, B. Pannetier and R. Rammal, Phys. Rev. Lett. 57, 1235-38 (1986).
[15] B. Derrida, L. De Seze and C. Itzykson, J. Stat. Phys. 33, 559 (1983); B. Derrida, C. Itzykson and J.M. Luck, Commun.Math.Phys. 94, 115-132 (1984)
[16] W.I. Newman, D.L. Turcotte and A.M. Gabrielov, Log-periodic behavior of a hierarchical failure model with applications to precursory seismic activation, Phys.Rev. E 52, 4827-4835 (1995)
[17] B. Kutjnak-Urbanc, Stefano Zapperi, S. Milošević and H. Eugene Stanley, Phys.Rev. E 54, 272-277 (1996).
[18] Bernasconi J. and W.R. Schneider, J.Phys.A15, L729-L734 (1983)
[19] D. Sornette, A. Johansen, A. Arnéodo, J.-F. Muzy and H. Saleur, Phys. Rev. Lett. 76, 251-254 (1996).
[20] A. Arneodo, F. Argoul, J. F. Muzy, M. Tabard, and E. Bacry, Fractals, 1 (1993) 629.
[21] Y. Huang, G. Ouillon, H. Saleur and D. Sornette, Physical Review E, in press (1997).

[22] G. Ouillon, D. Sornette, A. Genter and C. Castaing, J. Phys. I France 6, 1127-1139 (1996)
[23] J.-C. Anifrani, C. Le Floc'h, D. Sornette and B. Souillard, J.Phys.I France 5, 631-638 (1995).
[24] M. Sahimi and S. Arbabi, Phys. Rev. Lett. 77, 3689-3692 (1996).
[25] R.C. Ball and R. Blumenfeld, Phys. Rev. Lett. 65, 1784-1787 (1990); - Reply, Phys. Rev. Lett. 68, 2254-2254 (1992).
[26] D. Sornette and C.G. Sammis, J.Phys.I France 5, 607-619 (1995)
[27] H. Saleur, C.G. Sammis and D. Sornette, Nonlinear Processes in Geophysics 3, No. 2, 102-109 (1996)
[28] A. Johansen, D. Sornette, H. Wakita, U. Tsunogai, W.I. Newman and H. Saleur, Japan, J.Phys.I France 6, 1391-1402 (1996).
[29] D.J. Varnes and C.G. Bufe, Geophys. J. Int.124, 149-158 (1996)
[30] Bowman, D.D., and C.G. Sammis, EOS Trans. Am. Geophys. U., 77, page F486, 1996.
[31] Smith, S.W., and C.G. Sammis, EOS Trans. Am. Geophys. U., 77, page F480, 1996.
[32] Y. Huang, H. Saleur, C. G. Sammis, D. Sornette, Precursors, aftershocks, criticality and self-organized criticality, Phys. Rev. Lett. preprint cond-mat/9612065.
[33] D. Sornette, A. Johansen and J.-P. Bouchaud, J.Phys.I France 6, 167-175 (1996).
[34] Feigenbaum, J.A., and P.G.O. Freund, Int. J. Mod. Phys. 10, N27, 3737-3745 (1996).
[35] D. Sornette and A. Johansen
Physica A (in press) (cond-mat/9704127).
[36] M.J. Feigenbaum, J. Stat. Phys. 19, 25 (1978); 21, 669 (1979); P. Collet and J.P. Eckmann, Iterated maps of the interval and dynamical systems (Birkhauser, Boston, 1980).
[37] B. Derrida, J. Physique Colloques C5, 49 (1978); B. Derrida, A. Gervois and Y. Pomeau, Ann. Inst. H. Poincaré AXXIX, 305 (1978).; J. Phys. A 12, 269 (1979).
[38] F. Argoul, A. Arnéodo, P. Collet and A. Lesne, Europhysics Letters 3, 643-651 (1987).
[39] J. Vollmer and W. Breymann, Europhysics Letters 27, 23-28 (1994).
[40] D. Stauffer and A. Aharony, "Introduction to percolation theory", second edition, Taylor and Francis (1992).
[41] Aharony A., Critical properties of random and constrained dipolar magnets, Phys. Rev. B 12, 1049-1056 (1975)
[42] Chen J.-H.and T.C. Lubensky, Mean field and ϵ-expansion study of spin glasses, Phys. Rev. B 16, 2106-2114 (1977)
[43] D.E. Khmelnitskii, Impurity effect on the phase transition at T=0 in magnets. Critical oscillations in corrections to the scaling laws, Phys. Lett. A67,59-60 (1978).
[44] Boyanovsky D. and J.L. Cardy, Critical behavior of m-component magnets with correlated impurities, Phys. Rev. B 26, 154-170 (1982)

[45] A. Weinrib and B.I. Halperin, Critical phenomena in systems with long-range-correlated quenched disorder, Phys. Rev. B 27, 413-427 (1983)
[46] M. Mézard, Parisi G. and Virasoro M.A., "Spin glass theory and beyond", World Scientific, Singapore (1987).
[47] R. Rammal, G. Toulouse and Virasoro M.A., Rev.Mod.Phys.58, 765 (1986).
[48] Schlesinger M.F. and B.J. West, Phys.Rev.Lett. 67, 2106-08 (1991)
[49] B.J. West and W. Deering, Phys. Rep. 246, 1-100 (1994)
[50] E.A. Novikov, Dokl.Akad.Nauk SSSR 168/6, 1279 (1966); E.A.Novikov, Phys.Fluids A 2, 814-820 (1990).
[51] Anselmet F., Y. Gagne, E.J. Hopfinger and R.A. Antonia, J. Fluid Mech.140, 63 (1984)
[52] U. Frisch, Turbulence, the legacy of A.N. Kolmogorov (Cambridge University Press, 1995), p.130-131.
[53] F. Graner and B. Dubrulle, Astron. Astroph. 282, 262-268 (1994); 282, 269-276 (1994).
[54] B. Derrida, J. P. Eckmann, A. Erzan, J. Phys. A16 893 (1983)
[55] K.J. Falconer, J. Theoret. Prob. 7, 681 (1994).

LECTURE 19

Scale Relativity

Laurent Nottale

DAEC, Observatoire de Paris-Meudon, CNRS and Université Paris VII, 92195 Meudon cedex, France

1. INTRODUCTION

In his conference given at the St Louis congress in 1904, Henri Poincaré wrote [1]: "Comme les astres proprement dits, [les atomes] s'attirent ou se repoussent et cette attraction (...) ne dépend que de la distance. La loi suivant laquelle cette force varie en fonction de la distance n'est peut-être pas la loi de Newton, mais c'est une loi analogue; au lieu de l'exposant –2, nous avons probablement un exposant différent, et c'est de ce changement d'exposant que sort toute la diversité des phénomènes physiques. (...) Telle est la conception primitive dans toute sa pureté. (...) Néanmoins, il est arrivé un jour où la conception des forces centrales n'a plus paru suffisante. Que fit-on alors ? On renonça à pénétrer dans le détail dans la structure de l'Univers, (...) et l'on se contenta de prendre pour guides certains principes généraux: (...) [conservation de l'énergie et de la masse, principe de Carnot, égalité de l'action et de la réaction, principe de relativité et principe de moindre action]. L'application de ces cinq ou six principes généraux aux différents phénomènes physiques suffit pour nous en apprendre ce que nous pouvons raisonnablement espérer en connaître[1]."

---

[1] "As for stellar bodies, atoms are attracted or repelled to each other and this attraction (...) depends only on the distance. The law, according to which this force varies as a function of distance, is perhaps different from Newton's law, but is analogous to it; instead of the exponent –2, we probably have a different exponent, and it is from the change of the exponent that emerges all the diversity of physical phenomena. (...) Such is the primordial conception in all its purity. (...) Nevertheless, a day came when the concept of central forces did not appear sufficient anymore. What was done then? The

This quotation from Poincaré anticipates in a remarkable way some of the questions that are posed to us nowadays, nearly one century later, concerning scale invariance (and beyond), even though we are now concerned with understanding scale laws in more complex systems than those considered at that time. Indeed the simplest scale invariant equations have power-law solutions, and during some time a large part of the research on this subject was devoted to the calculation of the numerical values of the exponents. But we now know that we must go beyond such an approach, even if it has been a necessary step, and look for the fundamental principles that may allow us to really understand "where the physics is".

This is precisely the aim of the theory of scale relativity. As we shall see, we attempt to construct and generalize scale laws by using first principles, namely the principles of relativity and of covariance once extended to scale transformations [2,3], and the stationary action principle. Namely, we generalize Einstein's formulation of the principle of relativity, by requiring *that the laws of nature be valid in any reference system, whatever its state*. Up to now, this principle has been applied to changes of state of the coordinate system that concerned the origin, the axes orientation, and the motion (measured in terms of velocity, acceleration, ...).

In scale relativity, we assume that the space-time resolutions are not only a characteristic of the measurement apparatus, but acquire a universal status. They are considered as essential, intrinsic variables, inherent to the physical description. We define them as characterizing the "state of scale" of the reference system, in the same way as the velocity characterizes its state of motion, and we work in an extended phase space including coordinates, velocities and resolutions. The theory of scale relativity consists of applying the principle of relativity to such a scale-state. Its mathematical translation is *scale-covariance* [2,3], requiring that the equations of physics keep their simplest form under resolution transformations (dilations and contractions). The use of scale covariance for constraining scale laws has also more recently been proposed by Pocheau [4] in the context of turbulence.

The domains of application of this theory are typically the asymptotic domains of physics, small length-scales and small time-scales Δx, $\Delta t \rightarrow 0$ (microphysics), but also large length and time-scales Δx, $\Delta t \rightarrow \infty$ (cosmology).

Initially, the theory of scale relativity was mainly an attempt at refounding quantum mechanics on first principles [2,3]. We have demonstrated that the main axioms of quantum mechanics can be recovered as consequences of the principle of scale relativity, and that the quantum behavior can be recovered as the various manifestations of the non-differentiability and fractality of space-time at small scales [3,5,6]. Moreover, the theory allows to generalize standard quantum mechanics. Indeed, we have shown that the usual laws of scale (power law, self-similar, constant

---

goal of unravelling the details of the structure of the Universe was abandonned and replaced by certain general principles acting as guiding lines: (...) [conservation of energy and mass, Carnot's principle, equality between action and reaction, relativity principle and least action principle]. The application of these five or six general principles to the different physical phenomena is enough to teach us what we can reasonably hope to know."

fractal dimension) have the status of "Galilean" scale laws, while a full implementation of the principle of scale relativity suggests that they could be medium scale approximations of more general laws which take a Lorentzian form [7]. In such a "special scale relativity" theory, the Planck length- and time-scale becomes a minimal, impassable scale, invariant under dilations and contractions, which replaces the zero point (since owning all its physical properties) and plays for scales the same role as played by the velocity of light for motion. A similar proposal can be made at large scales: the scale of the cosmological constant can be reinterpreted as a maximal, impassable length-scale, invariant under dilations [3,6,9]. In this new framework, several still unsolved problems of fundamental physics may find simple and natural solutions: new light is brought on the nature of the Grand Unification scale, on the origin of the electron scale and of the electroweak scale, on the scale-hierarchy problem, and on the values of coupling constants [3,6,7]; moreover, one can construct a scale-relativistic interpretation of gauge invariance that yields new insights on the nature of the electric charge, and allows to predict new mass-charge relations for elementary particles [6,8].

2. TOWARD A NONDIFFERENTIABLE SPACE-TIME

Since more than three century, physics relies on the implicit assumption that space-time coordinates are a priori differentiable. However, it was demonstrated by Feynman [10] that the typical paths of quantum mechanical particles are continuous but non-differentiable. Now, one of the most powerful avenue for reaching a genuine understanding of the laws of nature has been to construct them, not from setting additional hypotheses, but on the contrary by attempting to give up some of them, i.e., by going to increased generality. An example of such an approach is Einstein's explanation of the nature of gravitation as the various manifestations of the Riemannian, i.e. non-flat geometry of space-time.

However, in the light of the above remark, Einstein's principle of relativity is not yet fully general, since it applies to transformations that are continuous and at least two times differentiable. The aim of the theory of scale-relativity is to look for the laws and structures that would be the manifestations of still more general transformations, namely, continuous ones (that can be differentiable or not). In such a construction the standard theory will be recovered as a special case, since differentiable spaces are a particular subset of the set of all continuous spaces.

In that quest, the first step consists of realizing that a continuous but non-differentiable space-time is necessarily fractal. More precisely, one can demonstrate [3,5,11] that a continuous but nondifferentiable function is explicitly resolution-dependent, and that its length $\mathcal{L}$ tends to infinity when the resolution interval tends to zero, i.e. $\mathcal{L} = \mathcal{L}(\varepsilon)_{\varepsilon \to 0} \to \infty$. This theorem naturally leads to the proposal that the concept of *fractal space-time*, [12,13,3] is the geometric tool adapted to the research of such a new description based on non-differentiability. This leads to introduce new intrinsic scale variables in the very definition of physical quantities (among which the coordinates themselves), but also to construct the differential equations (in the "scale space") that would describe this new dependence.

We shall see that, since the new scale equations are themselves constrained by the principle of relativity, the new concepts fit well established structures. Namely, the so-called symplectic structure of most physical theories, i.e., the Poisson bracket / Euler-Lagrange / Hamilton formulation, can be also used to construct scale laws. Under such a viewpoint, scale invariance is recovered as corresponding to the "free" case (the equivalent of what inertia is for motion laws).

3. SCALE INVARIANCE AND SCALE RELATIVITY

Scaling laws have been discovered and studied at length in several domains of science. A power-law scale dependence is encountered in a lot of natural systems. It can be described geometrically in terms of fractals [14,15], and algebrically in terms of the renormalization group [16,17]. As we shall see now, such simple scale-invariant laws can be identified with a "Galilean" version of scale-relativistic laws.

In most present use and applications of fractals, the fractal dimension D is defined, following Mandelbrot, from the variation with resolution ε of the main fractal variable (e.g., the length $\mathcal{L}$ of a fractal curve –which plays here the role of a fractal curvilinear coordinate–, the area of a fractal surface, etc...). Namely, the scale dimension $\delta = D - D_T$ is defined as:

$$\delta = \frac{d \ln \mathcal{L}}{d \ln(\lambda/\varepsilon)} \ , \tag{1}$$

where D_T is the topological dimension ($D > D_T$ most of the time). When δ is constant, we obtain a *power-law* resolution dependence $\mathcal{L} = \mathcal{L}_0 \, (\lambda / \varepsilon)^{\delta}$. Here the scale λ appears for dimensional reasons, but it remains arbitrary because there is a strict scale invariance. We shall see in Sec. 3.2 that it takes its full meaning once a scale symmetry breaking is introduced. The Galilean structure of the group of scale transformations that corresponds to this law can be verified in a straightforward manner from the fact that it transforms in a scale transformation $\varepsilon \to \varepsilon'$ as

$$\ln \frac{\mathcal{L}(\varepsilon')}{\mathcal{L}_0} = \ln \frac{\mathcal{L}(\varepsilon)}{\mathcal{L}_0} + \delta(\varepsilon) \, \ln \frac{\varepsilon}{\varepsilon'} \quad ; \quad \delta(\varepsilon') = \delta(\varepsilon) \tag{2}$$

The relativity of scales is now apparent from the fact that λ has disappeared from this transformation. It has exactly the structure of the Galileo group, as confirmed by the law of composition of dilations $\varepsilon \to \varepsilon' \to \varepsilon''$, which writes

$$\ln\rho'' = \ln\rho + \ln\rho' \ , \tag{3}$$

where $\rho = \varepsilon'/\varepsilon$, $\rho' = \varepsilon''/\varepsilon'$ and $\rho'' = \varepsilon''/\varepsilon$ are the three dilations

3.1 Least action principle for scale laws

We are then naturally led, in the scale-relativistic approach, to reverse the definition and the meaning of variables. The scale dimension δ becomes, in general, an essential, fundamental *variable*, that remains constant only in very particular situations (namely, in the case of scale invariance, that corresponds to "scale-freedom"). It plays for scale laws the same role as played by time for motion laws (at a given instant). The log-resolution is a measure of scale in the same way as velocity is a measure of motion, and it is now defined as a derived quantity in terms of the fractal coordinate $\ln\mathcal{L}$ and of the scale dimension δ:

$$\mathbb{V} = \ln(\lambda/\varepsilon) = \frac{d\ln\mathcal{L}}{d\delta} . \tag{4}$$

Our identification of the standard fractal behavior as a Galilean scale-law can now be fully justified. We assume that, as in the case of motion laws, scale laws can be constructed from a Lagrangian approach. A scale Lagrange function $\mathbb{L}(\ln\mathcal{L}, \mathbb{V}, \delta)$ is introduced, from which a scale-action is constructed:

$$\mathbb{S} = \int_{\delta_1}^{\delta_2} \mathbb{L}(\ln\mathcal{L}, \mathbb{V}, \delta)\, d\delta . \tag{5}$$

The action principle, applied on this action, yields a scale-Euler-Lagrange equation:

$$\frac{d}{d\delta}\frac{\partial\mathbb{L}}{\partial\mathbb{V}} = \frac{\partial\mathbb{L}}{\partial\ln\mathcal{L}} . \tag{6}$$

The simplest possible form for the Lagrange function is the equivalent for scales of what inertia is for motion, i.e., $\mathbb{L} \propto \mathbb{V}^2$ and $\partial\mathbb{L} / \partial\ln\mathcal{L} = 0$ (no scale "force", see Sec. 4.1). The Lagrange equation writes in this case:

$$\frac{d\mathbb{V}}{d\delta} = 0 \Rightarrow \mathbb{V} = cst. \tag{7}$$

The constancy of $\mathbb{V} = \ln(\lambda/\varepsilon)$ means here that it is independent of the scale-time δ. Then Eq. 4 can be integrated in terms of the usual power law behavior, $\mathcal{L} = \mathcal{L}_0 (\lambda / \varepsilon)^{\delta}$. This reversed viewpoint has several advantages:

(i) The scale dimension takes its actual status of "scale-time", and the logarithm of resolution $\mathbb{V}$ its status of "scale-velocity", $\mathbb{V} = d\ln\mathcal{L} / d\delta$. This is in accordance with its scale-relativistic definition, since it characterizes the state of scale of the reference system, in the same way as the velocity $v = dx/dt$ characterizes its state of motion.

(ii) This leaves open the possibility of generalizing our formalism to the case of four independent space-time resolutions, $\mathbb{V}^{\mu} = \ln(\lambda^{\mu} / \varepsilon^{\mu}) = d\ln\mathcal{L}^{\mu} / d\delta$.

(iii) Scale laws more general than the simplest self-similar ones can be derived from more general scale-Lagrangians (see Sec. 4 below).

3.2 Scale-symmetry breaking

An important point concerning the scale symmetry, which is highly relevant to the present study, is that, as is well-known from the observed scale-independence of physics at our own scales, the scale dependence is a spontaneously broken symmetry [2,3,7]. Let us recall the simple theoretical argument that leads to this result.

In the general framework of a continuous space-time (not necessarily differentiable), we expect a generalized curvilinear coordinate to be explicitly resolution-dependent, i.e. $\mathcal{L} = \mathcal{L}(\varepsilon)$. We assume that this new scale dependence is itself solution of a differential equation *in the scale space*. The simplest scale differential equation one can write is a renormalization-group-like, first order equation in which the scale variation of $\mathcal{L}$ depends on $\mathcal{L}$ only, i.e., $\mathrm{d}\mathcal{L}/\mathrm{d}\ln\varepsilon = \beta(\mathcal{L})$. The function $\beta(\mathcal{L})$ is a priori unknown but, still taking the simplest case, we consider a perturbative approach and we take its Taylor expansion. We obtain the equation:

$$\frac{\mathrm{d}\mathcal{L}}{\mathrm{d}\ln\varepsilon} = a + b\,\mathcal{L} + c\,\mathcal{L}^2 + \ldots \tag{8}$$

Disregarding for the moment the quadratic term, this equation is solved in terms of a standard power law of power $\delta = -b$, broken at some scale λ (integration constant):

$$\mathcal{L} = \mathcal{L}_0\,[\,1 + \left(\frac{\lambda}{\varepsilon}\right)^{\delta}\,]\,. \tag{9}$$

Depending on the sign of δ, this solution represents either a small-scale fractal behavior (in which the scale variable is a resolution), broken at larger scales, or a large-scale fractal behavior (in which the scale variable ε would now represent a changing window for a fixed resolution λ), broken at smaller scales.

The symmetry between the microscopic and the macroscopic cases can be seen from the properties of Eq. 8. Let us indeed transform the two variables $\mathcal{L}$ and ε by *inversion*, i.e. $\mathcal{L} \rightarrow \mathcal{L}' = 1/\mathcal{L}$ and $\varepsilon \rightarrow \varepsilon' = 1/\varepsilon$, we find that Eq. 8 becomes:

$$\frac{d\mathcal{L}'}{d\ln\varepsilon'} = c + b\,\mathcal{L}' + a\,\mathcal{L}'^2 + \ldots \tag{10}$$

This is exactly the same equation up to the exchange of the constants a and c. In other words, Eq. 8 is covariant (i.e. form invariant) under the inversion transformation, which transforms the small scales into the large ones and reciprocally, but also the upper symmetry breaking scale into a lower one. Hence the inversion symmetry, which is clearly not achieved in nature at the level of the observed structures, may nevertheless be an exact symmetry at the level of the fundamental laws. This is confirmed by directly looking at the solutions of Eq. 8 while keeping now the quadratic term. We get a scaling behavior which is broken toward both the small and large scales, in accordance with most real fractal systems:

$$\mathcal{L} = \mathcal{L}_0\;[\,1 + (\lambda_1/\varepsilon)^{\delta}\,]\,/\,[\,1 + (\lambda_2/\varepsilon)^{\delta}\,]\;. \tag{11}$$

There is another way to obtain the scale symmetry breaking, that allows to elucidate its origin. Consider a pure scale-invariant law for a "fractal coordinate" $\mathcal{L} = \mathcal{L}_0\,(\lambda/\varepsilon)^{\delta}$. We expect it to be also translation-invariant, i.e. $\mathcal{L}$ can be replaced by $\mathcal{L} - \mathcal{L}_1$. We then recover the broken law

$$\mathcal{L} = \mathcal{L}_0\,[1 + (\lambda'/\varepsilon)^{\delta}\,]\,, \tag{12}$$

where $\lambda' = \lambda\,(\mathcal{L}_0/\mathcal{L}_1)^{1/\delta}$. This law becomes scale-independent for $\varepsilon >> \lambda'$ when $\delta > 0$, and for $\varepsilon << \lambda'$ when $\delta < 0$. This means that the scale symmetry breaking results from the effect on the scale symmetry of the translation symmetry.

3.3 Special scale-relativity

It is well known that the Galileo group of motion is only a degeneration of the more general Lorentz group. The same is true for scale laws. Indeed, if one looks for the general linear scale laws that come under the principle of scale relativity, one finds that they have the structure of the Lorentz group [32,7]. Namely, it has been shown by Levy-Leblond [32] that the Lorentz group can be obtained from requiring linearity, group law and reflection invariance (the 4-dimensional Euclidean solution is easily excluded since unphysical). More recently, we have demonstrated [7] that in two dimensions, only three axioms were needed (linearity, internal composition law and reflection invariance). Therefore, we have suggested to replace the Galilean laws of dilation $\ln\rho'' = \ln\rho + \ln\rho'$ by the more general Lorentzian law:

$$\ln\rho'' = \frac{\ln\rho + \ln\rho'}{1 + \ln\rho\,\ln\rho' / \mathbb{C}^2} \quad . \tag{13}$$

This expression is yet incomplete, since under this form the scale relativity symmetry remains unbroken. Such a law corresponds, at the present epoch, only to the null mass limit. It is expected to apply in a universal way during the very first instants of the universe.

In Eq. 13, there appears a universal, purely numerical constant $\mathbb{C} = \ln\mathbb{K}$, where $\mathbb{K}$ plays the role of a maximal possible *dilation*. However, the effect of the spontaneous scale symmetry breaking which arises at some scale λ_0 is to yield a new law in which the invariant is no longer a dilation $\mathbb{K}$, but becomes a *length-time scale* $\mathbb{A}$. In other words, there appears in the theory a fundamental scale that plays the role of an unpassable resolution, invariant under dilations [7]. Such a scale of length and time is an horizon for scale laws, in a way similar to the status of the velocity of light for motion laws. The new law of composition of dilations and the scale-dimension now write respectively (in the scale-dependent domains, i.e. only below the transition scale in microphysics and beyond it in the cosmological case):

$$\ln\frac{\varepsilon'}{\lambda_0} = \frac{\ln(\varepsilon/\lambda_0) + \ln\rho}{1 + \dfrac{\ln\rho\,\ln(\varepsilon/\lambda_0)}{\ln^2(\mathbb{A}/\lambda_0)}} \quad ; \quad \delta(\varepsilon) = \frac{1}{\sqrt{1 - \ln^2(\lambda_0/\varepsilon)\,/\,\ln^2(\lambda_0/\mathbb{A})}} \quad . \tag{14}$$

A fractal curvilinear coordinate becomes now scale-dependent in a covariant way, namely $\mathcal{L} = \mathcal{L}_0\,[1 + (\lambda_0/\varepsilon)^{\delta(\varepsilon)}]$. One of the main new feature of special scale relativity with respect to the previous scale-invariant approach is that the scale-dimension δ, which was constant, is now explicitly varying with scale, and it even diverges when resolution tends to the new invariant scales. In the microphysical domain, the invariant length-scale is naturally identified with the Planck scale, $\mathbb{A}_{\mathbb{P}} = (\hbar G/c^3)^{1/2}$, that now becomes impassable and plays the physical role that was previously devoted to the zero point [3,7]. The same is true in the cosmological domain, with once again an inversion of the scale laws. We have identified the invariant maximal scale with the scale of the cosmological constant, $\mathbb{L} = \Lambda^{-1/2}$. The consequences of this new interpretation have been studied in [3,6,9]. We have found from various estimates of the cosmological constant that the value of $\mathbb{K} = \mathbb{L} / \Lambda_{\mathbb{P}}$ is about 5×10^{60} [3,6]. Moreover, from the identification of the cosmological constant with the gravitational self-energy density of vacuum fluctuations [33], we have been able to propose an explanation for the Eddington-Dirac large number coincidence, which can be now written as $\mathbb{K} \approx (m_{\mathbb{P}} / m)^3$, where m is a typical elementary particle mass [6,9].

Special scale-relativistic laws have also recently been considered by Dubrulle [18] and Dubrulle and Graner [19] for the description of turbulence, with a different interpretation of the variables and a different choice for the required symmetries. Namely, since we are interested here in the internal fractal structure of space-time trajectories, the fractal dimension is unique while there are four space-time resolutions, so that we are naturally led to consider the couple $(\ln\mathcal{L}, \delta)$ as the variables to transform and the resolution $\ln\varepsilon$ as the parameter of the transformation. On the contrary, Dubrulle and Graner are interested in the set of exponents that characterize the various moments of a statistical distribution in turbulence, so that the role of δ and $\ln\varepsilon$ is reversed in their approach. This reversal also led them not to use the axiom of reflection invariance, but to keep that of the existence of a symmetric element. In this case, the solution is a generalization of the Lorentz group defined by two constants instead of one [19].

It is also noticeable that recent developments in string theories have reached conclusions that are extraordinarily similar to those of scale relativity. One finds that there is a smallest circle in string theory (whose radius is about the Planck length), and that strings are characterized by duality symmetries. Therefore it has recently been argued by Castro [20] that scale relativity is the right framework in which the newly discovered string structures will take their full physical meaning.

4. GENERALIZED SCALE LAWS

4.1 Scale "dynamics"

The whole of our previous discussion indicates to us that the scale invariant behavior corresponds to freedom in the framework of a scale physics. However, in the same way as there exists forces in nature that imply departure from inertial, rectilinear uniform motion, we expect most natural fractal systems to also present distorsions in their scale behavior respectively to pure scale invariance. This means taking non-linearity in scale into account. Such distorsions may be, as a first step, attributed to the effect of a scale "dynamics", i.e. to a "scale-field". (Caution: this is only an *analog* of "dynamics" which acts on the scale axis, on the internal structures *of a given point* at this level of description, not in space-time. See Sec. 4.3 below for first hints about the effects of coupling with space-time displacements). In this case the Lagrange scale-equation takes the form of Newton's equation of dynamics:

$$\mathbb{F} = \mu \frac{d^2 \ln \mathcal{L}}{d\delta^2} \ , \tag{15}$$

where μ is a "scale-mass", which measures how the system resists to the "scale-force", and where $\Gamma = d^2 \ln \mathcal{L} / d\delta^2 = d \ln(\lambda/\varepsilon) / d\delta$ is the 'scale-acceleration'.

4.1.1 Constant scale-force

Let us first consider the case of a constant scale-force. The potential is $\varphi = \mathbb{F} \ln \mathcal{L}$, and Eq. 15 writes

$$\frac{d^2 \ln \mathcal{L}}{d\delta^2} = \mathbb{G} \ , \tag{16}$$

where $\mathbb{G} = \mathbb{F} / \mu$ = cst. It is easily integrated in terms of a parabolic solution:

$$\mathbb{V} = \mathbb{V}_0 + \mathbb{G} \ \delta \quad ; \quad \ln \mathcal{L} = \ln \mathcal{L}_0 + \mathbb{V}_0 \, \delta + \frac{1}{2} \mathbb{G} \ \delta^2 \ . \tag{17}$$

However the physical meaning of this result is not clear under this form. This is due to the fact that, while in the case of motion laws we search for the evolution of the system with time, in the case of scale laws we search for the dependence of the system on resolution, which is the directly measured observable. We find, after redefinition of the integration constants:

$$\delta = \delta_0 + \frac{1}{\mathbb{G}} \ln \left(\frac{\lambda}{\varepsilon}\right) \quad ; \quad \ln \frac{\mathcal{L}}{\mathcal{L}_0} = \frac{1}{2\mathbb{G}} \ln^2 \left(\frac{\lambda}{\varepsilon}\right) \ . \tag{18}$$

The scale dimension δ becomes a linear function of resolution (the same being then true of the fractal dimension $1+\delta$), and the ($\log \mathcal{L}$, $\log \varepsilon$) relation is now parabolic rather than linear as in the standard power-law case. There are several physical situations where, after careful examination of the data, the power-law models were clearly rejected since no constant slope could be defined in the ($\log \mathcal{L}$, $\log \varepsilon$) plane. In the several cases where a clear curvature appears in this plane (e.g., turbulence, sand piles, ...), the physics could come under such a "scale-dynamical" description. In these cases it might be of the highest interest to identify and study the scale-force responsible for the scale distorsion (i.e., for the deviation to standard scaling).

4.1.2 Harmonic oscillator

Another interesting case of scale-potential is that of a repulsive harmonic oscillator, $\varphi = -(\ln \mathcal{L} / \delta)^2 / 2$. It is solved as

$$ln \frac{\mathcal{L}}{\mathcal{L}_0} = \delta \sqrt{\ln^2 \left(\frac{\lambda}{\varepsilon}\right) - \delta^{-2}} \ . \tag{19}$$

For $\varepsilon << \lambda$ it gives the standard Galilean case $\mathcal{L} = \mathcal{L}_0 \, (\lambda / \varepsilon)^{\delta}$, but its large-scale behavior is particularly interesting, since it does not permit the existence of resolutions larger than a scale $\lambda_{max} = \lambda \ e^{1/\delta}$. Since the gauge symmetry group of

QCD, SU(3), is the dynamical symmetry group of the 3-dimensional isotropic harmonic oscillator, and since gauge invariance can be interpreted in scale relativity as scale invariance on space-time resolutions (see Sec. 4.3), such a behavior could provide a model of confinement in QCD [L.N., in preparation].

More generally, we shall be led to look for the general non-linear scale laws that satisfy the principle of scale relativity. Such a generalized framework, in which scale covariance can be fully implemented, implies working in a five-dimensional fractal space-time (the scale dimension plays the role of the fifth variable). The development of such a "general scale-relativity" lies outside the scope of the present contribution and will be considered elsewhere.

4.2 Discrete scale invariance and log-periodic behavior

Another correction to pure scale invariance is potentially important, namely the log-periodic correction to power laws that is provided, e.g., by complex exponents or complex fractal dimensions [21]. Sornette et al. (see Sornette's contribution to this school and refs. therein) have shown that such a behavior provides a very satisfactory and possibly predictive model of some earth-quakes and market krachs. This may be a first step toward a general description of the time evolution of crises. Let us show how one can recover log-periodic corrections from requiring scale covariance. Consider a scale-dependent "field" $\Phi(\varepsilon)$. The scale variable is identified with the interval $| t - t_c |$, where t_c is the date of crisis. Assume that Φ satisfies a renormalization-group first order differential equation,

$$\frac{d\Phi}{d\ln\varepsilon} - D\,\Phi = 0\,, \tag{20}$$

whose solution is a power law $\Phi(\varepsilon) \propto \varepsilon^D$. Now looking for corrections to this law, we remark that simply jumping to a complex exponent D would lead to large log-periodic fluctuations rather than to a controlable correction to the power-law. So let us assume that the right-hand side of Eq. 20 actually differs from zero

$$\frac{d\Phi}{d\ln\varepsilon} - D\,\Phi = \chi\;. \tag{21}$$

We can now apply the *scale-covariance* principle and require that the new function χ be solution of an equation which keeps the same form as the initial equation

$$\frac{d\chi}{d\ln\varepsilon} - D'\,\chi = 0\;. \tag{22}$$

Setting $D' = D + \delta$, we find that Φ must be solution of a second-order equation

$$\frac{d^2\Phi}{(d\ln\varepsilon)^2} - (2\,D + \delta)\,\frac{d\Phi}{d\ln\varepsilon} + D\,(D+\delta)\,\Phi = 0\;. \tag{23}$$

It writes $\Phi(\varepsilon) = a\;\varepsilon^D\,(\,1 + b\,\varepsilon^\delta\,)$, and finally, the choice of an imaginary exponent $\delta = i\omega$ yields a solution whose real part includes a log-periodic correction:

$$\Phi(\varepsilon) = a\ \varepsilon^{D}\ [\ 1 + b \cos(\omega \ln\varepsilon)\ ]. \tag{24}$$

4.3 Scale-motion coupling and gauge invariance

The theory of scale relativity also allows us to get new insights about the physical meaning of gauge invariance [8,6]. In the previous sections, only scale transformations at a given point were considered. But we must also wonder about what happens to the structures in scale of a scale-dependent object when it is displaced. Consider anyone of these structures, lying at some (relative) resolution ε (such that $\varepsilon < \lambda$, where λ is the fractal /nonfractal transition) for a given position of the particle. Under a translation, the relativity of scales implies that the resolution at which this given structure appears in the new position will *a priori* be different from the initial one. In other words, ε is now a function of the space-time coordinates, $\varepsilon = \varepsilon(x,t)$, and we expect the occurrence of *dilatations of resolutions induced by translations*, which read:

$$e\ \frac{d\varepsilon}{\varepsilon} = -\ A_{\mu}\, dx^{\mu}\ , \tag{25}$$

where a four-vector A_μ must be introduced since dx^{μ} is itself a four-vector and $d\ln\varepsilon$ is a scalar (in the case of a global dilation). This behavior can be expressed in terms of a new scale-covariant derivative:

$$e\ D_{\mu} \ln(\lambda/\varepsilon) = e\ \partial_{\mu} \ln(\lambda/\varepsilon)\ +\ A_{\mu}\ . \tag{26}$$

However, if one wants the "field" A_μ to be physical, it must be defined whatever the initial scale from which we started. Starting from another scale $\varepsilon' = \rho\varepsilon$ (we consider only Galilean scale-relativity here), we get $e\ d\varepsilon' / \varepsilon' = -A'_{\mu}\, dx^{\mu}$, so that we obtain:

$$A'_{\mu} = A_{\mu} + e\ \partial_{\mu} \ln\rho\ , \tag{27}$$

which depends on the relative "state of scale", $\mathbb{V} = \ln\rho = \ln(\varepsilon/\varepsilon')$. If one now considers translation along two different coordinates (or, in an equivalent way, displacement on a closed loop), one may write a commutator relation which defines a tensor field $F_{\mu\nu} = \partial_{\mu} A_{\nu} - \partial_{\nu} A_{\mu}$. This field is, contrarily to A_μ, independent of the initial scale. One recognizes in $F_{\mu\nu}$ the analog of an electromagnetic field, in A_μ that of an electromagnetic potential, in e that of the electric charge, and in Eq. 27 the property of gauge invariance which, in accordance with Weyl's initial ideas [22,23], recovers its initial status of scale invariance. However, Eq. 27 represents a progress compared with these early attempts and with the status of gauge invariance in today's physics. Indeed the gauge function, which has, up to now, been considered as arbitrary, is now identified with the logarithm of internal resolutions.

Another advantage with respect to Weyl's theory is that we are now allowed to define four different and independent dilations along the four space-time resolutions instead of only one global dilation. The above U(1) field is then expected to be embedded into a larger field (in accordance with the electroweak theory) and the charge e to be one element of a more complicated, "vectorial" charge [6,8]. Moreover, when combined with the Lorentzian structure of dilations of special scale relativity, our interpretation of gauge invariance yields new relations between the charges and the masses of elementary particles [6,8].

5. DISCUSSION AND OPEN PROBLEMS

In the present contribution, we have mainly considered the question of pure scale laws. We have shown how they can be constructed and generalized by using the standard tools of classical physics (Euler-Lagrange equations / action principle), once applied not only to displacements in space-time, but to displacements in the scale-space. In this regard our tool can be thought of as the theoretical equivalent of what are wavelets in fractal and multifractal data analysis (see e.g. [24,25]). It allows one to construct conservation laws and invariant quantities through a "scale mechanics" [3,7], quite analog to the case of motion.

There is another well-developed domain of application of scale relativity that we did not mention here for lack of place, namely, the study of the induced effect of scale laws on motion laws. These effects can be described in terms of a complex, scale-covariant time derivative [3,5,6]. When written in terms of this new derivative, the equations of classical mechanics take a quantum-like form [3,6,11,26,27]. This new approach provides a new formulation of quantum mechanics [3], allows to obtain solutions to the Schrödinger equation by numerical simulations without using it [28; see also 29], and leads to new proposals concerning the formation and evolution of gravitational structures at all scales [11,30,31].

Several problems remain open in the theory. Only the minimal effects of nondifferentiability have been up to now taken into account. One of the main study to be developed in the near future is special scale-relativity in the case of four independent space-time resolutions. This can be carried out by working in a five-dimensional fractal space-time, where the role of the fifth dimension is played by the scale exponent. In the end, through a proper covariant representation, scale transformations and space-time transformations could be treated on the same footing. This would open the road toward a generalized theory of scale relativity, including non-linear scale transformations and "scale-accelerations" in its description.

Acknowledgements. The author acknowledges very interesting discussions with Drs. Bérangère Dubrulle and Didier Sornette.

References

[1] Poincaré H., Bull. des Sciences Mathém. 2è série, t. XXVIII (1904) 302.
[2] Nottale L., *Int. J. Mod. Phys.* **A4** (1989) 5047.
[3] Nottale L., Fractal Space-Time and Microphysics: Towards a Theory of Scale Relativity (World Scientific, London, 1993).
[4] Pocheau A., *Phys. Rev.* **E49** (1994) 1109, and these proceedings.
[5] Nottale L., *Chaos, Solitons and Fractals*, **4** (1994) 361.
[6] Nottale L., *Chaos, Solitons and Fractals*, **7** (1996) 877.
[7] Nottale L., *Int. J. Mod. Phys.* **A7** (1992) 4899.

[8] Nottale L., in "Relativity in General", (1993 Spanish Relativity Meeting, Salas, Spain), Ed. J. Diaz Alonso, (Editions Frontières, 1994), pp.121-132.
[9] Nottale L., in "Clustering in the Universe", proceedings of 15th Moriond Astrophysics meeting, eds. S. Maurogordato et al, (Frontières, 1995), p. 523.
[10] Feynman R., Hibbs A., Quantum Mechanics and Path Integrals (MacGraw-Hill, 1965).
[11] Nottale L., *Astron. Astrophys.* (1997) in press.
[12] Nottale L. & Schneider J., *J. Math. Phys.* **25** (1984) 1296.
[13] Ord G.N., , *J. Phys. A: Math. Gen.* **16** (1983) 1869.
[14] Mandelbrot B., Les Objets Fractals (Flammarion, 1975).
[15] Mandelbrot B., The fractal geometry of nature (Freeman, 1983).
[16] Wilson K., *Rev. Mod. Phys.* **47** (1975) 774.
[17] Wilson K., *Sci. Am. (August)* **241** (1979) 140.
[18] Dubrulle B., *Phys. Rev. Lett.* **73** (1994) 959, and these proceedings.
[19] Dubrulle B. & Graner F., *J. Phys. (Fr.)* **6** (1996) 797.
[20] Castro C., *Foundations of Physics*, (1997) in press.
[21] Sornette D. & Sammis C.G., *J. Phys. I France* **5** (1995) 607.
[22] Weyl H., in "*The Principle of Relativity*", (Dover publications, 1918), p. 201.
[23] Dirac P.A.M., *Proc. Roy. Soc. Lond.* **A 333** (1973) 403.
[24] Argoul F., Arnéodo A., Grasseau G., Gagne Y., Hopfinger E.J. & Frisch U., *Nature* **338** (1989) 51.
[25] Arneodo A., Grasseau G. & Holschneider M., *Phys. Rev Lett.* **61** (1988) 2281.
[26] Nottale L, & Pissondes J.C., (1997) submitted for publication.
[27] Pissondes J.C., (1997) submitted for publication.
[28] Hermann R., *J. Phys. A*, (1997) in press.
[29] Ord G.N., *Ann. Phys.* **250** (1996) 51.
[30] Nottale L, *Astron. Astrophys. Lett.* **315** (1996) L9.
[31] Nottale L, Schumacher G., & Gay J., *Astron. Astrophys.* **322** (1997) 1018.
[32] Levy-Leblond J.M., *Am. J. Phys.* **44** (1976) 271.
[33] Zeldovich Ya.B., *JETP Lett.* **6** (1967) 316.

LECTURE 20

Scaling in Turbulent Flows

L. Biferale

University of Rome "Tor Vergata", Department of Physics and INFM, Via della Ricerca Scientifica 1, 00133 Rome, Italy

1. INTRODUCTION

Turbulent flows appear in nature in many different contests. Incompressible fluid evolution is described by Navier-Stokes (NS) equations. By denoting with V_0 the large-scale velocity field and with L_0 the scale of the box where the fluid is confined, one can rewrite Navier-Stokes equations in terms of dimensionless variables, $x \to x/L_0$ and $v \to v/V_0$:

$$\partial_t v + (v \cdot \partial_x)v = -\partial_x P + Re^{-1}\partial_{xx} v \tag{1}$$

where P is the pressure and where with Re we denote the Reynolds number: the only free parameter entering in the equations. The inverse of the Reynolds number multiply the linear-dissipative term of the equations and can, therefore, be understood as the ratio between non-linear and linear terms in the NS eqs. At fixed external geometry, i.e. fixed V_0 and L_0, the Reynolds number is inversely proportional to the viscosity: $Re \sim \nu^{-1}$.

Turbulent flows are described by the time evolution of the NS eqs at "high" Reynolds numbers, i.e. for numbers much larger then the critical value at which the flow shows a transition from a smooth and laminar evolution to a complex and spatio-temporal chaotic behaviour. One speaks about "Fully Developed Turbulence" (FDT) in the ideal limit of $Re \to \infty$.

In the limit of FDT the NS eqs are invariant under a scaling transformation:

$$x \to \lambda x, \;\; v(x,t) \to \lambda^h v(\lambda x, \lambda^{1-h} t), \;\; t \to \lambda^{1-h} t, \tag{2}$$

where h is a general scaling exponent.
Due to the high chaoticity of the flow, symmetries can be satisfied only in a statistical sense. Both experiments and standard phenomenological arguments focused on scaling properties of turbulent flows. A lot of attention has been put in measuring scaling properties of moments of velocity differences at the scale r:

$$S_p(r) =< |v(x+r) - v(x)|^p >=< |\delta v(r)|^p > \tag{3}$$

where $< \cdots >$ stands for ensemble average and v is the velocity component parallel to r. At high Reynolds $S_p(r)$ show a power law behaviour

$$S_p(r) \propto r^{\zeta(p)} \tag{4}$$

for $L_0 > r >> \eta_k$ where $\eta_k = (\nu^3/\epsilon)^{1/4}$ is the dissipative (Kolmogorov) scale, ϵ is the mean energy dissipation rate, ν the kinematic viscosity. The range of length $L_0 > r >> \eta_k$, where the scaling relation (4) is observed, is called the inertial range. The Kolmogorov (K41) theory [1] predicts $\zeta(p) = p/3$, but experimental and numerical results show that $\zeta(p)$ deviates substantially from the linear law. Using mainly dimensional analysis one can link this phenomenon to the intermittent behavior of the energy dissipation. Indeed, it is natural to generalize K41 theory by rewriting eq.(4) in the following way:

$$S_p(r) \propto< \epsilon_r^{p/3} > r^{p/3} \propto r^{\tau(p/3)+p/3} \tag{5}$$

where ϵ_r is the average of the local energy dissipation $\epsilon(x)$ on a volume of size r centered on a point x. A comparison of eq.(3) and eq.(5) leads to the conclusion that the scaling exponents $\tau(p/3)$ of the energy dissipation are related to those of S_p by $\zeta(p) = \tau(p/3) + p/3$.

Intermittent corrections to K41 can also be interpreted as the appearance of a statistically "local scale invariance" at the place of a "global scale invariance" as it would be if $h \equiv 1/3$ globally in the fluid.

Many phenomenological models, [2], [3] [4], [5] have been suggested to describe the behavior of the $\zeta(p)$. However, it turns out that the $\zeta(p)$ may be not universal in non homogeneous, anisotropic flow and may depend on the location where measurements are done. Specifically, they may have different values if one measures either far away from boundaries, where turbulence is almost homogeneous and isotropic, or in locations of the flow where a strong mean shear is present. The $\zeta(p)$ depend also on the way in which turbulence is produced, for example 3D homogeneous turbulence, boundary layer turbulence, thermal convection and MHD. Thus there is the fundamental question of understanding in which way all these parameters influence the scaling laws. Furthermore all the above mentioned models assume the existence of two well defined intervals of lengths that are the inertial range and a dissipation range. According to idea of multiscaling these two ranges may eventually be connected by an intermediate region where the viscosity begins to act [6]. However this idea of a well defined inertial range, where viscosity does not act at all, and the idea of multiscaling turns out to be incompatible with the recently introduced new form of scaling, which has been named Extended Self Similarity (ESS) [7] [8].

The main discrepancy arises concerning the modelization of viscous effects which is clearly an important point in order to safely use large eddy simulations in real applications.

In this paper we review some recent ideas about "robust" scaling properties of turbulent flows. We study a generalized form of ESS (G-ESS) which has been checked in many different flows.

Our interpretation of ESS and of this generalized scaling suggest that there is no sharp viscous cut-off in the intermittent transfer of energy.

2. EXTENDED SELF SIMILARITY

Extended Self Similarity (ESS) is a property of velocity structure functions of homogeneous and isotropic turbulence [7, 8]. It has been shown using experimental and numerical data [10] that the structure functions present an extended scaling range when one plots one structure function against the other, namely:

$$S_n(r) \propto S_m(r)^{\beta(n,m)} \tag{6}$$

where $\beta(n,m) = \zeta(n)/\zeta(m)$. The details of ESS have been reported elsewhere [8].

ESS has been also checked for the temperature and velocity fields in Rayleigh-Benard convection [12]. In this case, due to the active feedback played by the temperature on the velocity time evolution a new lenght scale appears in the system: the so-called Bolgiano scale. The presence of the Bolgiano scale allow two different dimensional ansatze and therefore two different scaling behaviours, depending on the importance of boyancies effects on the energy transfer dynamics. It turns out that ESS is a very useful tool in order to distinguish between Kolmogorov and Bolgiano scaling [12]. In the case of the Bolgiano scaling it has been found that $\zeta(3) = 2.08$ which is clearly very different from the Kolmogorov value $\zeta(3) = 1$. In spite of this large difference between the values of the exponent $\zeta(3)$, using ESS one discovers [11] that the ratio $\zeta(p)/\zeta(3)$ in the case of Bolgiano are equal to those of homogeneous and isotropic turbulence. The same property is observed for the $\zeta(p)$ obtained from measurements done on the solar wind [13].

Let us stress that this kind of universality, observed in different flows, disappears if the system is influenced by the presence of a strong mean shear. In this case ESS does not work, because an extended range of scaling is not present when S_n is drawn as a function of S_3.
Violation of ESS has been observed experimentally in boundary layer turbulence [9] and in the shear behind a cylinder [14].

In a recent letter [5] She and Leveque have proposed a new phenomenological model to explain the anomalous scaling exponents of velocity structure functions. The theory yields a prediction

$$\zeta(p) = ph_0 + d_0(1 - \beta^{p/3})$$

which is in very good agreement with available experimental data [8]. The parameters h_0 and d_0 describe the singular exponent and the fractal codimension of the most intermittent structure present in the flow, and β is connected to intermittency along the energy cascade. She-Leveque took $h_0 = 1/9$, $d_0 = 2$ because they considered vortex filaments as responsible of the strong intermittent burst and $\beta = 2/3$ from phenomenological dimensional analysis.

The model in itself has nothing more then other previous proposals of the same kind. The interesting aspect, in our view, stems from the physical interpretation of the free parameters h_0, d_0.

In [11] we have proposed a new form of scaling which should be exactly satisfied at all resolvable scales: from the integral scale down well inside the viscous subrange. The main idea was to use the phenomenological interpretation of the three free parameters entering in the log-Poisson description h_0, d_0, β in such a way to incorporate all non-universal and viscosity dependency in two of them h_0, d_0 while keeping the other β constant for all flows and at all scales. The main phenomenological motivation was that h_0 and d_0 being connected to the statistics of filaments could be strongly affected from possible anisotropies in the flow (at both small and large scales) and from viscosity at small scales.

Our claim is that most of the *fragile* aspects of intermittency and scaling in turbulence are only due to non-universal properties of the most singular scaling structure, i.e. we can imagine that h_0, d_0 become functions of r/η by approaching the viscous cut-off and could be strongly dependent on anisotropic effects while β stays highly universal at all scales and independently of the isotropic properties of the flow.
In this way we disentangle robust from fragile characteristic of intermittency in fluidynamics.

3. GENERALIZED-ESS

For this purpose we introduce the dimensionless structure function

$$G_p(r) = \frac{S_p(r)}{S_3(r)^{p/3}} \tag{7}$$

According to Kolmogorov theory (7) should be a constant both in the inertial and in the dissipative range, although the two constant are not necessarily the same.

Let us now study the self scaling properties of $G_p(r)$ or, equivalently:

$$G_p(r) = G_q(r)^{\rho(p,q)} \tag{8}$$

where we have by definition:

$$\rho(p,q) = \frac{\zeta(p) - p/3\ \zeta(3)}{\zeta(q) - q/3\ \zeta(3)} \tag{9}$$

$\rho(p,q)$ is given by the ratio between deviation from the K41 scaling. It will play an essential role in our understanding of energy cascade. Indeed, it is easy to realize that this quantity can stay constant along all the cascade process: from the integral to the sub-viscous scales. It is reasonable to imagine that the velocity field becomes laminar in the sub-viscous range, $S_p(r) \propto r^p$, still preserving some intermittent degree parameterized by the ratio between corrections to K41 theory. In [11], relation (8) has been tested in many different experimental set-up [14], [12], done at different Reynolds numbers and for some direct numerical simulation with and without large scale shear and inside boundary layers, i.e. in cases where ESS is not observed. In all cases a very good scaling behavior where detected.

4. A THEORETICAL INTERPRETATION

The aim of this section is to discuss a possible theoretical interpretation of the experimental and numerical results previously discussed. Our starting point is to revise the concept of scaling in fully developed turbulence.

Let us consider three length scales $r_1 > r_2 > r_3$ and our basic variables to describe the statistical properties of turbulence, namely the velocity difference $\delta v(r_i)$. We shall restrict ourselves to those statistical models of turbulence based on random multiplier.

Thus we shall assume that there exists a statistical equivalence of the form:

$$\delta v(r_i) = a_{ij} \delta v(r_j) \tag{10}$$

where $r_i < r_j$ and a_{ij} is a random number with a prescribed probability distribution P_{ij}.

By definition, we have:

$$a_{13} = a_{12} a_{23} \tag{11}$$

Equation (11) is true no matter which is the ratio $\frac{r_1}{r_2}$ and $\frac{r_2}{r_3}$. Now we ask ourselves the following question: what is the probability distribution P_{ij} which is functionally invariant under the transformation (11)? This question can be answered by noting that equation (11) is equivalent to write:

$$\log a_{13} = \log a_{12} + \log a_{23} \tag{12}$$

(we assume $a_{ij} > 0$). Thus our question is equivalent to ask what are the probability distribution stable under convolution. For independently distributed random variables a solution of this problem can be given in a complete form [15]. If the variables are correlated the situation becomes much more difficult to solve, as it is well known from the theory of critical phenomena. For the time being we shall restrict ourselves to independent random variables.

In this case, for instance, the gaussian and the Poisson distributions are well known examples of probability distribution stable under convolution. These two examples correspond to two turbulent model proposed in literature, namely the log-normal model and the log-Poisson model [5], [17], [18]. A more general description can be found in [15, 16] a.

We can have a different point of view on our question which is fully equivalent to the above discussion. A simple solution to our question is given by all probability distribution P_{ij} such that:

$$\langle a_{ij}^p \rangle \equiv \prod_{k=1}^{n} \left(\frac{g_k(r_j)}{g_k(r_i)} \right)^{\gamma_k(p)} \tag{13}$$

for any functions $g_k(r_i)$ and $\gamma_k(p)$ ($\langle \cdots \rangle$ represents average over P_{ij}). Indeed we have:

$$\langle a_{13}^p \rangle = \langle a_{12}^p a_{23}^p \rangle = \prod_{k=1}^{n} \left(\frac{g_k(r_1)}{g_k(r_2)} \right)^{\gamma_k(p)} \prod_{k=1}^{n} \left(\frac{g_k(r_2)}{g_k(r_3)} \right)^{\gamma_k(p)} = \prod_{k=1}^{n} \left(\frac{g_k(r_1)}{g_k(r_3)} \right)^{\gamma_k(p)} \tag{14}$$

We want to remark that equation (13) represents the most general solution to our problem which is independent of the scale ratio r_i/r_j.
Let us give a simple example in order to link equation (13) to the case of probability distribution stable under convolution. Following [5], [17] and [18] let us consider the case of a random log-Poisson multiplicative process, namely:

$$a_{ij} = A_{ij} \beta^x \tag{15}$$

where x is a Poisson process $P(x = N) = \frac{C_{ij}^N e^{-C_{ij}}}{N!}$ and A_{ij} are constants which satisfies $A_{ij} A_{jk} = A_{ik}$
By using (15) we obtain:

$$\langle a_{ij}^p \rangle = A_{ij}^p \exp(C_{ij}(\beta^p - 1)) \tag{16}$$

Equation (16) is precisely of the form (13) if we write

$$A_{ij} = \frac{g_1(r_j)}{g_1(r_i)} \qquad \exp C_{ij} = \frac{g_2(r_i)}{g_2(r_j)} \tag{17}$$

In order to recover the standard form of She-Leveque model we need to assume that (23):

$$g_1(r_i) \sim {r_i}^h \tag{18}$$

$$g_2(r_i) \sim r_i^2 \tag{19}$$

This example highlights one important point in our discussion, i.e. the general requirement of scale invariant random multiplier (13) does not necessary imply a simple power law scaling as expressed by the equations (18- 19). Moreover, the general expression (13) is compatible only with infinitively divisible distribution. For instance, previous random multiplier model for turbulence, such as the β-random model or the p-model, cannot be expressed in the general form (13) independently of the ratio r_i/r_j.

It is worthwhile to review the multifractal language at light of the previous discussion. In the multifractal language for turbulence, the two basic assumptions are:

I) The velocity difference on scale r shows local scaling law with exponent h, i.e. $\delta v(r) \sim r^h$;

II) the probability distribution to observe the scaling $\delta v(r) \sim r^h$ is given by $r^{3-D(h)}$.

In the multifractal language, therefore, there are two major ansatz: one concerns power law scaling of the velocity difference (assumption I) and the other one concerns a geometrical interpretation (the fractal dimension $D(h)$) of the probability distribution to observe a local scaling with exponent h. How is it possible to generalize the multifractal language in order to take into account equation (13)?

As we shall see, the theory of infinitively divisible distribution is the tool we need to answer the previous question. All published model of turbulence based on infinitively divisible distribution are equivalent to write $D(h)$ in the form:

$$3 - D(h) = d_0 f\left[\frac{h - h_0}{d_0}\right] \tag{20}$$

where d_0 and h_0 are two free parameters while the function $f(x)$ depends only on the choice of the probability distribution. For instance for log-normal distribution $f(x) = x^2$. Equation (20) allows us to write the structure functions as an average over all possible local scaling behaviours weighted by the probability to meet a aprticular local scaling exponent:

$$\langle \delta v(r)^p \rangle = \int d\mu(h) r^{hp} r^{3-D(h)} = r^{h_0 p + d_0 H(p)} \tag{21}$$

where

$$H(p) = \inf_x (px + f(x)) \tag{22}$$

We can see that equation (21) is equivalent to a random multiplicative process given by:

$$\langle a_{ij}^p \rangle = \left(\frac{r_j}{r_i}\right)^{h_0 p} \left(\frac{r_j}{r_i}\right)^{d_0 H(p)} \tag{23}$$

Equation (23) can be generalized to the form (13) by allowing h_0 and d_0 to depend on r, i.e.

$$\langle a_{ij}^p \rangle = \frac{\left(r_j^{\bar{h}_0(r_j)}\right)^p}{\left(r_i^{\bar{h}_0(r_i)}\right)^p} \left[\frac{r_j^{\bar{d}_0(r_j)}}{r_i^{\bar{d}_0(r_j)}}\right]^{H(p)} \tag{24}$$

where:

$$\overline{h}_0(r) = h_0 s_h(r) \qquad \overline{d}_0(r) = d_0 s_d(r) \tag{25}$$

Equation (24) is equivalent to (13) by using:

$$g_1(r_i) = r_i^{\overline{h}_0(r_i)} \qquad g_2(r_i) = r_i^{\overline{d}_0(r_i)} \tag{26}$$

$$\gamma_1(p) = p \tag{27}$$

$$\gamma_2(p) = H(p) \tag{28}$$

The same results can be obtained by (21), i.e. we have

$$\langle \delta v(r)^p \rangle = r^{\overline{h}_0(r)p + \overline{d}_0(r)H(p)} \tag{29}$$

Note that the saddle point evaluation of (21) is not spoiled by the dependence of h_0 and d_0 on r.

We have seen that (13) can be reformulated in terms of multifractal language for infinitively divisible distribution whose function $D(h)$ can be rewritten as in (20). We can ask the following question: what is the physical meaning of (13) or its multifractal analogous (24- 29)? It is precisely the multifractal language which allows us to answer this question. Indeed, the two basic assumption for the multifractal language can now be replaced in the following way:

I) the velocity difference on scale r behaves as

$$\delta v(r) \sim g_1(r) g_2(r)^x; \tag{30}$$

II) the probability distribution to observe I is $g_2(r)^{f(x)}$.

Then we have

$$\langle \delta v(r)^p \rangle = \int d\mu(x) g_1(r)^p g_2(r)^{px+f(x)} = g_1(r)^p g_2(r)^{H(p)} \tag{31}$$

by employing a saddle point integration. The most clear physical interpretation of (31) is that the probability to observe a given fluctuation of the velocity difference has no more a geometrical interpretation linked to the fractal dimension $D(h)$. The probability distributions are controlled by a dynamical variable $g_2(r)$ which at this stage we still need to understand. An insight on the dynamical meaning of $g_2(r)$ can be obtained by the following considerations.

Let us define $\epsilon(r)$ the average of the energy dissipation on a scale r. We can define the typical time scale, i.e. the eddy turnover time, $\tau(r)$ on scale r as:

$$\frac{\delta v^2(r)}{\tau(r)} \sim \epsilon(r) \tag{32}$$

In [11] we have seen that all experimental and numerical data suggest that the following relation is always valid:

$$\frac{\epsilon(r)}{\langle \epsilon \rangle} =^s \frac{\delta v^3(r)}{\langle \delta v^3(r) \rangle} \tag{33}$$

where $=^s$ means that all moments on the r.h.s. are equal to l.h.s. By using (32-33) we obtain the definition of length $L(r)$:

$$L(r) \equiv \delta v(r)\tau(r) = \frac{\langle \delta v^3(r) \rangle}{\epsilon} \tag{34}$$

$L(r)$ cannot be regarded as a real length scale in the physical space. Rather, $L(r)$ should be considered as a dynamical variable entering into the statistical description of turbulence. This is precisely the idea behind ESS which formulate the scaling properties of turbulence in terms of $L(r)$. Indeed in order to obtain ESS from (31) it is sufficient to state that, within the range of scales where ESS is observed, $g_1(r)^{1/h_0} \sim g_2(r)^{1/d_0} \sim L(r)$. The physical meaning of ESS is strictly linked to (34) and in particular to (33) which is a generalization of Kolmogorov Refined Similarity Hypothesis.

Let us summarize all our previous findings:

(i) we have introduced the idea of scale invariant random multiplier satisfying equation (13);

(ii) we have shown that infinitively divisible distributions are all compatible with(13);

(iii) we have shown that the multifractal language specialized for the case of infinitively divisible distribution gives equation (13) (with $n = 2$ and $\gamma_1(p)$ linear in p) and it is equivalent to scale invariant random multiplier;

(iv) finally we have argued that the correct scaling parameter to describe the statistical properties of small scale turbulent flows is not directly linked to a simple geometrical interpretation, rather it should be considered a dynamical variable.

Our finding (i)-(iv) enables us to have a unified theoretical interpretation of the experimental and numerical results presented at the beginning of this paper. Indeed equation (29) or (31) tells us that the anomalous part of the structure functions:

$$G_p(r) \equiv \frac{\langle \delta v^p(r) \rangle}{\langle \delta v^3(r) \rangle^{p/3}} \tag{35}$$

satisfies the scaling properties:

$$G_p(r) = G_q(r)^{\rho_{p,q}} \tag{36}$$

where $\rho_{p,q} \equiv \frac{\zeta_p - p/3}{\zeta_q - q/3}$. According to our analysis of the experimental and numerical results, the scaling (36) is observed down to the smallest resolved scale. We have shown that, in the theoretical framework so far exposed, we recover the ESS when $g_1(r)^{1/h_0} \sim g_2^{1/d_0} \sim L(r)$. If $g_1(r)^{1/h_0} \neq g_2(r)^{1/d_0}$ we lose ESS, but its generalized version (36) is still valid.

5. MULTISCALING

It is generally argued that the anomalous scaling can be observed for scale larger than a given viscous cutoff. The physical interpretation of this statement is that non linear, intermittent, transfer of energy is acting only for scale larger than the viscous scale. Below such a scale the structure function are supposed to show a simple (regular) scaling $\langle \delta v^p(r) \rangle \sim r^p$.

Usually the viscous cutoff is introduced as the scale at which the local Reynolds number is of order one, namely:

$$\frac{\delta v(r) r}{\nu} \sim 1 \tag{37}$$

This condition can be obtained by the requirement that the local energy transfer $\epsilon(r) \sim (\delta v^3(r)/r)$ becomes equal to the energy dissipation $\nu(\delta v^2(r)/r^2)$:

$$\nu \frac{\delta v^2(r)}{r^2} \sim \frac{\delta v^3(r)}{r} \tag{38}$$

which gives equation (37). There is a well defined prediction [6], called Multiscaling, based on (37), using the multifractal language. Indeed for any exponent h one can introduce the h-dependent viscous cutoff given by:

$$r_d^{h+1} \sim \nu \tag{39}$$

where $\delta v(r) \sim r^h$.
It follows that $r_d(h)$ is a fluctuating quantity. There are two consequences of this theory. The first one predicts that for the structure functions $\langle \delta v^p(r) \rangle$ there exists a cutoff scale r_p dependent on p and moreover $r_p < r_q$ for $p > q$. The second prediction concerns the moment of the velocity gradients Γ which are:

$$\langle \Gamma^p \rangle \sim \langle r_d(h)^{(h-1)p} r_d(h)^{3-d(h)} \rangle \sim Re^{-z(p)} \tag{40}$$

with $z(p) = \sup_h \left[\left((h-1)p + 3 - D(h) \right) / (1+h) \right]$.
Between the two predictions the first one is the strongest consequence of (39). In particular, the first prediction states that between the end of the inertial range (i.e. the region where anomalous scaling of $\langle \delta v^p(r) \rangle$ with respect to r is detected) and the dissipation cutoff r_d, the local slope is controlled by $D(h)$. The second prediction is somehow weaker because present experimental data do not distinguish among several models, so far proposed, for the Re-dependence of $\langle \Gamma^p \rangle$.

In order to compare the multiscaling in the dissipation range with our experimental and numerical data, we have produced two synthetic turbulence signals (signals A and B hereafter) [11]. Signal A is defined such as to reproduce G-ESS exactly: no sharp cut-off between inertial and viscous range should be detectable. Signal B, is defined such as the transition among inertial and viscous ranges is introduced by a Multiscaling effect, i.e. we let the iscous cutoff fluctuate according (39).

Since experimental data do not show any deviation from G-ESS behaviour we want to test how much relevant could be the effects introduced by Multiscaling.

By studying the compensated slope, i.e. $G_p(r)/(G_q(r))^{\rho(p,q)}$ at varying r for signal A and B we detected [11] a strong scaling violation (order 40%) in the Multiscaling-signal for scales of the order of the Kolmogorov scale. On the other hand, by definition, signal A shows a pure scaling at all scales within numerical precision which is of the order of a few per cent.

The above discussion rules out the effect of multiscaling on the viscous cutoff (39).

Previous claims on the validity of multiscaling effects should be considered either wrong or affected by experimental errors. On the other hand our model, used to implement the synthetic signal A, should be considered a very accurate model even for scale close to the regular region where $\delta v(r) \sim r$.

There is, however, the theoretical possibility that Multiscaling indeed acts somewhere at very small scales (much smaller then the usual Kolmogorov scale) and that it therefore should be taken into account in the G-ESS scenario if one would really like to reach, by using multiplicative models, scales well inside the viscous sub-range.

6. CONCLUSIONS

In this paper we have reviewed several new results concerning the scaling behaviour of small scale statistical properties of turbulence. 1) We have reviewed the main results on ESS and in particular we have stressed that in homogeneous and isotropic flows in turbulence, Rayleigh Benard convection and solar wind magnetohydrodynamics, the ratio $\zeta(p)/\zeta(3)$ seems to have an universal behaviour. This is a rather striking and unexpected result which implies that anomalous violation of dimensional scaling may be explained in an universal way. We do not know any simple phenomenological explanation for our finding.

2) We have mentioned that ESS is not observed when relatively strong shear flows are present. A phenomenological analysis, based on the Kolmogorov equation, shows the relevance of a length scale based on the mean energy dissipation and the shear strength.

3) We have reviewed a recently proposed generalization of ESS. This generalization is supported both by experimental and numerical data and it seems to be not affected by viscous cutoff.

4) We have described a theoretical interpretation which unifies the previous points. The theory is based on the assumption that the probability distribution is infinitively divisible and predicts the existence of the generalized ESS. The theory can also be used to generate artificial signals which displays all the scaling features observed in real data.

5) We have shown that the original proposal on the multiscaling for the viscous cutoff is incompatible with the turbulence data. The theory formulated

in this paper removes this incompatibility and suggests that multiscaling is acting at much smaller scales than previously proposed. The new point on the theory is a change of view in the probability distribution of the original multifractal model which is not directly linked to a geometrical interpretation in terms of fractal dimensions.

Acknowledgments

The author wish to thank R. Benzi, S. Ciliberto, M.V. Struglia, R. Tripiccione and M. Pasqui for continuous collaborations on these subjects

References

[1] A. N. Kolmogorov, *C. R. Acad. SCI USSR* **30** (1941) 299.
[2] R. Benzi, G. Paladin, G. Parisi, A. Vulpiani, *J. Phys. A* **17** (1984) 3521.
[3] C. Meneveau and K. R. Sreenivasan, *Phys. Rev. Lett.* **59** (1987) 1424.
[4] Y. Gagne and B. Castaing, *C.R. Acad. Sci. Paris* **312** (1991) 441.
[5] Z. S. She, E. Leveque, *Phys. Rev. Lett.* **72** (1994) 336.
[6] U. Frish, M. Vergassola, *Europhys. Lett.* **14** (1991) 439.
[7] R. Benzi, S. Ciliberto, R. Tripiccione, C. Baudet, S. Succi, *Phys. Rev. E* **R29** (1993) 48.
[8] R. Benzi, S. Ciliberto, C. Baudet, G. Ruiz Chavarria, Physica D, 80, 1995, 385.
[9] G. Stolovitzky, K. R. Sreenivasan, *Phys. Rev. E* **48** (1993) 32.
[10] M. Briscolini, P. Santangelo, S. Succi, R. Benzi, *Phys. Rev. E* **50** (1994) 1745.
[11] R. Benzi, L. Biferale, S. Ciliberto, M.V. Struglia and R. Tripiccione, *Physica D* **96** (1996) 162.
[12] R. Benzi, R. Tripiccione, F. Massaioli, S. Succi, S. CIliberto, *Europhys. Lett.* **25** (1994) 331.
[13] R. Grauer, *Physics Lett. A* **195** (1994) 335.
[14] R. Benzi, S. Ciliberto, C. Baudet, G. Ruiz Chavarria, R. Tripiccione, *Europhys. Lett.* **24** (1993) 275.
[15] E.A. Novikov, *Phys. Rev. E* **R3303** (1994) 50.
[16] B. Dubrulle, Scale Invariance and Beyond (eds. B. Dubrulle & F. Graner & D. Sornette, Springer, Heidelberg, 1997) p. 12.
[17] B. Dubrulle, *Phys. Rev. Lett.* **73** (1994) 959.
[18] Z.-S. She and E.C. Waymire, *Phys. Rev. Lett.* **74** (1995) 262.
[19] R. Benzi, L. Biferale, A. Crisanti, G. Paladin, M. Vergassola and A. Vulpiani, *Physica D* **65** (1993) 352.

LECTURE 21

Statistical Scale Symmetry Breaking

B. Dubrulle

CNRS, CEA/DSM/DAPNIA/service d'Astrophysique, L'Orme des Merisiers, 709, 91191 Gif-sur-Yvette, France

1. INTRODUCTION

Many physical systems are characterized by fluctuating quantities over a wide range of scale. During this school, we met several examples: exchange rate in economics, velocity increments in turbulence, density fluctuations in cosmology, intensity of earthquakes, size of disordered domains in Ising models... In this lecture, I shall focus on the description of fluctuations in a scale invariant system, i.e. occuring in a system described by laws which keep the same shape at any scale. Mathematically, this property is referred to as "scale covariance" of the physical laws. Examples of such scale covariant laws are given in the lecture of Pocheau. It is now natural to wonder how this scale covariance is reflected in the random processes solutions of the laws. An easy answer, known by everybody, is when the random process keeps the symmetry, and is itself scale invariant. It is then characterized by power laws- the computation of the corresponding scaling exponents has become a whole industry in physics. This nice situation is however almost never reached in real systems: often, boundary conditions, finite size effects or forces spoil the system and induce a *symmetry breaking*. The solution is not scale invariant anymore, there are no exact power laws (see the lecture by Castaing and [3, 4, 5, 7]). This is a bit of a mess, then: if you do observe power laws, you have well defined tools and a well defined langage to describe to the guy next door your experiment. If he is interested and wants to compare his measurements to yours, he knows exactly what you are speaking about, and what quantities-namely exponents, and prefactors!- he should measure in his experiment, even though its experimental set up is quite different from yours. When scale symmetry is broken, what can you measure? What can you compare? What is likely to replace exponents and prefactors? The goal of this lecture is to present the tools and the "langage" which can be built to study the scale symmetry breaking.

2. SCALE INVARIANT CASE

Let me first consider an ideal scale invariant solution, to introduce the concepts and the notations I shall need in the sequel. From my fluctuating quantity, I build a scale dependent random process ϕ_ℓ ([1]), e.g. by taking increments over a distance ℓ like in turbulence, or using a wavelet decomposition, as explained in the lecture by Arneodo. Scale invariance of this solution means that, upon rescaling of scales by a factor λ, the rescaling of the process by a factor λ^h, where h is any number in a given interval $[h_{min}; h_{max}]$, leads to a process having the same statistical properties than the original one, that is:

$$\lambda^h \phi_{\lambda\ell} \approx \phi_\ell, \tag{1}$$

where the equality $\approx$ is meant in a statistical sense. The moments of ϕ_ℓ are then power laws, and are characterized by a set of scaling exponents $\zeta(n)$ and prefactors A_n

$$<\phi_\ell^n>= A_n \ell^{\zeta(n)}. \tag{2}$$

If you are an experimentator, you then have two sets of characteristic numbers to describe your system: the exponents, which can be found by fitting power laws to the moments of your field , and the prefactors, which you can obtain by one measurement of each moment at a given scale (e.g. the largest one). The exponents $\zeta(n)$ can be related to another set, made of "local" exponent h (1) by a Legendre transform property:

$$\begin{aligned} \zeta(n) &\equiv \frac{d\ln <\phi_\ell^n>}{d\ln \ell}, \\ &= \min_h (nh - f(h)), \\ h(n) &= \frac{d\zeta(n)}{dn}, \\ &= \frac{d^2 \ln <\phi_\ell^n>}{dnd\ln \ell}, \end{aligned} \tag{3}$$

where $f(h)$ characterizes the probability of occurence of the exponent h. This is the multifractal description. Note that in the present case, $\zeta(n)$ and, then h are scale independent numbers. I'll develop in Section 3 and 4 a formalism to be able to deal with situations where these quantities are allowed to vary with scale (no more power law moments!).

([1]) In what follows, I only consider positive processes, to be able to deal with its logarithm or non-integer power.

The shape of $\zeta(n)$ is completely determined by the function $f(h)$, called the multifractal spectrum. It can be mathematically shown to be a "convex" function (with a negative second derivative). The simplest example is given by the "paraboloid function" $f(h) = -(h - h(0))^2/(2\sigma^2)$, which can be seen as the expansion of $f(h)$ around its maxima $h(0)$, the most probable exponent. In such case, one gets $\zeta(n) = nh(0) - n^2\sigma^2/2$. This corresponds to the case where the statistics of the random process ϕ_ℓ is log-normal. By essence -think about regular expansions of f around $h(0)$-, this statistics is an asymptotic limit of many multifractal statistics, and is used in various areas of physics.

3. BUILDING NEW TOOLS

3.1. The Composition Rule

In fact, any multi-fractal shape for $\zeta(n)$, can also be obtained via a "composition rule" for the exponents $h(n) = d\zeta/dn$. For example, the log-normal case obeys the very simple "additive rule" $h(n+p)-h(0) = h(n)-h(0)+h(p)-h(0)$. This property is in fact not so surprising, if one looks at the structure of the set of possible exponents h and order of moments n. Both are subsets of the space of real numbers, and are characterized by minimal and maximal values: h_{min} and h_{max}, the smallest and the largest "singularity" exponent, and n_{min} and n_{max}, the order of the lowest and the largest converging moment[2]. There is a powerful theorem, used in special relativity, which states that any group law in an interval of the space of real numbers is isomorph to the composition law:

$$V \otimes V' = \frac{V + V' - VV'\left(C_+ + C_-\right)/C_-C_+}{1 - VV'/C_-C_+}, \tag{4}$$

where C_+ and C_- are two constants determined by the boundary of the interval: starting from any number V in the interval $]C_-;C_+[$ and composing it with itself (forward) or with its inverse (according to the composition law) (backward) recursively, one reaches any number in the interval $]C_-;C_+[$. [3] So, I just have to use the theorem to know that I can always find two bijections $F:]h_{min}, h_{max}[\to]F(h_{min}), F(h_{max})[$ and $G:]n_{min}, n_{max}[\to]G(n_{min}), G(n_{max})[$ so that the function $V(n) \equiv (F \circ h \circ G)(n)$ obeys the composition rule:

$$V(n\tilde{\otimes}p) = V(n) \otimes V(p), \tag{5}$$

where $\tilde{\otimes}$ and $\otimes$ are two composition rule of the shape (4) with respective constants $\tilde{C}_- = G(n_{min})$ and $\tilde{C}_+ = G(n_{max})$ and $C_- = F(h_{min})$ and $C_+ = F(h_{max})$. For example, in the log-normal case, G is the identity, F is a translation by $h(0)$, and $h_{min} = n_{min} = -\infty$ and $h_{max} = n_{max} = +\infty$. A slightly

---

([2]) I can always exclude the case of moment divergencies induced by diverging prefactor of the power law by a suitable renormalization of the random process, via multiplication by a scale independent process.

([3]) The same property also holds for union of open interval $]-\infty, C_-[\sqcup]C_+, \infty[$.

more complicated case can be obtained by letting h_{min} take a finite value, while keeping all the order parameters as in the log-normal case. One gets [10]:

$$h(n) = h(0) + h_{min}\left(1 - \beta^n\right), \quad \beta = \frac{h(1) - h(0)}{h_{min}},$$
$$\zeta(n) = n\left(h(0) + h_{min}\right) + \frac{h_{min}}{\ln \beta}\left(1 - \beta^n\right). \tag{6}$$

This is the log-Poisson case, which has recently attracted much attention in turbulence (see lectures by Arneodo or Biferale). Other log-infinitely distributions can be found with simple choices of the parameter [10].

Because of the apparent freedom on F and G, this remark does not appear as very useful, maybe. It is not, as long as one ignores our main joker, the underlying scale invariance. Taking the latter into account, one is however able to select or understand the role of F and G via my next trump, the scale transformations.

3.2. The Scale Transformations

Even more amazing: associated with the composition group law (4) is a set of "scale transformation" defined as follows. I introduce a log-scale coordinate characterizing the measurement of a scale with respect to any scale unit ℓ_0, via:

$$T = \frac{\ln(\ell/\ell_0)}{\ln(K)},$$

where K is the "unit" (the basis) of the logarithm. In a same way, I introduce a log-coordinate characterizing the measurement of the random process ϕ_ℓ, at this scale, in a given realization with respect to any "field unit" ϕ_0 (possibly random and or scale dependent [8])

$$\begin{aligned} X &\equiv \frac{d\ln < (\phi_\ell/\phi_0)^s >}{ds}, \\ &= \frac{\langle(\phi_\ell/\phi_0)^s \ln(\phi_\ell/\phi_0)\rangle}{\ln(Q)\,\langle(\phi_\ell/\phi_0)^s\rangle} \equiv \frac{\langle\ln(\phi_\ell/\phi_0)\rangle_{\mathcal{R}_s}}{\ln(Q)}, \end{aligned} \tag{7}$$

where Q is another unit for the logarithm. The notation $<>_{\mathcal{R}_s}$ is used to recall the the average is taken with respect to the weighted probability distribution $\phi^s/ < \phi^s >$, which plays the role of a "reference frame" (see e.g. [6, 19]). My definition is chosen so that, for $\phi_0 = 1$, the derivative dX/dT is just the local multifractal exponent $h(s)$ (3), within a factor $\ln(K)/\ln(Q)$. When ϕ_0 is not constant, this derivative is a function of h (e.g. a translation). It is the quantity $F \circ h$ we met earlier. For clarity, I denote it $V_{\phi/\mathcal{R}_s,K,0}$, to recall that it depends both on s, the units K and Q and the references ϕ_0 and ℓ_0. Note that I do not impose *a priori* that this number is constant, i.e. independent of the scale. This freedom will be the cornerstone of the extension towards symmetry breaking.

I am now going to relate these two measurements X, T, performed in a given "reference frame" $(s, K, 0)$, in a given realization, to similar measurements performed in another "reference frame" $(s', K', 0')$, in another realization X', T', that is varying ℓ to ℓ' and s to s'. I stress here that my notations are purposively ambiguous because the same numbers X' and T' can be obtained by independent variations of different quantities: either ℓ and s, or ℓ_0 and ϕ_0 (this is the point of view adopted in [17, 18, 8]), or even by varying K and Q (as discussed in [11]), or of course by varying all of them together. I shall come back on the meaning of this later, but I stress that they are equivalent points of view as far as the final results is concerned. I now state the *result*, proved in [8, 10]. To obtain the composition rule for the exponents V, I have to take the link between sets of log-coordinate (X,T) and (X',T') according to the linear transformation

$$\begin{aligned} \begin{pmatrix} CT' \\ X' \end{pmatrix} &= \Gamma(V) \begin{pmatrix} 1-2\Lambda V/C & (\Lambda^2-1)V/C \\ -V/C & 1 \end{pmatrix} \begin{pmatrix} CT \\ X \end{pmatrix} \\ \Gamma(V)^{-2} &= 1-2\Lambda V/C+(\Lambda^2-1)(V/C)^2. \end{aligned} \tag{8}$$

The transformation (8) depends on three parameters: Λ, C and V. The two former depend only on the random process itself: Λ breaks the symmetry between small and large scales, C is a characteristic exponent. The latter, V, characterizes the scaling exponent of the first coordinate system, with respect to the second, $V = V_{\mathcal{R}_{s,K,0}/\mathcal{R}_{s',K',0'}}$. Applying (8) successively from (ϕ_ℓ to $\phi'_{\ell'}$), then from ($\phi'_{\ell'}$ to $\phi''_{\ell''}$), one obtains a group structure, where a composition law for V is (4), with $C_+ = C/(\Lambda-1)$ and $C_- = C/(\Lambda+1)$. Even nicer: starting from a realization of a random field such that $X = hT$ (i.e. $\phi_\ell \sim \ell^h$, one obtains another realization so that $X' = h'T'$ (i.e. $\phi'_\ell \sim \ell'^{h'}$) with $h = h' \otimes V$. So, the transformation really maps one scaling exponents onto another one, via a clever change of scale...and of realization (it is a statistical transfomation!).

So, now, we see more clearly where appear the two functions F and G introduced in Section 3.1. The first one is associated with the normalization of ϕ_ℓ. The second one comes from the link between $s, K, 0$ and $s', K', 0'$. What this means practically is that if your data correspond to a multifractal, there is a clever way to organize and analyze your data (your are the one who choose $K, 0, ..!$) so that the local exponent dX/dT obeys the simple relation (5) with simple functions F and G. For example, in the log-normal case, it is easy to show that if you normalize your data with $\ln(\phi_0) = <\ln(\phi_\ell)>$, then your local scaling exponent obeys the simple addition rule $V'' = V + V'$.

3.3. Analogy with Special Relativity

There is an interesting analogy between the transformations I just defined and the Lorentz transforms, i.e. between scale symmetry and special relativity.

To my knowledge, Nottale [12, 15] was the first to mention this analogy, and to try to exploit it. In his lecture, he presents an analogy to be used for describing the scale variation of deterministic fields. In this lecture, I presented a different analogy, involving random fields. Here, X and T are variables describing the value of the log of the random field and the value of the log of scale. They are our coordinates. The "reference frame" here is made of the chosen units ϕ_0 and ℓ_0 and of the resolution K and Q (the basis of the logarithm). Changes of frame are achieved by changes of units and of resolution. There is a postulate of relativity which tells you that values of exponents, like velocities in special relativity, are relative (to the choice of your reference frame): of course, the value of $d\ln_K \phi_\ell / d\ln_Q \ell$ depends of the choices of the basis of the logarithm! When applied to deterministic fields, this principle does not lead to any interesting result. For random fields, it is not trivial because $< \phi_\ell^n >$ and $< \phi_\ell >^n$ are not equivalent! (these different moments are controlled by different part of the distribution).

So, for example, one might just as well *derive* the shape of the "scale transfomration" (8) using this relativity principle and the requirement of scale symmetry. This symmetry implies, in a broad sense, homogeneity in our log-coordinates (consider eq. (1) when you take the logarithm. This is a translation invariance!). This homogeneity automatically selects linear transformations. Then, the principle of exponent relativity imposes to consider a group structure for the transformations, and then selects the shape (8) [8], in analogy with special relativity [12, 11].

4. SCALE SYMMETRY BREAKING

I have now at my disposal all the tools required to study the *scale symmetry breaking*. Practically, this means that you have given me data which in an ideal situation obeys power laws, with certain multifractal exponents, that I can manipulate cleverly to obtain the underlying composiyion law. But the data you gave me a real data, coming from finite size experiments, and the local exponent dX/dT is not scale independent anymore...So, you ask me: what can you tell me in such case? How can you describe this new situation? The practical consequence of scale symmetry breaking is that i.e. the derivative d^2X/dT^2 is not zero anymore. In perfect analogy with special relativity, the symmetry breaking can be described via the introduction of *"forces"*, created by the boundary conditions, or finite size effects, or anisotropies...To describe the action of these "forces", a theoretical framework is needed. In the sequel, I consider a Lagrangian framework which is conceptually very fruitful: in any Lagrangian description, there are conservation laws associated with symmetries. Using them, one may then hope to find new scalings generalizing the self-similar scaling (in which dX/dT is conserved along the scale). An example of such new scaling is given in the lecture of Biferale. The explaination within the Lagragian framework will be provided in Section 4.2.2.

4.1. Lagrangian Framework

I consider only a restricted class of Lagrangians: those compatible with the group structure described in Section 3, i.e. compatible with the "scale symmetry". In the present case, there is only one space dimension, so no curl. The only possibility is an "electro-magnetic" theory, with no "magnetic field" [11]. The Lagrangian takes the form [11]

$$\mathcal{L} = -\frac{1}{\Gamma(\dot{X})}\left[MC^2 + e\dot{X}_i A^i\right], \tag{9}$$

where the vector $A^i = (\Phi, A)$, $\dot{X}^i = \left(\Gamma, \Gamma\dot{X}/C\right)$ and the metric used for the scalar product is given in appendix. Possible signification of the scalars M, e and the potentials A, Φ, depend on the physical quantity I want to describe. The associated Euler-Lagrange equation is

$$MC\frac{d}{dT}\left(\Gamma\left[\Lambda - (\Lambda^2 - 1)\frac{\dot{X}}{C}\right]\right) = eE, \tag{10}$$

where I note

$$E = -\partial_X\Phi - \Lambda\partial_{CT}\Phi + \Lambda\partial_X A + (\Lambda^2 - 1)\partial_{CT}A. \tag{11}$$

the analog of an electric field.

Symmetry constraints on the field E, or equivalently on charge distribution, are far less severe than on the tri-dimensional full electro-magnetism [13]. In fact, in such a one-dimensional space, the magnetic field being necessarily zero, Maxwell's equations reduce to:

$$\begin{aligned}\frac{\partial E}{\partial X} &= \rho,\\ \frac{\partial E}{\partial T} &= -J.\end{aligned} \tag{12}$$

The permittivity is taken equal to 1. Here ρ is the analog of the charge density and $J^i = (C\rho, J)$ is a bi-vector with zero divergence:

$$\partial_i J^i = 0.$$

If eE/MC is given, one can solve (10) to determine Γ as a function of the log-scale T, then obtain X. An example relevant to turbulence is treated in the next subsection.

4.2. An Example: The Log-Normal Case

To illustrate the procedure, I now treat an explicit example. The simplest, non trivial case is given by the log-normal case. This case is relevent to hydrodynamical turbulence, see the lecture by Arneodo. As I discussed earlier, the log-normal case corresponds to F being a translation by $h(0)$, G being identity, and the composition law being additive. I thus fall back on the analog of the "Newtonian mechanics", which is easier to deal with.

4.2.1. Dynamics

I then define the "clever" log-coordinates as:

$$T = \ln_K(\ell/\ell_0), \quad X_n = \frac{d}{dn}\ln\left[\frac{<\phi_\ell^n>}{e^{n<\ln\phi_\ell>}}\right], \tag{13}$$

where I have used $ln(\phi_0) =< \ln(\phi_\ell) >$ to define my X_n coordinate, in the reference frame $\mathcal{R}_n$. The equation of motion for X_1 is then:

$$\frac{d^2X_1}{dT^2} = \frac{e}{M}E. \tag{14}$$

To find the equation for the nth component, I have to remember the composition law and its associated scale transformation. Here, the composition law is "Newtonian" ($V(n+1) = V(n) + V(1)$), and the scale transformation is Galilean:

$$X_{n+1} = X_n + \dot{X}_1 T. \tag{15}$$

Using that, I find for the nth component X_n:

$$\frac{d^2X_n}{dT^2} = \frac{ne}{M}E. \tag{16}$$

To find this equation, one may proceed by recurrence, using the following observation: by the composition rule, the motion of the nth component X_n is the motion of the $n-1^{th}$ component X_{n-1} in the accelerated frame attached to X_1. The force acting on X_n is then the force acting on X_{n-1} plus the inertial force taking into account the acceleration of X_1. By definition, the inertial force is the acceleration of X_1, i.e. eE/M.

The quantities X_n are known once the three quantities are given:

• its coupling constant nE/M.

• its initial values of X_n and $\dot{X}_n$ at, say, $T = 0$. They are fixed by the experimental conditions (boundary , size,..). By the Galilean composition law, I know that for consistence the derivative at $T = 0$ should obey: $\dot{X}_n(0) = n\dot{X}_1(0)$.

• and of course the electric field itself. It cannot be derived from mere symmetry considerations and should be deduced from experimental conditions, or computed using another systematic theory. I give an example in section 4.2.3. Before, I show that an interesting conclusion can be reached when the electric field is such that only the X-translation symmetry is broken.

4.2.2. *General Scaling*

The introduction of a "force" in our scale equation can break two symmetries: the X-translation symmetry and/or the T-translation symmetry. The former means invariance of the physical laws with respect to changes of amplitude of the field ϕ_ℓ; the latter means invariance of the physical laws with respect to changes of scale. In general, it is the T transaltion symmetry which is easier to break: it is sufficient to introduce a larger physical size in the system (e.g. by putting it in a box) to priviledge one scale. So let me for one moment break this T translation symmetry, and maintain the X-translation symmetry. This is in some sense the simpler realistic scale symmetry breaking one can imagine. Associated with the X translationnal symmetry, there is a conservation law: a generalized impulsion, $\mathcal{P} = \partial\mathcal{L}/\partial\dot{X}$ is conserved along the scale. This conservation gives rise to a new scaling.

Indeed, let me consider the relation (16), when E is independent of X. Using the initial conditions, it can be integrated into:

$$X_n - X_n(0) = n\,(X_1 - X_1(0)) \tag{17}$$

Integrating (17) with respect to n, substracting $n/3$ times the result for $n = 3$, and repeating the procedure for p, I get a result which can be put under the form of a new scaling, extending throughout the whole range of scale:

$$\ln\left(\frac{<\phi_\ell^n>}{<\phi_\ell^3>^{n/3}}\right) = \frac{n^2-3n}{p^2-3p}\ln\left(\frac{<\phi_\ell^p>}{<\phi_\ell^3>^{p/3}}\right). \tag{18}$$

The property (18) is described in the lecture of Biferale (where $\phi_\ell = |\delta u_\ell|$), and was observed in turbulence by Benzi and collaborators [2]. It was named "general scaling" because it is valid from the injection scale down to the smallest scale resolved into the system, and thus generalizes the self-similar properties of the structure functions in the inertial range. I have here shown that it stems directly from the conservation along the scales of the generalized impulsion, in the log-normal case. It is not clear whether this property still holds for different statistics than log-normal. Checking this requires a non-trivial "general relativity" type of computation, see [11]. Even if it were not rigorously true, one expects to see (18) to hold only approximatively with a slightly different proportionality constant, if the turbulence is nearly log-normal, as seems to be the case (see the lectures by Arneodo and Biferale).

4.2.3. *Ohmic Case*

I have just shown that, with the Lagrangian formalism, one can find new scalings from new conservation laws. It is also possible to get "generic " (simplest) symmetry breaking. There are of course other ways to get such generic solutions, like a mere normal form approach [7]. The Lagrangian solution is a little bit more elegant and systematic, since it can be used for any type of multifractal statistics.

For simplicity, I still consider the log-normal statistics, and the case where the electric field is X-independent. The General Scaling property means that the variations of a given X_n are completely determined by the variations of X_1 in a fixed reference frame. These variations can be found by solving the equations:

$$\frac{d^2X_1(T)}{dT^2} = \frac{e}{M}E(T). \tag{19}$$

The electric field itself is linked to the current J, via the Maxwell equation [11]:

$$\frac{\partial E}{\partial T} = -J. \tag{20}$$

In a given experiment, physical conditions select the shape of J, and then the shape of the structure functions via (19). The major challenge is now to relate J and E to physical quantities characterizing an experiment, i.e. to find a closure relation complementing Maxwell's equation (20).

In the case where X translation symmetry holds, E and J are functions of T only. Thus the theorem of implicit functions implies that one can write J as a function of E, i.e. $J = J(E)$. In a perfectly scale invariant system (infinite Reynolds number turbulence), $\dot{X}_1$ is constant, and E is zero. If the system is sufficiently close to the scale invariant situation (in the vicinity of the inertial range, or when the Reynolds number is large enough), E is small. Thus one can expand J as a function of the parameter E, a standard procedure in electricity [14]. When E is zero, J is also zero as a consequence of (20). The leading term in the expansion is therefore *linear*:

$$J = \sigma E. \tag{21}$$

This is the linear, or "ohmic" case, and σ is the analog of a conductivity. Different scale symmetrical systems are expected to be characterized by different "conductivities " , and to belong to different classes of solutions, possibly not even linear. I show below that the linear case is by itself very rich.

The system of equations (19, 12, 21) is closed. They are two integration constants $E_0 = E(0)$ and $V_1 = \dot{X}_1(0)$. Solving the system is simple: when σ is different from 0, I have the "generic case":

$$\begin{aligned} E(T) &= E_0\, e^{-\sigma T}, \\ \dot{X}_1(T) &= V_1 + \frac{eE_0}{M\sigma}\left(1 - e^{-\sigma T}\right), \\ X_1(T) - X_1(0) &= \left(V_1 + \frac{eE_0}{M\sigma}\right)T + \frac{eE_0}{M\sigma^2}\left(e^{-\sigma T} - 1\right). \end{aligned} \tag{22}$$

When the conductivity is zero, I have the insulating case:

$$\begin{aligned} E(T) &= E_0, \\ \dot{X}_1(T) &= \frac{eE_0}{M}T + V_1, \\ X_1(T) - X_1(0) &= \frac{eE_0}{2M}T^2 + V_1 T. \end{aligned} \tag{23}$$

There are 3 important physical parameters in the solution: the coupling constant eE_0/M, the initial derivative V_1 and the conductivity σ ($X_1(0)$ only contributes to a prefactor in the moments of ϕ_ℓ). The symmetry arguments presented here can give no informations about the *value* of these constants, or given about the variations of these constants with the external parameters (like the size of the system, or the control parameter). This is a well known pitfall of symmetry arguments, which only give you shapes and not numbers! These numbers should be computed directly from the equations describing the systems, or merely fitted from the data. For example, in [9], this procedure was applied to turbulence data. It was found that indeed σ varies with the turbulent configuration: it is zero for turbulence behind a cylinder, but non zero for turbulence in a Helium tank, in a wind tunnel, or computed numerically on a periodic grid. A more detailed study, looking at variations with external parameters such as the Reynolds number, is in progress [1].

I can now comment on the "symmetry breaking" which was just found: as expected, the presence of an "electric" field broke the self-similarity (which would be given by $X_1 - X_1(0) \propto T$. This self-similarity is only recovered for $eE_0/M = \sigma = 0$. These two parameters are thus our "symmetry breaking" parameters. Note that the exponential term found in the generic solution was already found using two different approaches: a statistical approach by Castaing [3, 4, 5] (see his lecture), and a normal form approach [7]. The insulating solution was also found in [7] via the normal form approach. This shows that these solutions are really "generic", and independent on the methods found to obtain it.

5. CONCLUSION

In the past, it was the rule to search for power laws whenever considering a physical system where scale symmetry is thought to hold. In my opinion, it is a biased approach, similar to that developped by the man searching his key car under the street light "because it is the only place where he can see enough". Exact scale symmetry is probably very rare, and it is the scale symmetry breaking which is the common situation. New tools are being built to study this symmetry breaking, as can be found in the lectures by Castaing, Nottale, Pocheau, and here. These are probably only little flash lights, for the time being, but they might be quite enough to help the lost man find his keys in a number of situations!

Acknowledgments

The works of Castaing, Nottale and Pocheau were sources of constant inspiration in the development of the ideas presented here. Finally, I must thank especially F. Graner, who was my main collaborator on the subject described in this lecture.

Appendix A

The analogy presented in Section 3 introduces for the (CT, X) space a constant metric tensor $\bar{\bar{g}}$, i.e. all Christoffel symbols being identically zero [13]. The space-time is flat, and I am in special relativity. The only possible metric, left invariant by all similarity transformations (8):

$$\bar{\bar{S}}^t(V).\bar{\bar{g}}.\bar{\bar{S}}(V) = \bar{\bar{g}}, \tag{A1}$$

is, up to a multiplicative constant:

$$\bar{\bar{g}} = (g_{ij}) = \begin{pmatrix} 1 & -\Lambda \\ -\Lambda & \Lambda^2 - 1 \end{pmatrix}; \tag{A2}$$

I keep the solution with determinant -1.

This metric g_{ij} reflects the fact that the relation $g_{ij}R^iR^j = 0$ is invariant under any similarity transformation. Its matrix $\bar{\bar{g}}$ is diagonalizable and invertible:

$$\bar{\bar{g}}^{-1} = (g^{ij}) = -\begin{pmatrix} \Lambda^2 - 1 & \Lambda \\ \Lambda & 1 \end{pmatrix}$$

References

[1] Babiano A. *et al* (50 authors), in preparation.
[2] Benzi R., Biferale L., Ciliberto S., Struglia M., Tripiccione R., *Physica D* **96** (1996) 162.
[3] Castaing B., *J.Phys. II France* **50** (1989) 147.
[4] Castaing, B., Gagne Y., Marchand M. *Physica D* **68** (1993) 387.
[5] Castaing B. *J. Phys. II France* **6** (1996) 105.
[6] Dubrulle B., *Phys. Rev. Letters* **73** (1994) 959.
[7] Dubrulle B. *J. Phys. II France* **6** (1996) 1825.
[8] Dubrulle B., Graner F. *J.Phys. II France* **6** (1996) 797.
[9] Dubrulle B., Graner F., Submitted to Phys Rev. E, 1996.
[10] Dubrulle B., Breon F-M., Graner F., Pocheau A., in preparation.
[11] Graner F., Dubrulle B., Submitted to Phys Rev. E, 1996.
[12] Nottale L., *Int. Journal of Modern Physics* **A7** (1992) 4899.
[13] Landau L., Lifschitz L., Field Theory (Mir, Moscow, 1989) p. 1.
[14] Landau L., Lifschitz L., Electrodynamics of continuous Media (Mir, Moscow, 1989) p. 1.
[15] Nottale L., Fractal space-time and microphysics (World Scientific, Singapour, 1993) p. 1.
[16] Parisi G., Frisch U., Turbulence and Predictability in Geophysical Fluid Dynamics (Proceed. Intern. School of Physics 'Enrico Fermi', North Holland, Amsterdam, 1983) p. 84.
[17] Pocheau A., *Phys. Rev. E* **49** (1994) 1109.
[18] Pocheau A., *Europhys. Lett.* **35** (1996) 183.
[19] She Z. S., Leveque E., preprint 1997.

Imprimé en France. – JOUVE, 18, rue Saint-Denis, 75001 PARIS
N° 252853X. Dépôt légal : Décembre 1997

GPSR Compliance

The European Union's (EU) General Product Safety Regulation (GPSR) is a set of rules that requires consumer products to be safe and our obligations to ensure this.

If you have any concerns about our products, you can contact us on ProductSafety@springernature.com

In case Publisher is established outside the EU, the EU authorized representative is:

Springer Nature Customer Service Center GmbH
Europaplatz 3
69115 Heidelberg, Germany

Batch number: 10704711

Printed by Printforce, the Netherlands